T0295430

The Arrival of the Fittest

The Arrival of the Fittest

BIOLOGY'S IMAGINARY
FUTURES, 1900–1935

Jim Endersby

The University of Chicago Press   CHICAGO AND LONDON

The University of Chicago Press, Chicago 60637
The University of Chicago Press, Ltd., London
© 2025 by The University of Chicago
Published 2025
Printed in the United States of America

34 33 32 31 30 29 28 27 26 25   1 2 3 4 5

ISBN-13: 978-0-226-83754-3 (cloth)
ISBN-13: 978-0-226-83756-7 (paper)
ISBN-13: 978-0-226-83755-0 (e-book)
DOI: https://doi.org/10.7208/chicago/9780226837550.001.0001

Published with support of the Susan E. Abrams Fund.

Library of Congress Cataloging-in-Publication Data

Names: Endersby, Jim, author.
Title: The arrival of the fittest : biology's imaginary futures, 1900–1935 /
    Jim Endersby.
Description: Chicago : The University of Chicago Press, 2025. |
    Includes bibliographical references and index.
Identifiers: LCCN 2024026532 | ISBN 9780226837543 (cloth) |
    ISBN 9780226837567 (paperback) | ISBN 9780226837550 (ebook)
Subjects: LCSH: Science fiction, English—History and criticism. |
    English fiction—20th century—History and criticism. | Science fiction,
    American—History and criticism. | American fiction—20th century—
    History and criticism. | Evolution (Biology) in literature. | Utopias—
    Great Britain. | Utopias—United States. | Science in mass media—
    History—20th century. | Evolution (Biology)—Public opinion.
Classification: LCC PR888.E95 E54 2025 | DDC 823/.08762090912—dc23/
    eng/20240724
LC record available at https://lccn.loc.gov/2024026532

For Jim Secord

Contents

About This Book

The Arrival of the Fittest examines how various early twentieth-century publics made creative use of new theories of heredity (particularly the now largely forgotten mutation theory of Hugo de Vries). Science fiction writers, socialists, feminists, and utopians are among those who seized on the amazing possibilities of rapid, potentially controllable, evolution. De Vries's highly respected scientific theory only briefly captured the attention of the scientific community, but its many fans appropriated it for their own, often wildly imaginative ends. Writers from H. G. Wells and Edith Wharton to Charlotte Perkins Gilman, J. B. S. Haldane, and Aldous Huxley, created a new kind of imaginary future—the biotopia—which took the utopian and dystopian possibilities of biology and presented them in ways that still influence the public's understanding the sciences of life. Jim Endersby recovers the fascinating, long-forgotten origins of ideas that have informed fictions from *Brave New World* to the *X-Men* movies while reflecting on the lessons—positive and negative—that this period might offer us.

Introduction

ARRIVALS

Evolution Talk

In 1909, *Scribner's Magazine* published "The Debt," a new short story by Edith Wharton, which began with an unidentified narrator asking: "You remember—it's not so long ago—the talk there was about Dredge's 'Arrival of the Fittest'? The talk has subsided, but the book of course remains: stands up, in fact, as the tallest thing of its kind since—well, I'd almost said since 'The Origin of Species.'"[1] Like the *Origin*, Dredge's fictitious book had offered the public sensational new ideas about evolution, and the comparison between Dredge and Darwin was a topical one; Wharton's story appeared in the year that marked both the fiftieth anniversary of the *Origin of Species* and the centenary of its author's birth. Yet even as most of the scientific community were celebrating Darwin, his theory of natural selection was being challenged by scientists who claimed it did not actually explain how species originated. Improvements to existing species might well come about through natural selection, but many could not see how genuine evolutionary novelty arose from the slow, gradual elimination of the less fit. As one writer put it, "Natural selection may explain the survival of the fittest but it cannot explain the arrival of the fittest."[2] Wharton's use of the same phrase evoked the wider debates and reflected her interest—and self-taught expertise—in the latest biology.[3] Her story hints at just how widespread such interest was. As we shall see in later chapters, during the twentieth century's first decades scientific fact and fiction were being freely blended by everyone from agricultural entrepreneurs to socialist revolutionaries, from elite scientists to the pioneers of pulp science fiction—all of whom interpreted the challenge to

1. Wharton 1909.
2. Harris 1904, 195.
3. For Wharton and biology, see Lewis 1975, 56; Bender 1996, 314; Hayes 1996, 181–82; Saunders 2009; Ohler 2014.

SCRIBNER'S
FOR MARCH NOW READY

Everybody is talking of
"THE HOUSE OF MIRTH," by
Edith Wharton in Scribner's
Are you reading it?

FIGURE 1.1. Edith Wharton's novel *The House of Mirth* was publicized effectively by *Scribner's*, helping to make her into one of America's best-known contemporary writers.

Darwinism as opening up the possibility of a new, rapid evolution that might eventually allow humans to remake life itself.

The debates over Darwinism in the decades around 1900 were part of a broader reconfiguration; thanks in part to the newly rediscovered work of Gregor Mendel and a renewed interest in Francis Galton's science of eugenics, biology would soon be recognized as one of the most potent—and personal—of sciences. That story is well-known to historians of science, but this book will focus on a third element—the all-but-forgotten mutation theory of Hugo de Vries—to illustrate a broader narrative: of how biology was taken out of the hands of specialists and transformed into mass, public culture—into a set of tools with which to imagine the future.[4] As Wharton's story illustrates, somewhat abstruse scientific arguments were becoming matters of public interest; the public not only *took* an interest in the latest biology but assumed it was a matter *of* public interest, which gave them the right to debate and interpret the latest scientific ideas.[5] If we push the views of scientists to one side for a moment and examine the history of biology from the perspectives of its various publics, one of the many surprises that emerges is that it was primarily the supposedly failed mutation theory that changed the way science in public was done and understood. And, as this volume's conclusion argues, the impact of that change is still apparent when twenty-first-century publics respond to the latest news about such topics as genetically modified crops, cloning, or "designer babies."

Tracing what happens when scientific ideas and practices escape the world of science and the scrutiny of its experts (including its historians) reveals a different history, which complements existing specialist accounts.[6] It shows, for example, why classics of early twentieth-century fiction, most notably Aldous Huxley's *Brave New World* (1932), are still being cited today as epitomizing biology's newly unleashed power—and its threat. Utopian and dystopian fictions (of which Huxley's is only the

4. The claim that Darwinism had suffered an eclipse was first made by Julian Huxley (1942). His version was largely accepted by specialist historians of science in the later twentieth century (e.g., Provine 1971; Hull 1973; Bowler 1983 [1992]; Smocovitis 1996; Sapp 2003, 63–72). More recently, the standard account has been challenged, but largely in relation to scientific communities and their work (e.g., Largent 2009; Bowler 2017a).

5. See Thomas Broman's (1998; 2012) analyses of Jurgen Habermas's notion of the eighteenth-century public sphere and the role of nonspecialist publications in redefining it.

6. See note 4 above. See also Paul and Kimmelman 1988; Aldridge 1996; Thurtle 2007; Carlson 2011; Radick 2023.

best known) are a further reminder that it was not the biologists alone who made theirs into the century's dominant science. A decisive role was played by those who interpreted—and appropriated—the new biologies and put them to imaginative new uses.

Novelists, journalists, autodidacts, and science enthusiasts of every kind collaborated haphazardly—as both readers and writers—and drew on sources as diverse as technical monographs, popular journalism, and college textbooks to create a new language with which to discuss the new biologies.[7] Early twentieth-century evolution talk was often characterized by an optimistic anticipation of the future and its possibilities. It boosted biology's rise to public prominence by encouraging those Gillian Beer characterized as "unforeseen readers" to begin debating rival definitions and their significance.[8]

Wharton's story gives a sense of how broad interest in the Darwinian debates was becoming, and it offers a tangible clue as to how biology caught the public's imagination. Despite being fictitious, Dredge was introduced as a professor of *experimental evolution*, a new area of study that was not only real but the focus of intense interest both within and beyond the scientific community. *Experimental evolution* was a vague term (and thus ripe for interpretation) that was coined to describe the potential application of several new ideas about heredity, but at its heart was mutation. Natural selection was proverbially slow and gradual, which meant that evolution could never be observed, but mutations were rapid, suggesting that evolution was sometimes fast enough to be studied in laboratories and scientific gardens. And if it could be studied, it could be manipulated. Many experimenters hoped to further accelerate evolution and persuade it to make larger leaps. That would allow new species of plants to be created to order—and created quickly—perhaps to be followed by improved animals and eventually even radically remodeled humans.

A key reason why nineteenth-century thinkers had assumed evolution was slow was that it was held in check by the conservative force of heredity. Under the models of inheritance which prevailed at the time, it seemed inevitable that organisms with novel features (however advantageous) would be swamped by the burden of the past—the overwhelming preponderance of the organism's original, unimproved form. One reason Galton proposed eugenics was that his own, widely accepted, law of ancestral inheritance seemed to prove that regression to the mean—the survival

7. Bakhtin 2010, 1076, 1078; Holquist 2002, 18–19, 20–22.
8. Beer 1999b, 1.

of the average—was heredity's most common outcome.[9] Helping the fittest to arrive was part of his original eugenic dream, but in Galton's day it was almost impossible to see how it might be achieved—his own work suggested that any improvement that might raise the average would be rapidly diluted until it disappeared. As a result, eugenics tended to focus on eliminating the unfit, to at least prevent them further lowering the human average, a strategy whose horrifying consequences are all too well-known.[10] However, as we shall see in later chapters, experimental evolution persuaded a minority of eugenicists to refocus their work on increasing the numbers of supposedly fit people.

Histories of modern biology, including those of eugenics, tend to focus on 1900, when Mendel's work was supposedly rediscovered and gradually developed into modern genetics.[11] For example, Gregory Radick's *Disputed Inheritance* is the most up-to-date, scholarly account of how and why genetics emerged when it did.[12] Radick's focus is on the scientific community—its ideas, individuals, and institutions—and as a result he barely mentions the mutation theory, because scientists rapidly lost interest in it. By contrast, this book focuses on science in public, largely ignoring the laboratories and plant-breeding stations, the scientific journals, and the private arguments of scientists. The mutation theory's public career was decisive in launching experimental evolution and transformed the public's perception of biology's possibilities. By contrast, Mendel's ideas were too mathematical and complicated to ignite widespread public interest, which significantly delayed their wider impact.

De Vries's theory (also announced in 1900) not only persuaded the public that evolution had become experimental and could be controlled, it offered a sense that time itself was changing—which transformed the way people began to imagine the future (a shift to which early twentieth-century physics also contributed).[13] Experimental evolutionists believed they would soon be able to do quickly what unaided evolution might

9. Radick 2023, 132–35.

10. For a brief introduction to the scholarship on eugenics, see Searle 1976; Paul 1984; Kevles 1986; Mazumdar 1992; Cuddy and Roche 2003; Black 2004; Currell and Cogdell 2006; Bashford and Levine 2010.

11. Keller 2000 (2002). Various scholars have shown that the narrative of "rediscovery" is misleading and oversimplifies key aspects of the story. Nevertheless, it was widely accepted as an accurate term in the period under discussion. For more recent historiography, see Zirkle 1968; Brannigan 1979; Theunissen 1994a; Wolfe 2012; Stamhuis 2015.

12. Radick 2023.

13. Canales 2010; 2015.

only achieve after thousands of years—if it ever got there at all. Thanks to mutation, heredity became a kind of time machine that could accelerate today's plants, animals, and people into an unimaginably better tomorrow.[14] Instead of being pictured as the burden of the past, holding back innovation, heredity increasingly became a tool with which to speculate about a seeming limitless future.

The ways in which new words and concepts have reshaped the history of science are, of course, well-known.[15] A key thinker who influenced much later work is Reinhart Koselleck, whose arguments for the link between conceptual shifts and new sense of time have proved particularly useful for thinking about the issues covered in this book. Koselleck argued that among the defining experiences of late eighteenth-century and early nineteenth-century modernity was that the "space of experience" (the past that could be known and assessed) was increasingly overshadowed by future possibilities, the expansive "horizon of expectation."[16] Modernity came to combine the prospect of an apparently limitless future with the sensation that time itself was speeding up.[17] A similar—and perhaps even more intense—transition occurred in the early twentieth century, which H. G. Wells argued was marked by "a discovery of the future"—a shift from "things accomplished to things to come."[18] As increasing numbers of people found themselves not only living *in* a new time but having a new sense *of* time, the word *progress* became ever more ubiquitous. Progress implied an unpredictable future that bore little relation to past experience, thus transforming the balance between experience and expectation, past and present.[19] Following the changing meanings of words not only helps us track new or modified concepts, but also shifting understandings of time.[20] In later chapters we will witness biology's contribution to this wider process, as words like *Darwinism* and *evolution* were modified or redefined by radically new concepts, including mutation. Claiming that

14. The idea of a novel scientific technique becoming a form of time travel is explored in Radin and Kowal 2017 (8–9); Radin 2017.

15. See, e.g., Daston and Galison 1992; Daston 2001; Latour 1993; Müller-Wille and Rheinberger 2007; Müller-Wille and Brandt 2016b.

16. Koselleck 1985, xxiii–xxiv, 267–70. See also Radin 2017, 9–12.

17. *Zeitschichten* (layers of time) were central to Koselleck's *Begriffsgeschichte* (history of concepts), as for example when he argued that when the concept of *Volk* (people/race) became a future-oriented idea in the early nineteenth century, it changed the relationship of past to future and gave a different sense to each. See Motzkin 1996, 41; Zammito 2004; Jordheim 2012.

18. Wells 1902b, 327.

19. Koselleck 1985, 280–81; Motzkin 1996, 41; Zammito 2004, 126–28.

20. Dixon 2008a, 35–37.

mutation was the key to evolution not only changed the pace of evolutionary change, it helped change the public's sense of time as—almost overnight—evolution became less about the past and more about the future, a future that seemed closer than anyone had imagined.

The combination of new languages for thinking about biology and a new sense of time helped create new genres, most obviously science fiction (chapter 8). However, even the seemingly staid genre of the school or college textbook was undergoing a transformation. Perhaps the best example is *The Science of Life* (1929–1930), a biology textbook that aimed to be as authoritative as it was accessible and was initially published in fortnightly magazine-like parts, with attractive color covers, in order to reach the widest possible audience. Its authors reflected the hybrid breadth of its ambitions: H. G. Wells collaborated on the book with two accredited university biologists: Wells's son G. P. "Gip" Wells and Julian Huxley. As chapter 6 shows, *The Science of Life* was not the only textbook whose style reflected the way biology was changing the balance between past experience and the widening horizon of expectation. Genres both reflect and shape readers' expectations; they guide us toward what to read but also how to read it. However, as Beer notes, "although each genre establishes expectations, these expectations cannot be enforced." Readers bring their own knowledge and questions to each text and hence discover possibilities that "lie latent in the work's terms and forms, waiting for the apt and inappropriate reader."[21] Many other theorists have also stressed the key role that readers play in creating and sustaining genres.[22] Those who read and wrote about new biological theories felt that new genres were required to do justice to them, but the boundaries of even the newest genres were constantly being challenged and expanded, all of which helped make once-outlandish predictions seem plausible.[23]

Biology became known as the "science of life" after World War I, a name that created a clear contrast with the "sciences of death," such as chemistry and physics, which were often blamed for the mechanized slaughter of the war. While the chemists were perfecting poison gas, the biologists were making life longer and healthier (for example by creating synthetic vitamins, which helped end painful childhood diseases like

21. G. Beer 1999b, 186–87.

22. Koselleck's contemporary Hans Robert Jauss first used the term *horizon of expectation* to refer to genre in 1959, when he and Koselleck were both taught and influenced by Hans-Georg Gadamer, whose philosophy underpinned the idea. See Keith Tribe, "Translator's Introduction," in Koselleck 1985, viii, xiii–xv. See also Dubrow 1982, 31; Wolfreys 2001, 151–55, 282–84.

23. Esposito 2017, 2.

rickets).[24] Such contrasts shaped the increasing optimism that characterized the genres in which biology was explored during the inter-war period, as its tools became more powerful and the discipline became increasingly influential. By the 1930s, biologists—particularly in Britain and the USA—were confident that genetics could be combined with Darwinian natural selection to produce what would become known as the modern synthesis, which made evolution mathematically precise and rigorous (and thus more prestigious).[25] Although the biologists quickly abandoned their initial enthusiasm for de Vries's ideas, they—thanks in large part to their audiences—retained experimental evolution's often hyperbolic confidence in biology's ability to improve life.

Alongside the Darwinian debates and the impact of experimental evolution, the third decisive factor in this story was the changing nature of the mass media, which created new readerships who often developed into new interpretative communities.[26] One result was that the boundaries between science and fiction became particularly porous. This phenomenon is illustrated by another Wharton story, "The Descent of Man" (1904), which reflected on what she called the "reverberating age," the epoch of mass media being created by ever-growing and diversifying numbers of publications and readers.[27] The story offered a fictionalized account of what happened to science when it left the safety of the laboratory to appear in public, and found itself admired, interpreted, and judged by readers who had not been trained in the customs and manners of the scientific community (the story is analyzed in more detail in the next chapter). Wharton's tale reflected the way that she (and many of her readers) learned about the latest science—from newspapers, magazines, or popular books—from which she developed her own interpretations of its significance. Her stories are just one example of the variety of fictions that were built on science. Wharton's were firmly rooted in everyday reality, but other writers used biology to imagine utopias that sometimes included new, improved people who would be built in the same laboratories that would turn out new, improved plants and animals.

24. Wells, Huxley, and Wells 1929–1930; Abir-Am 1982.

25. For the history of the synthesis, see Huxley 1942; Adams 1970; Largent 2009; Provine 1978 and 1992; Mayr and Provine 1998; Smocovitis 1996; Stoltzfus and Cable 2014.

26. For interpretive communities, see Fish 1980; Leiserowitz 2007; Hine et al. 2013.

27. The story appeared in *Scribner's* 25, no. 32 (March 1904): 313–32, but page references are from the eponymous collection of her stories which appeared in the same year (Wharton 1904).

The growth of new print media helped spread and enrich the language with which biology was discussed. As titles proliferated and circulations grew, a dizzying variety of interpreters joined conversations. As a result, early twentieth-century biology began to resemble what some media theorists have dubbed "participatory culture"—characterized by bricolage (exemplified by the ways that fans of TV science fiction franchises refashion the media products they enjoy by enriching them with new meanings).[28] In recent decades, numerous scholars have begun to analyze recent mass media in terms of the *uses* to which its audiences put it and the ways in which they participate in remaking it. Such approaches make readers and viewers into active makers of meaning, as is also evident in much recent work on genre (particularly in film), which acknowledges the role of audiences in defining and sustaining genres. One result is that the consumption of culture (even of mass-produced, commodified culture) is no longer seen as a largely passive process.[29] In the chapters that follow, some of these approaches are utilized to understand how science became such a pervasive and influential part of twentieth-century culture. Obviously, science's influence has long been recognized as a defining feature of the period, but the ways in which it came to exercise that influence have conventionally been described with terms such as *popularization, dissemination,* or *education*—all of which assume that something that is unambiguously "science" is concentrated somewhere called "the scientific community," whence it is transmitted to a nonscientific "general public" (who possess little or no science). This diffusion seems modeled on osmosis—a passive, chemical process in which unequal concentrations simply even themselves out.[30]

Patterns of participation in science changed during the twentieth century as a more structured scientific profession emerged (clearly demarcated by formal qualifications, peer-reviewed journals, and institutional bases). Nevertheless, as later chapters show—that emergence did not

28. Jenkins 1992 (2013). See also Certeau 1984 (1988); Bacon-Smith 1992; Penley 1997; Kilgore 2003; Milburn 2010; Carrington 2016.

29. Altman 1999; Vint and Bould 2009.

30. By contrast, recent work on public science in the nineteenth century and earlier has demonstrated that scientific communities had fluid boundaries that allowed active participation in the making of scientific knowledge. For nineteenth-century examples, see Secord 1994; Secord 2000; Fyfe and Lightman 2007; Lightman 2007; Karpenko and Claggett 2017. For early-modern examples, see Stewart 1992; Fara 1996; Golinski 1999; P. H. Smith 2004 and 2009; Long 2011.

instantly create a stable distinction between scientists and the public.[31] Like today's TV and mass-media fans, the early twentieth-century science "fans" considered in this book were seldom quiescent consumers. There are, of course, significant differences, of which the most important was that early twentieth-century biology was not a finished product (in the way a TV series is) but was still taking shape. The definition of *experimental evolution*—which referred to both theories and practices—was still being negotiated, and its audiences played an important role in an always-unfinished dialogue around what the term would eventually mean. As we shall see in chapter 8, there are parallels with the emergence of pulp science fiction in the 1920s, whose very name (originally "scientifiction") was one of many features of the new genre that emerged through continuing dialogue between readers, writers, editors, and publishers (often carried out through the letters pages of the new magazines). Like many of the contemporary sciences upon which it drew, the science fiction genre was (and still is) a work in progress. And, as later chapters show, experimental evolution remained an unruly, undisciplined, public activity whose enthusiasts were anything but passive.

Biology's fans included scientists themselves, of course, who also speculated about the applications of this optimistic new biology, but the vast majority of those considered here had no formal scientific posts or qualifications. They included journalists, novelists, editors, science fiction writers, autodidacts, socialists, feminists, and many others who have traditionally been considered outside the scientific community (except, occasionally as recipients of its findings). They nevertheless helped shape the language around biology, its broader influence, and occasionally the discipline itself. The audiences' roles are particularly apparent in Britain and the USA, two countries whose inhabitants read many of the same books and magazines and shared a fascination with the future and how evolution might shape it. Their responses were also shaped by the fact that de Vries's key work, *Die Mutationstheorie* (1901–1903) was not translated into English for almost a decade, during which time many Anglophone readers had to rely on interpreters and intermediaries in order to understand the new theory. There were also crucial differences between Britain and the USA (including the manner in which contrasting ways of imagining biological futures helped shape their contrasting, national scientific traditions), but on both sides of the Atlantic throughout these decades,

---

31. Peter Bowler makes this point in the introduction to his recent book, *Science for All* (2009), which provides an invaluable overview of elite scientists who continued to write for broad audiences in the twentieth century.

evolution and new understandings of heredity defined possible futures, from eugenic improvements to better crops or cures for disease.[32] Biology even promised to redefine human nature itself, according to some of its enthusiasts. The possibilities of such power being wielded by an unaccountable elite frightened many, but whether science's interpreters were inspired or appalled by the looming future, they encouraged a diverse range of readers to take a keen interest in biology's implications and often to become writers themselves.

The Arrival of the Fittest

The ways biology was shaped by its audiences were epitomized by Dredge's fictional, biological bestseller *The Arrival of the Fittest*. Wharton gave no hint as to the nature of the new evolutionary theories that her imaginary professor of experimental evolution had proposed. However, the book's title tells us a great deal about the Darwinian debates and the way readers like Wharton became writers about the latest biology, borrowing ideas and phrases to construct new arguments, which characterized the numerous different communities of interpreters discussed in later chapters.[33]

As noted, by the time Wharton's story appeared, many claimed that Darwin's key idea—natural selection—failed in its most important objective: it did not explain the first appearance of new forms of life. Darwin had argued that organisms varied and competition between them would ensure that the best-adapted organisms survived. That argument was widely accepted as an explanation of apparent improvement (successive generations gradually became better adapted to their environments), but it seemed unable to explain the origins of variation itself, particularly the origin of really radical novelties. The botanist James Arthur Harris (who had argued in 1904 that natural selection explained the survival but not the arrival of the fittest), acknowledged that Darwin had convinced everyone (at least, everyone "whose opinions are most worth consideration") of the *fact* of evolution, but debate continued as to the precise *method* by which it occurred.

The claim that Darwinism could not explain the "arrival of the fittest" lay at the heart of the (often ferocious) arguments about whether Darwinism needed revising, updating, radically expanding, or abandoning. Historians have described these arguments in terms of Darwinism being

32. Van Dijck 1998, 12–14; Jasanoff and Kim 2015; Rees and Morus 2019.
33. Certeau 1984 (1988), xviii. See also Claude Lévi-Strauss's comparison of western science and "primitive" mythic thought in *La Pensée sauvage* (1962 [1974]).

eclipsed, but that claim is based on the perspectives of scientists, not least on the anachronistic assumption that the way Darwinism came to be defined after the modern synthesis was correct. However, as numerous historians have shown, the histories of specific words can allow us to trace conflicts over their shifting meanings. Treating such words as *Darwinism* as the names of concepts assumes they have been clearly defined, but their importance at specific periods suggests that their very meaning was being contested.[34]

When considering science in public, historians have to acknowledge that the public has always possessed Humpty Dumpty's power of using a word to mean what they "choose it to mean—neither more nor less." Almost none of the participants in these early debates would have endorsed modern historians' judgments as to which ideas were "non-Darwinian," and making anachronistic judgments can obscure the larger point—that practically everyone involved in the debates claimed to be a Darwinian (and usually a better Darwinian than any of their opponents). When, for example, historians of ideas argue that the prevailing concept of evolution in these decades owed at least as much to the ideas of Herbert Spencer as it did to Darwin, it is worth noting that most contemporary readers neither noticed nor cared that their version of evolution was really "Spencerian" (and the same could be said about "Lamarckian" ideas, also discussed later).[35] Tracing the phrase *arrival of the fittest* allows historians to trace debates over rival meanings of key terms such as *Darwinian* and *evolution*. As scientific rivals attempted to attach the prestige of Darwin's name to their interpretations, their arguments created opportunities for all kinds of interpreters to join the conversation.

To understand the Darwinian debates, a quick reminder of Darwin's ideas may be useful. He had assumed that evolution by natural selection was too slow to be observed directly, and so had turned to the activities of plant and animal breeders (whose methods he referred to as "artificial selection") to prove how much change humans could effect in a

34. Kenny 2004, 1–8.

35. Richards 1987, 2009. The pre-Darwinian French biologist Jean-Baptiste de Lamarck argued that changes in an organism's body (such as the effects of using, or failing to use, its muscles) were inherited by its offspring, which explained how a species became better adapted and thus progressed. This mechanism became known as Lamarckian inheritance, or the inheritance of acquired characteristics. This interpretation of Lamarck was widely held, particularly in the nineteenth century, but is neither a complete nor an accurate account of Lamarck's thinking. For a fuller account of Lamarck and his legacy, see Jordanova 1984; Burkhardt 1995; Gliboff 2011; Gissis and Jablonka 2011.

comparatively short space of time. A key example was the common rock pigeon, which Victorian pigeon fanciers had transformed into a startling array of ornamental breeds by simply selecting birds that possessed a particular feature.[36] In nature, competition does the selecting; organisms vary at random and those with any kind of advantage in the struggle for existence (especially the struggle to reproduce) are more likely to survive and pass on that advantage (while the reverse happens to those that possess disadvantageous features). Darwin then argued by analogy that, given the almost infinite time during which natural selection could act, even the most dramatic transformations could emerge from this process. However, as Harris had noted, many still questioned the validity of the analogy between artificial and natural selection. Breeders could undoubtedly improve a feature of an animal or plant (as was demonstrated by the dramatic improvements in the sugar content of sugar beet that had been made in just a few generations). However, it seemed impossible for selection to put sugar into a sugar-free plant—or to create any other major evolutionary novelty.

As noted, de Vries's massive book *Die Mutationstheorie* did not appear in English for many years, but even after it was translated, its length and highly technical nature ensured that most British and American readers encountered de Vries's radical claims in reviews or from articles in newspapers or general-interest magazines. Such interpretations were usually created outside the conventionally defined scientific community but were nevertheless the main route by which the mutation theory became the most widely discussed contribution to the Darwinian debates. Chapter 1 considers de Vries's theory in more detail, demonstrating how experimental evolution was largely created by his eye-catching claim that the fittest arrived as mutations, radically new forms that appeared suddenly. He thus challenged orthodox Darwinism's claim about the tempo of evolution (that it was a slow, incremental process), because mutation changed the mode (evolution occurred in large leaps, not small steps).[37] According to Darwin, species changed slowly because natural selection worked by gradually accumulating slight, favorable variations over many generations. Each step was small, but—given the earth's enormous age—any degree of change was theoretically possible. Yet even some of his supporters struggled to see how real novelties could emerge from this process: no matter how much time evolution was granted, how could such small changes

36. Secord 1985.

37. For a wider discussion of these terms and their history, see Simpson 1944; Ayala, Fitch, and Ayala 1995.

confer sufficient benefit on their possessors to increase their chances of survival? How could a wingless species acquire wings in small steps? And, if evolution were going on all around us, why had nobody ever seen a new species evolve? Such questions framed the problem of explaining the arrival of the fittest.

Darwin had, of course, anticipated such objections and spent many pages in the *Origin* and elsewhere explaining how such complex structures as the vertebrate eye could have arisen. He acknowledged that the claim that a series of essentially random variations could be built into something so sophisticated must initially seem "absurd in the highest possible degree," yet he continued, "reason tells me, that if numerous gradations from a perfect and complex eye to one very imperfect and simple, each grade being useful to its possessor, can be shown to exist"—and he believed they could be—then the problem of evolving an eye "though insuperable by our imagination, can hardly be considered real."[38] He offered examples of creatures with eyes of varying complexity, from simple light-sensitive patches of tissue to fully formed vertebrate eyes, showing how each step conferred a small but significant benefit on the organism. Yet Darwin's explanation still did not convince everyone. His examples still seemed to demonstrate the *improvement* of existing structures, rather than the creation of wholly new ones (how did that first light-sensitive patch arise?). Some of Darwin's contemporaries argued that perhaps such novelties were generated by larger, more abrupt leaps, epitomized by the aberrant forms that plant and animal breeders referred to as sports, saltations (from the Latin *saltare*, to leap), or simply monstrosities. Darwin had considered these cases and had accepted that such forms sometimes gave rise to new domestic breeds. However, he concluded that such cases were rare; breeders usually selected small variations over numerous generations to create new varieties, which was the analogy he adopted for the long, slow process of producing new species by (natural) selection.

Darwin's skepticism about the role of sports partly resulted from what was known as the "swamping problem." This problem was best exemplified in an 1867 review of the *Origin* by Fleeming Jenkin, who argued that a single improved individual could never transform the population within which it emerged. Jenkin used the image of a shipwrecked white man stranded among a darker-skinned tribe to explain his objection. As most of his contemporaries would have done, he assumed that the white man was greatly superior to the indigenous people, yet he argued that the lone survivor would be forced to breed with darker-skinned women and that

38. Darwin 1859, 186–88.

their offspring would also have to find darker-skinned mates (to avoid incest). After a few generations the average color of the population's skin would be as a dark as ever, because as Jenkin put it, even "a very highly-favoured white cannot blanch a nation of negroes."[39] Pale skin did not, Jenkin acknowledged, confer any advantage on its possessor, but served as a marker of the other, invisible, traits which he presumed explained white dominance (such as supposedly superior intelligence or courage)— these would also be diluted. His example seemed particularly persuasive because human skin color is a trait that exhibits "blending" (parents with differently colored skins tend to produce offspring whose skin color is intermediate between those of their parents). During the nineteenth century, blending was assumed to be a common feature of heredity, an assumption that exacerbated the problem of swamping. However, even without blending inheritance, it was difficult to see how large, dramatic changes could play a major role in evolution, simply because they were so rare.[40] Even though an individual organism might possess a significant improvement, its rarity would force it to interbreed with the much more numerous "unimproved" members of its species—so the improvement would be quickly swamped.

The swamping problem was a key issue that had led Darwin to conclude that evolution's tempo must be slow and its mode gradual (which also explained why nobody had actually seen a species evolve). He summarized his view in the *Origin*:

> As natural selection acts solely by accumulating slight, successive, favourable variations, it can produce no great or sudden modification; it can act only by very short and slow steps. Hence the canon of "Natura non facit saltum," ["nature makes no leaps"] which every fresh addition to our knowledge tends to make more strictly correct.[41]

Darwin's vision of the gradual accumulation of favorable variations meshed well with his contemporaries' assumptions about the nature of historical change, which is one reason why (contrary to popular myth), his theories were accepted comparatively rapidly in Victorian Britain.[42] Bernard Lightman has argued that Victorian readers' expectations were

39. [Jenkin] 1867 (1973), 291.
40. For more on swamping and saltationist evolution, see Gillham 2001, 259–61; Bulmer 2004; Radick 2023, 32.
41. Darwin 1859, 471.
42. Endersby 2009.

shaped by a genre he christened the "evolutionary epic," a narrative of endless, gradual progress driven by natural law that was best exemplified by the anonymous bestseller *The Vestiges of the Natural History of Creation* (1844).[43] Many of the *Origin's* early readers interpreted it as the latest (and most reputable) addition to this existing genre.[44] Both books were interpreted through the shared Victorian assumption that evolution was a form of stadial history (the belief that history moved through a predictable series of progressive stages), which had become widespread in the eighteenth century. This view was in many respects a secularized version of Christian sacred history, and it inherited aspects of its teleology (history had a goal and end). This legacy led many Victorians to imagine evolution as progressive: organisms, machines, religions, human institutions, and whole societies were envisaged as evolving through a predictable series of stages, from lower to higher, becoming ever more sophisticated and complex.[45] It was a vision that united many Victorian political philosophers, from Herbert Spencer to Karl Marx, and numerous theories of biological evolution expounded a similar vision (unsurprisingly in the case of Spencer, whose philosophy was partly inspired by the *Vestiges*).[46]

Stadial evolution became a common feature of the way time was embedded in Victorian thinking. Everything from barnacles to butter knives was imagined as part of a tale of onward and upward improvement, a series of small steps leading to an ever-brighter future. Darwin's vision was directly influenced by his friend the geologist Charles Lyell, whose arguments for slow geological change underpinned Darwin's. And of course, for many Victorians (particularly the wealthier ones—who included Darwin himself) slow, gradual change was more acceptable than rapid, abrupt change—revolution. (As we shall see in chapter 7, the contrast between evolution and revolution became crucial to interpreting new biological ideas.) Nineteenth-century narratives of stadial evolution (whether social, economic, or biological) reassured their audiences that the future would arrive predictably—and at a pace which humans would be able to comprehend and adjust to. The mutation theory challenged these prevailing

43. Wharton's idea for Dredge's book may have been partly inspired by *The Vestiges of the Natural History of Creation* (1844), which decisively shaped Victorian views of evolution. My analysis of the communities of early twentieth-century readers of new biological ideas is deeply indebted to James A. Secord's (2000) study of the *Vestiges'* reception.

44. Lightman 2007. See also J. Secord 2000.

45. Meek 1976; Roemer 1976, 15–33; Williams 1976 (1983), 118–20; Koselleck 1985, 2–20; Harrison 2017, 62; Mandler 2000; Kelley 2003.

46. Taylor 2007, 35–38.

assumptions and also claimed to have solved some of Darwinism's major problems, notably swamping.

De Vries claimed his plant experiments had revealed the existence of what he called "mutants" (translations of his work introduced the noun into English), which were different from sports or saltations because they were usually sterile when fertilized with the parental pollen. Mutants were thus new, stable forms—perhaps even new species—created in a single jump. Their limited fertility ought to have condemned them to rapid extinction, but de Vries was convinced that each species occasionally entered what he called a "mutation period" during which numerous mutations were produced simultaneously, with each variety appearing many times. The mutants were briefly common enough to be able to breed with each other, ensuring their survival. Finally, mutation periods were rare, and species were normally stable, which explained why nobody had noticed mutation before. Taken together, these claims comprised the core of the mutation theory (although, as we shall see in later chapters, the precise meaning of the word *mutation* became another topic for debate). The original theory attracted considerable scientific interest because de Vries, who had been trained in the most advanced German laboratory techniques, had spent decades proving his claims experimentally; *Die Mutationstheorie* presented a lengthy, detailed account of his work with plants, particularly a species of evening primrose, *Oenothera lamarckiana*, which provided the best evidence of mutation. Large leaps between generations changed the mode of evolution (it was discontinuous rather than gradual), which implied an increase in its tempo (no need to wait hundreds or thousands of generations for a new type). However, what really fired the imaginations of those who heard about de Vries's theory was the claim that further experiments would reveal what caused a species to enter a mutation period; once that was known, it would be possible to induce mutations artificially, which opened up the dramatic prospect of controlling evolution.

Harris was one of many who thought that the mutation theory might finally explain the arrival of the fittest, but—as he had acknowledged in his review—he had not coined the phrase but was quoting it. It had been first been used by Jacob Gould Schurman, a professor of Christian ethics and moral philosophy, in his book *The Ethical Import of Darwinism* (1887): "The survival of the fittest, I repeat, does not explain the arrival of the fittest."[47] He repeated the phrase almost verbatim in *Agnosticism and Religion* (1896), and it was regularly used by Christian critics as a stick with which

47. Schurman 1887, 78.

to beat their evolutionary opponents.[48] Numerous religious thinkers used the scientific contention over the precise meaning of *Darwinism* to argue that evolution itself had been debunked (the role of mutation theory and experimental evolution in the well-known struggles to ban the teaching of evolution in US schools are analyzed in chapter 6). Writers of many denominations concurred with the German natural philosopher Eberhard Dennert when he claimed in 1904 that "Darwinism will soon be a thing of the past"; it was on "its death-bed, while its friends are solicitous only to secure for it a decent burial."[49] And Christian writers, particularly in the USA, regularly used the arrival-of-the-fittest problem in attempts to discredit evolution.

Nevertheless, de Vries himself (perhaps unaware of its original context) borrowed the phrase from Harris and used it to describe his key breakthrough. When he gave his first extended English-language lectures on mutation at the University of California, he concluded them by quoting from "a friendly criticism of my views" by Arthur Harris: "Natural selection may explain the survival of the fittest, but it cannot explain the arrival of the fittest." De Vries was confident that he had finally solved this problem: "the origin of species by mutation instead of continuous selection" explained how the fittest arrived.[50] His lectures were published as *Species and Varieties* (1905), the first extended English-language discussion of his theory, so the originally *anti*-Darwinian phrase "arrival of the fittest" became part of de Vries's argument for mutation as a central part of Darwinism. (He assumed mutation was essentially random while natural selection retained its key function of determining whether the mutants survived or perished.) Thanks to de Vries's use of it, "arrival of the fittest" was regularly invoked to summarize his solution to Darwinism's major problem. So, by the time the phrase made its debut in fiction (thanks to Wharton), it had already migrated from a theological context to a specifically biological one, and had made the move via a book review, to de Vries's public lectures and finally into *Species and Varieties*, which was—as we shall see in later chapters—read and interpreted by everyone from expert men of science to socialist agitators.[51] Meanwhile, Wharton's story was only one of many examples of a recondite biological debate

48. See, for example, Kneeland 1892 and Andrews 1894.
49. Dennert 1904, 28.
50. De Vries 1905c, 825–26.
51. De Vries's book was among those owned by John Edwin Peterson, a Swedish migrant who worked for the Pullman car company. He was a member of Industrial Workers of the World (IWW) and the Socialist Party of America (Peterson 1986). For socialists and mutation, see chapter 7.

migrating from science to fiction—gaining new readers and generating fresh interpretations in the process.

Experimental evolution and the Darwinian debates were two factors that made rather esoteric biological ideas into topics of mass-media interest during these decades. They coincided with an apparent revolution in plant-breeding, which promised to transform the world's food supply with seemingly new plants. These three factors led the influential general-interest American magazine the *Atlantic Monthly* to publish a lengthy overview of recent biology in the summer of 1908 (the year before "The Debt" appeared). They reviewed recent books on evolution and heredity, noting that their sheer number was evidence of "the world's perennial interest in the topic"—an interest that was increasing sharply as science revealed how living creatures might be made into "something else." Biologists claimed they were acquiring the power to transmute life's elements into new kinds of animals and plants; soon even new types of people might be produced in biological laboratories.[52] Alongside works on Darwinism and theoretical biology, the *Atlantic*'s reviewer considered de Vries's most recent book, *Plant-Breeding* (1907), praised its accessibility, and described its author as "the world's first authority in his field."[53] *Plant-Breeding* described de Vries's own ideas alongside the work of the celebrated Californian plant breeder, Luther Burbank, whose extraordinary fame as a "plant wizard" also reverberated through the press at this time and played a crucial role in shaping US interest in experimental evolution (see chapters 3 and 4).[54] Burbank not only seemed to promise new foods and flowers (and substantial profits to those who grew them), he also exemplified another meaning of *bricolage* (which in vernacular French has the same meaning as the English expression "do-it-yourself"); Burbank's homely appeal included the promise that ordinary people could both understand and participate in the new biology, enhancing the feeling that the future was close and everyone could play a part in creating it.

Reviews like the *Atlantic*'s brought de Vries and his theory to new audiences, but Wharton (a regular contributor to the magazine), would

---

52. In this review, Brewster used the term *bionomics* to describe the novel approach, as did other writers. However, bionomics was rarely used and has since acquired a much narrower, technical meaning (referring to the ecology of a particular species of organism). I have preferred the broader and more widely used term *experimental evolution*.

53. Brewster 1908, 121.

54. Bowler (2017b, 178) touches on this connection briefly but does not explore it.

FIGURE I.2. The early twentieth-century explosion in printing brought cheap magazines and newspapers onto every American city street.

already have known of them.[55] She was a voracious reader with a keen interest in contemporary biology, who read several of the books the *Atlantic* reviewed. These included Robert Heath Lock's *Recent Progress in the Study of Variation, Heredity, and Evolution* (1906), which described the mutation theory as a response to doubts about "whether the selection of continuous variations . . . can ever lead to the development of a permanent new race."[56] The magazine also considered Vernon Kellogg's widely reviewed *Darwinism To-Day* (1907). Wharton read that too, and annotated Kellogg's discussion of mutation, which included the phrase "arrival of the fittest." She underlined his conclusion that the origin of variation "is the basic problem of evolution, for it is the problem of beginnings."[57] The *Atlantic* also reviewed J. A. Thomson's *Heredity*, which also

55. Wharton published nine stories or poems in the *Atlantic* between 1889 and 1912; see "Edith Wharton," *Atlantic*, accessed February 5, 2024, https://www.theatlantic .com/author/edith-wharton/.

56. Lock 1906, 114. See also Lewis 1975, 228; Ramsden and Lee 1999, xxvi, 78.

57. Kellogg 1907, 89. Other magazines that reviewed Kellogg's book included the *Dial* (Jordan 1907, 163), the *Nation* ("Review of Kellogg, *Darwinism To-Day*" 1907), and Britain's *Athenaeum* ("Review of Kellogg, *Darwinism To-Day*" 1908), and many scientific journals (e.g., *Science* [Dean 1908] and *American Naturalist* [F.T.L. 1908]).

explained how the mutation theory claimed to explain "the arrival of the fittest."[58] Wharton could have learned the phrase from any of these (or several other sources); using it placed both her and her story at the heart of then-current biological debates.

Wharton's example encapsulates the way scientific claims migrated from specialist, technical publications to reach wider readerships. Most early twentieth-century readers left no detailed evidence of their reading, nor were their eventual interpretations of its significance published, but the evidence examined in later chapters suggests that countless other readers acquired a similar fascination with the new biology from general-interest publications. Many became enthusiastic devotees of experimental evolution whose scientific basis made its promised futures seem plausible—even unavoidable. And, despite being produced in the shadow of war and in the midst of the worst economic depression the world had ever seen, many of these imaginary futures were manifestly optimistic; as the *Atlantic*'s review commented, many of the writers whose works were considered shared a "living faith in the power of science to transform humanity and thereby to make men happy." In the future, it was promised, everyone would be able to "transform himself into the kind of man who will be happy amid his own handiwork," a world of transformed plants, animals, and perhaps even people. As one of the biologists quoted in the review put it, "the scientific philosopher must not think of existing human nature as immutable, but must try to modify it for the advantage of mankind."[59] It was that claim—that nature—including human nature—was simply raw material for the biological imagination to work with—that made this period so distinctive and extraordinary.

Biotopia

The transformation of biology into a shared, public culture began with intense public interest in the claim that biology could redesign nature, which made a new kind of utopia possible. The word *utopia* has been used to mean a perfected world ("an epoch of rest," in William Morris's phrase) with no problems left to solve, which can make the search for utopia seem a sterile project, more likely to produce imposed uniformity than an ideal world. However, the utopian impulse can also offer the paradoxical

58. Thomson 1908, 97.
59. Elie Metchnikoff, *The Prolongation of Life: Optimistic Studies* (1908, 235), quoted in Brewster 1908, 124.

opportunity to make an imaginative leap into an unimaginable future (hence its appeal to those who fear they won't survive long enough for sociopolitical reform to improve their lives).[60] Mutation theory seemed to offer a similar jump, circumventing the slow, intractable processes of both biological and social evolution (see chapter 7). Science's preoccupation with evidence and proof would seem to make it the antithesis of speculative, imaginative utopianism, yet the two have been linked since at least the seventeenth century, when Francis Bacon wrote *New Atlantis* (1627). J. Colin Davis has argued that true utopias (unlike Golden Ages or Arcadias) are created by human effort (or what Raymond Williams called a "willed transformation, in which a new kind of life has been achieved by human effort").[61] What is most significant about scientific utopias is, as Davis noted, that they assume "nature is deficient or unaccommodating and must be altered."[62] This belief in nature's *im*perfection puts this form of utopianism in conflict with many influential intellectual traditions in Western thought, which share a faith in nature's goodness (for example, natural theology and some forms of environmentalism). Part of what made the new biological utopia—which I will call a *biotopia*—distinctive was its refusal to accept what nature provided (including normative judgments), its willingness to use the latest biology to change nature, and its unwillingness to wait for the future to arrive.

Utopia is considered here primarily as a literary genre, one that encourages readers to imagine alternatives to the existing world.[63] Once European science emerged as an influential tool for investigating the world, biology and allied studies (such as natural history and anthropology) were often used to argue that social inequalities resulted from innate biological differences.[64] For many, the appeal of imaginary futures has been the opportunity to rethink the supposedly natural categories—such as race and gender—that three centuries of biology had helped create. Speculative

60. Moylan 1986; Baccolini and Moylan 2003; Levitas 2011; Nersessian 2017, 92.

61. R. Williams 1978, 203.

62. J. C. Davis 1984, 26–27. H. G. Wells (1939 [1982]) also regarded *New Atlantis* as the first modern utopia. Overviews of utopianism include Kateb 1963; Berneri 1950 (1982); Williams 1978; Manuel and Manuel 1979; Moylan 1986; Kumar 2003; Segal 2005. For alternatives to Davis's view, see Levitas 2011 and Goodwin 2001.

63. Suvin 1979 (2016); Ruppert 1986; Kumar 1987; Ferns 1999.

64. For examples, see Sander Gilman, "Black Bodies, White Bodies: Toward an Iconography of Female Sexuality in Late Nineteenth-Century Art, Medicine, and Literature," in Gates 1986, 223–61; Schiebinger 1989 and 1993; Pierson and Chaudhuri 1998; Bancel, David, and Thomas 2014.

fictions create counterfactual worlds, thought experiments that show that claims about certain people being "naturally" inferior are just another kind of science fiction. The possibility of challenging those hierarchies stimulated early twentieth-century socialists and feminists to appropriate the new biology as a tool for imagining alternatives. Mutation, experimental evolution, and the Darwinian debates—all of which originated within scientific communities—were all necessary conditions for biotopia, but none were sufficient to generate the range of excitement analyzed here. What I am identifying as the biotopian mode was created by dialogue and bricolage within what might be called biological fan communities—those who made use of science for their own ends.[65]

Biotopianism rejected the plants and animals that nature provided by seeking to improve both, and some of its proponents also rejected supposedly natural categories that they saw as elitist and divisive. As a result, biotopians often rejected the claim that nature offered ethical norms, either by rejecting the very category of "natural" or provocatively preferring the "unnatural" (especially in relation to sex, reproduction, and contraception). By contrast, earlier scientific utopias generally left nature well managed but otherwise unchanged, as they utilized physical sciences and technologies to imagine futures characterized by material abundance and universal leisure.[66] One of the most influential and widely read nineteenth-century examples was Edward Bellamy's *Looking Backward: 2000–1887* (1888), which foresaw a technologically advanced America being steadily transformed into an apparently egalitarian socialist state.[67] Bellamy's time-traveling protagonist, Julian West, is astonished by the news that such things as demagoguery and corruption have vanished in just over a century and comments that "human nature itself must have changed very much" (60), but he is assured that it has not changed at all. The compressed time frame, which helped to make the imagined future seem plausible, would not have allowed for such radical change. Moreover, imagining such a change would have made Bellamy's vision seem impractical (see chapter 5). And so the novel was firmly committed to the assumption that human nature was biologically ordained, inherited from the past, and could not change. These assumptions were vividly illustrated by the fact that the women in Bellamy's future utopia have freedom, education, and economic equality, but remain so similar to their ancestors that West falls in love with the great-granddaughter of the girl he left behind—and

65. Penley 1997.
66. Marx 1964 (2000); Segal 2005 and 1994.
67. Bellamy 1888 (1917); Roemer 1976, 2–3; 1981; Lipow 1982; Pittenger 1993, 68.

she returns his love because her behavior, morals, tastes, and personality are identical to those of the nineteenth-century ancestress from whom she inherited them.[68]

Bellamy's utopia shared a premise with earlier ones (whether religious, political, or legal), that nature—particularly human nature—was largely immutable; we could no more change ourselves than we could persuade flowers to bloom in the desert. A key source for that assumption was the Christian origins of the tradition (most utopian writers took it for granted that human nature was deeply flawed, as our fall and expulsion from Eden proved).[69] Although nineteenth-century writers reimagined original sin in secular forms, particularly as the Malthusian fear that overpopulation would make utopia impossible, they retained the belief that we were inherently flawed—because evolution had burdened us with an insatiable instinct to reproduce and compete.[70] Since most utopias assumed humans were flawed creatures, they usually focused on reforms intended to mitigate the impact of those failings. Among the solutions was the belief that civilization itself had damaged humanity's instinctive ability to live in harmony with nature's laws, which could be mitigated by a renewed trust in nature. Those who held (and still hold) such views are often accused by philosophers of committing the naturalistic fallacy, of confusing facts (the way the world *is*) with ethical values (the way the world *ought* to be).[71] As later chapters show, biotopians rejected naturalistic ethics and avoided the fallacy by treating nature simply as a set of raw materials, ready for human ingenuity to transform to meet human standards (whether ethical, aesthetic, or utilitarian).

Experimental evolution gave some early twentieth-century utopians a justification for rejecting both God's and Nature's guidance. Many of the new biologists believed humans might create a new Garden of Eden (like the one Bacon had described in *New Atlantis*), in which scientific plant-breeding would not only feed the world but transform it into a beautiful, abundant utopian garden.[72] Despite their European origins, these dreams became most popular in the USA, particularly in various writings about Burbank. In 1901, Charles Howard Shinn, Inspector of Experiment Stations at the University of California, described Burbank and his fellow

68. Hamlin 2014b, 161.

69. Roemer 1976, 15–33; Nersessian 2017, 93–94.

70. Hale 2014.

71. My thinking on this topic was initially prompted by Stephen Jay Gould's essay "Nonmoral Nature" (1983a, 32–45).

72. Koerner 1999, 113–39; Spary 2000.

pioneers of scientific plant-breeding as "explorers" whose discoveries "when rightly understood shall in due season release brain-tired men from gray city pavements, sending each one to his own well-watered, fruit-giving, life-supplying acre." This, he announced, was now possible as a result of the "marvellous gospel of plant-evolution."[73]

Burbank and his contemporaries also inspired popular US horticultural writer William Harwood to write *The New Earth* (1906), in which he argued that "the Old Earth was far from paradise ... [a] cheerless, desolate home, often untidy and usually cursed with food unfit to eat." Fortunately, science had come to our aid, and thanks to the "creation" of improved crop plants ("one of the master acts of the men of the New Earth") we would enjoy a future characterized by "broad acres, well kept and well stocked; ... a modern home with its good cheer, its books, its music, its culture; a close touch with progress; ... the pride of strong men and sensible women in a calling as old as the human race, but never until now come into its own,—these are the tokens of the New Earth."[74] It is easy to imagine that H. G. Wells might have written these words, and Harwood's language was often redolent of science fiction (SF), yet the genre barely existed at the time he wrote. However, Istvan Csicsery-Ronay has argued that SF is best understood not as a genre but as a mode of awareness, characterized by a tension between the threat and promise of imaginable and possible futures. SF's characteristic technoscientific trappings are often used to make a given future more plausible (exactly as Harwood did), but also serve to highlight the artificiality and arbitrariness of what might be created. And, of course, the use of science emphasizes the role of human agency in creating any future.[75] Much writing in the biotopian mode shared SF's tendency to remind its audiences that the future would not be determined by God or Nature, but by humans—for better or worse; we can imagine anything, so should be careful what we wish for. (Biotopianism's role in creating SF is explored in chapter 8.)

The creation of new plants was a key feature of biotopias, but even more important was the hope of creating new people to populate science's new garden—the distant prospect of repairing fallen human nature. When the Carnegie Institution founded its Station for Experimental Evolution in 1905, newspapers reported its work in unabashedly utopian terms as fulfilling "the dream of Bacon" (chapter 2). And the well-known writer and sociologist Charlotte Perkins Gilman imagined a women-only

73. Shinn 1901, 3–5.
74. Harwood 1906, 49, 1–2.
75. Csicsery-Ronay 1991, 387–88.

utopia, *Herland* (1915), whose inhabitants had used what they learned from scientific crop-breeding to perfect themselves (so that Herland's children were like "perfectly cultivated richly developed roses compared with—tumbleweeds").[76] Meanwhile, on the other side of the Atlantic, the Baconian garden was central to H. G. Wells's utopia, *Men Like Gods* (1923), from which "ill-bred weeds" had been eliminated. Wells's utopians were as orderly as their garden-like world, which (like *Herland*) had undergone "a great cleansing . . . from noxious insects."[77] And the British biologist John Burdon Sanderson (J. B. S.) Haldane imagined a future where deliberately modified algae fed the world and turned the oceans purple—a sight which had come to seem "so natural."[78] These and other examples will be analyzed in later chapters. Despite their differences they share characteristic themes that define biotopia: the world would become a scientifically cultivated garden; nonhuman nature would be changed to suit our needs; and eventually human nature itself would be transformed. Of course many biotopians showed no interest in the details of the new scientific ideas, but simply expressed confidence that evolution could be controlled and directed. Alongside mutation theory, the new science of Mendelian genetics was an important source of their confidence. Its proponents generally foresaw an increase in evolution's tempo (a precise, mathematical understanding of heredity would make progress faster), but assumed that the painstaking selection of small variants (its mode) would remain central to plant-breeding. By contrast, fans of the mutation theory claimed it changed both evolution's tempo *and* mode, promising rapid leaps into a better future.

Because biotopian writers rejected the idea that nature was harmonious or self-regulating, some reached conclusions that will strike many twenty-first-century readers as profoundly dystopian (but of course one person's utopia is always another's dystopia). As every writer on utopianism notes, the ambiguity began with Thomas More's original coinage (*utopia* sounds as if could mean either "good place," *eutopia*, or "no place," *outopia*). Ruth Levitas has argued that More's original coinage haunts utopian studies "like a familiar but rather troublesome ghost," but an inclusive, ambiguous definition is the only kind that will do justice to the

---

76. Gilman 2013, 98.

77. Wells 1923 (2002), 32, 85–86.

78. Haldane 1924, 61–62, 66. The idea of "genetic engineering" did not really exist at this time, but the idea was beginning to be discussed by geneticists. The US geneticist Albert Blakeslee may well have been the first to call himself a "genetics engineer" (Campos and Schwerin 2016, 402; Campos 2015).

diversity of the utopian tradition.[79] Certainly, for the purposes of this book, ambiguity was the most attractive aspect of utopianism; the lack of clarity is an open invitation to bricoleurs and would-be interpreters looking for opportunities to expound their own ideas. Many advocated the deliberate eradication of any nonhuman species that was deemed—from an anthropocentric perspective—harmful or merely useless (a biotopia is thus the opposite of an ecotopia). Similarly, the biotopians went well beyond earlier proposals to reform relationships, the family, or childcare and instead imagined babies in bottles and a complete end to human sexual reproduction.[80] The rejection of nature and naturalistic ethics was exemplified by the way some exponents of these new utopias embraced the term *perversion* as positive rather than pejorative (chapter 5).

The desire to interfere in the most seemingly natural (and personal) aspect of human life was shared by both biotopians and eugenicists, advocates for the supposed science of controlled human breeding. (As I found repeatedly while writing this book, if you say you are working on how early twentieth-century biology became popular and inspired mass enthusiasm for creating a biologically inspired utopia, everyone assumes that you are working on eugenics.) Eugenics had many fans during these decades, who believed their science could reverse the apparent degeneration of the Anglo-American stock who had risen to global dominance during the preceding century. Scientific and moral progress seemed to have had the paradoxical effect of making life too easy; improved medicine and an increasingly compassionate approach to society's less fortunate members had allowed many who were regarded as "unfit" to survive and breed.[81] Eugenicists argued that such people were poor because of their low intelligence or weak morals, which led them to breed uncontrollably, producing endless offspring as "unfit" as their parents. The eugenic solution was to take control of human breeding, primarily by discouraging or forcibly sterilizing those who insisted on multiplying their inferior stocks. It may be controversial to call such programs "utopian," but that was the firm belief of their advocates, who were confident that an altogether finer human species would emerge from the process. The horrifying legacy of eugenics has, of course, prompted considerable scholarly attention to its history, which has revealed that eugenics was popular among people with many different political views, which led to the formation of large and active fan clubs—the eugenic societies that formed in many countries.

79. Levitas 2011, 3, 207–8.
80. Haldane 1924; Gilman 2013; Squier 1994.
81. Among the first to make this argument was Greg (1868).

For historians of this period, eugenics is unavoidable, and it was undoubtedly a strand within biotopianism. Nevertheless, eugenics is not this book's main focus, partly because it is already well-studied, which has marginalized other important issues. Also, historians of eugenics tend to treat it as a stable concept whose meaning is evident. However, the writings of the overlapping communities of eugenics fans and those of experimental evolution suggest that the precise meaning of the word *eugenics* was being actively debated during these decades (just as *Darwinism* and *evolution* were).[82] Hints of those debates will emerge throughout the book, and their significance is analyzed in more detail in the conclusion. The eugenicists and biotopians clearly shared a willingness to use science to interfere in people's intimate lives, yet Galton and the early eugenicists were indebted to a form of naturalistic ethics that resulted from the nineteenth-century view of heredity. Eugenicists argued that well-intentioned but wrongheaded interference in natural selection's ruthless work of pruning the human stock was resulting in degeneration.[83] Eugenic policies were intended to take up the work that natural selection had previously done, thus turning to nature for guidance as to what was best. By contrast, the biotopians largely rejected any hint of naturalized ethics (see chapter 5).

Alongside the rejection of nature's normative role, biotopianism was also inspired by the early twentieth-century shift from a focus on the past to an emphasis on the future. Victorian thinkers like Galton assumed people were defined by their unalterable "nature": whatever they had been born with dominated their "nurture" (what happened after they were born). Despite the Victorian enthusiasm for various kinds of self-help, a majority seemed to accept that those with a good bloodline (or huge tracts of land) were destined for greatness. No amount of self-improvement could lift the burden of poor heredity or inherited disease, hence the recourse to eugenic measures.[84] For many biologists with an interest in heredity, 1900 was the tipping point when shared perceptions of what was, or might become, possible began to dominate their thinking. Expectation and experience are dynamically related—changing one changes the other—and so heredity, previously imagined as what was *un*changeable, was reimagined

82. Koch 2006.

83. Greg 1868. See also Paul 1995, 22–39.

84. I am indebted to Staffan Müller-Wille and Christina Brandt, whose work first made me recognize this point. They argue that the "view that sees (cultural as well as biological) inheritance as a common stock of dispositions seems to lie in the association of heredity with the future rather than the past," which they associate with the broad idea of progress; however, they don't offer a specific causal account of the shift (Müller-Wille and Brandt 2016a, 17).

as individuals and societies re-evaluated the past in the light of current events and future possibilities.[85]

Britain's conservative, class-bound culture meant the past weighed heavier there, while Americans were generally more excited by the future, but on both sides of the Atlantic, the early twentieth century saw the study of heredity change profoundly. Its earlier work of cataloguing and detailing the legacies of the past was increasingly replaced by speculation about possible futures. A focus on whatever one had inherited (which I will refer to as "ancestral heredity") emphasized the past, the space of experience.[86] Around 1900 it gave way to excitement over each organism's stock of exciting, as-yet-unrealized possibilities ("speculative heredity"), which emphasized the expanding horizon of expectation. Writers began to describe nature *and* nurture as equally susceptible to human intervention, thanks in part to experimental evolution. Nevertheless, there was more to biotopianism than experimental evolution's claims. They were often little more than grist to the imaginative mills of those who read about the way biology was being transformed thanks to the future-oriented mood of speculative heredity (one result being that the biotopians occasionally mocked the eugenicists for being out of touch with the latest biology). As later chapters show, biotopian optimism papered over the cracks between incompatible theories (old and new), making them feel coherent to their fans. As a result, the term *biotopia* may describe everything from new crop plants to feminist utopias to space colonies built by artificially altered humans. It promised unlimited profits to some of its enthusiasts and a scientific justification for socialist revolution to others. Yet despite their differences, what united these dreamers was optimism based on biology's newest theories, the rejection of a normative role for nature, and—above all—upon heredity reimagined not as the past's legacy but as life's future potential.

Reverberating Science

Biotopianism reflected the environment in which it was created—the world's first genuinely *mass* media. New theories about heredity could never have captured the public's imagination had they not appeared in the middle of an unprecedented media revolution. Like their nineteenth-century precursors, the early twentieth-century biological fantasies

85. Koselleck 1985, 275–76.
86. For a detailed analysis of the cultural impact of this view, see Shuttleworth 2010, 267–363.

considered here depended on a complex interdependence of new socio-economic circumstances and scientific ideas.[87] However their full impact depended on near-universal literacy, leading to dramatic rises in the numbers of both publications and readers, which in turn created new genres and formats. Deliberately challenging modernist novels and poems were as characteristic of the period as disposable, cheap fiction (both *bestseller* and *short story* were new coinages).[88] Detective stories, spy novels, and early SF were among the new genres that appeared, examples of what literary theorists call "paraliterature" (any and all writing that is not considered literature).[89] Samuel Delany has argued that the concept is useful for understanding a genre such as SF, arguing that "paraliterature exists to delimit literature" (in much the same way homosexuality came into existence during the same period to define heterosexuality, and to "provide it with an equally false sense of itself").[90] Paraliterature is defined in part by material practices such as the way it is marketed, which shape the ways it is read (SF, for example, might be defined simply as anything that can be read *as* SF).[91] Most of the sources considered in this book could be considered as paraliterature, and Delany's argument suggests a useful parallel between the literature/paraliterature split and the demarcation of science and so-called pseudoscience (that which appears in peer-reviewed science journals is science—because it is *read* as science—and vice versa). Analyzing the biologically inflected paraliterature—from fictions of all kinds to journalism, plant-growers' catalogues, and biology textbooks—shows how new theories succeed (or fail) to become accepted as science.

The new mass-media market brought the mutation theory to many different kinds of texts and readers, helping it to become what might be called an "undisciplined science," one that isn't yet (and may never be) part of a recognized discipline, with defined rules and properly qualified gatekeepers to decide who or what may join the accredited-sciences club. Undisciplined sciences force historians to question widespread assumptions, such as the existence of a stable divide between the scientific community and its audiences (or between expert and lay, elite and popular, science and pseudoscience). It was precisely the absence of such boundaries that made the new biology so enticing. When the *Atlantic Monthly*

87. For an overview of such arguments applied to nineteenth-century evolution, see Endersby 2009.

88. Ledger and Luckhurst 2000, xiii–xiv; Tattersdill 2016, 1–27.

89. Jameson 1991, 2–3. The term was used earlier but in a different sense by Rosalind E. Krauss (*The Originality of the Avant-Garde and Other Modernist Myths*, 1985).

90. Delany 1999, 205.

91. Delany 1994, 273–81.

article (mentioned above) wanted to illustrate the excitement the new biology was generating, it offered two quotes. The first was from British biologist Reginald C. Punnett, from *Mendelism* (1907), the second edition of his brief introduction to what would soon be called genetics. He listed various questions that geneticists had been unable to answer just a couple of years earlier, but about which they had been confident that "experiment would give us the solution" and that "our confidence has been justified."[92] This is precisely the kind of quotation that a historian of science might offer as evidence that the rediscovery of Mendel's ideas led to a rapid transformation of biology. Punnett had done exactly what such a historian would expect—written a "popularizing book for the curious" that would diffuse this new knowledge to the science-deficient public.[93] However, the *Atlantic*'s second quotation was from an "Illinois farmer writing in a farm paper," who claimed that, by applying Mendel's laws, a stock breeder could "obtain any character you desire from any breed and graft this character on to your favorite breed" (120–21). According to the reviewer, these quotes *together* proved that just "ten years ago, organic evolution was one of the speculative sciences," whereas today's farmer could demand such things as "his wheat must ripen by such and such a date"—and the modern scientific breeder "builds him the plant to order" (121). The opinion of an American farmer was being given equal weight to that of a prominent British scientist, each of whom reiterated experimental evolution's central claim—that living things could be built to order. Even more importantly, their claims appeared in a broad, general-interest publication, ensuring that the conversation was a public one.

The *Atlantic Monthly*'s review concluded that many of the biological innovations discussed "could be begun now. All of it will have to be done sooner or later," because the world would be inherited by "that nation which does it first" (124). The claim that nobody could afford to ignore biology's promised future embodied a rhetorical assumption about actions and possibilities.[94] By contrast with stadial evolution, experimental evolution embodied the compression of time and space foreshadowed by nineteenth-century technologies of travel and communication, which appeared to be annihilating space and time.[95] By the early twentieth century, this idea had become familiar; the future was now, urgently demanding our attention and action. Time-space compression underpinned demands

92. Brewster 1908, 121.
93. Radick 2023, 267.
94. Jack 2006, 54–55. See also Harvey 1990; Haraway 1997, 41–45.
95. Marx 1964 (2000), 194. See Schivelbusch 2014 for examples of this rhetoric.

for progress, as everyone from political leaders to advertising copywriters sold their audiences the claim that they would fall behind if they failed to consume the future on offer.[96] Abrupt, rapid change (such as the mutation theory offered) captured the look and feel of the future, while slow and steady change was increasingly associated with the past.

Interpreting Science

The new media markets were both a visible symptom and a driver of the twentieth century's rapid pace of change. The growing market for writing about science meant that, as we shall see in later chapters, experts like de Vries struggled to offer the public a correct interpretation of their science, not least because readers had become active consumers in the scientific marketplace and were deciding for themselves which intellectual products to trust. Although individual readers' opinions are usually irrecoverable, historians can identify communities of readers with shared assumptions and values, which makes it possible to analyze the range of possible interpretations that were acceptable within specific communities.[97] For example, chapter 7 examines socialist periodicals, which debated the mutation theory's ability to advance the socialist cause; some scientific experts might have been appalled, but many socialist readers would have regarded an overtly political interpretation of science as the most (or, indeed, the only) reliable one. However, a common political or religious affiliation was not the only foundation for a community of readers; they also formed around a shared interest in a specific genre of writing, such as SF. The marketing of particular kinds of writing to specific audiences helped shape such communities and, as chapter 8 explores, readers also shape their genres. SF soon developed a distinctive visual style, which promised would-be readers more of the kind of thing they liked, just as literary fiction by authors like Wharton was promoted in ways that made similar promises to a different set of readers, and the same was true of journalistic accounts of the latest science. As noted earlier, genres teach audiences both what and how to read. Genres thus have a regulatory aspect, similar to codes of etiquette; by helping readers decide how they ought to read, they also suggest how readers might think and act.[98] Genres thus provide another tool with which historians can identify the

96. Haraway 1997, 41–42; Jack 2006.
97. For reading communities and their roles, see Secord 2000. See also note 26 above on communities of interpretation.
98. Dubrow 1982, 31.

range of interpretative possibilities likely to be available within a specific community of readers.

Interpreter is a deliberately broad term, not least because, as André Carrington puts it, "every interpretive act is an act of authorship, and every act of authorship is an act of interpretation."[99] Since this book's focus is on public science, *interpreter* will be used to describe those who shared their interpretations, usually by publishing them (although lecturing, broadcasting, and filmmaking are briefly considered in later chapters). Interpretation usually began with translation—simply substituting a familiar term for a complex one (or using an established term, such as *sport*, in place of a novel one like *mutant*). However, even the most ostensibly objective translation altered the meaning of the original text. A familiar term like *sport* looked back to earlier meanings, while the novelty of *mutant* looked forward and might make the text more exciting or disconcerting. And, of course, translating scientific writing also entailed finding accessible analogies; interpreters of mutation sometimes compared the appearance of new kinds of plants to the transmutation of chemical elements, evoking unfamiliar ideas such as radioactivity.[100] Interpretation invariably entailed translation, explanation, and—as each interpreter constructed their own meanings—a degree of creativity. Novel ideas usually require some degree of interpretation before other scientific experts can understand them (never mind wider publics). In cases like experimental evolution, where there was no obvious community of experts to judge the validity of new ideas, the discoverer was often forced to become their own interpreter, as de Vries did when he decided to communicate his ideas to lecture audiences at the University of California by using Luther Burbank's work to provide evidence and examples (chapter 4).[101] The claims of really novel sciences are inevitably evaluated in public, and consensus emerges (if it emerges at all) through dialogue between different audiences.

The very process of explaining ideas to wider publics encouraged readers to express their own opinions, to join the conversation over the new science's claims to legitimacy. Journalists generally tried to extend their audiences through such techniques as finding arresting images or analogies with which to communicate the novelty of experimental evolution. Similar creative methods were employed by textbook writers, who needed to compete for the attention of school-textbook purchasing committees

99. Carrington 2016, 7, 9.

100. Campos 2015.

101. For example, de Vries used Burbank's experiments to illustrate his own concepts, such as elementary and compound species (De Vries 1905c, 116–18).

(particularly in the USA, where states would often mandate a specific textbook for every school in its jurisdiction; see chapter 6). Textbook publishers, editors, and authors competed to produce the most readable and up-to-date texts, which affected their choice of everything from specific terminology to the examples and illustrations used. Even in the sober world of the textbook, interpretation was always a creative process. Creativity became even more evident when interpreters appropriated new biological ideas to imagine radically different futures, as when the women of Gilman's *Herland* attributed their unique, parthenogenetic culture to "to the law of mutation" (chapter 7). As later chapters show, interpreters could be arranged along a spectrum, with attempts at objective translation at one end and unabashed creativity at the other. As a result, many early twentieth-century readers of biology probably thought of interpreters as more or less interesting, rather than as true or false. Experimental evolution emerged out of these conversations between equally diverse sets of interpreters and readers; early twentieth-century biology came to function as a shared public culture—a set of tools with which to imagine the future.[102]

The Arrival of the Fittest tells the largely unknown history of mutation theory's role in creating an enormously influential public culture around biology that shaped the early twentieth century. Analyzing the biotopian style of speculation and its lasting influence may help us to think about science's potential role in imagining our own future (if any). I will return to that possibility in the epilogue, but first we need to understand more about the mutation theory itself and examine its initial reception.

102. I am indebted to Jan Golinski (1999) for the idea of science as public culture.

1

Undisciplined Futures

The American family magazine the *Youth's Companion* was a typical early twentieth-century popular periodical—cheap, accessible, and appealing. On August 1, 1901, its "Nature and Science" column began with an item headed "New Light on the Origin of Species," which informed readers that Hugo de Vries had made a "momentous discovery"—that "new species appear suddenly by mutation," not gradually as Darwin had believed.[1] In a single, brief paragraph, de Vries's mutation theory received its first coverage in the US national press—not in a specialized scientific journal, but in a best-selling magazine aimed primarily at Christian families.[2] As we shall see, "New Light" was short on details and rather ambiguous; it was sketchy and unfinished, almost demanding that readers complete the story for themselves.

The media market in which mutation made its debut was crucial to understanding the theory's later career. In 1885, America's four best-selling monthlies had all cost between 25 and 35 cents and had a combined circulation of six hundred thousand; by 1905 there were twenty comparable magazines, but their price had fallen to 10–15 cents, while their combined circulation had passed 5.5 million.[3] By 1914, US printers were publishing fourteen billion publications every year—99 percent of which were newspapers and magazines.[4] And by 1920, every home in America received on

---

1. "New Light on the Origin of Species" (1901).

2. I have located a couple of brief mentions that appeared slightly earlier: *Sioux City Journal* (Sioux City, IA, March 7, 1901: 9); and *Arkansas Democrat* (Little Rock, Pulaski, AK, July 31, 1901: 5).

3. Mott 1957, 8.

4. US *Census of Manufactures,* cited in Thompson 2019. US literacy was 95.7 percent by 1930 (see Wilbur Schramm, ed., *Mass Communications,* 2nd ed. [Urbana-Champaign: University of Illinois Press, 1960]: 112). While 48 percent of US citizens read books, 69 percent read magazines and 82 percent newspapers (Angus Campbell

FIGURE 1.1. "Why Is a Neighbor's Newspaper Always More Attractive Than Your Own?" As the number of competing periodicals grew, they became more specialized, leading readers to worry whether they were missing out, no matter what they read.

average at least one newspaper.[5] Similar transformations were underway in Britain, where new titles such as the *Daily Mail* (f. 1896) opened up new markets (particularly by attracting more women readers) and were soon developing a new language with which to address them.[6] As one contemporary commentator noted, magazines were being born at a rate "to make Malthus stare and gasp," yet their indecent growth continued as ever-larger numbers of words were being put into ever more hands at ever-decreasing prices.[7]

Scribner's magazine (with over two hundred thousand readers) was typical of the new magazines that emerged during these decades.[8] It published many of Edith Wharton's stories, including "The Descent of Man," which reflected explicitly on how the public made sense of science in the age of mass-market periodicals. Wharton's story concerned

and Charles Metzner, *Public Use of the Library and Other Sources of Information* [Ann Arbor: University of Michigan, Institute for Social Research, 1950]: 1–14). Schramm and Campbell and Metzner are cited in Cheng 2012 (64).

5. Aurora Wallace, *Newspapers and the Making of Modern America: A History*, 1st ed. (Westport, CT: Greenwood Press, 2005): 3, quoted in Chalaby 1998 (183–91).

6. LeMahieu 1988; Matheson 2000, 558–59; Bowler and Morus 2005.

7. "Some Magazine Mysteries" 1895, 342.

8. Mott 1957, 11, 717–32.

a fictitious Professor Linyard, a naturalist and evolutionary theorist who was nostalgic for the days when the audience for "the man of science" had consisted only of "fellow-students." He was appalled by the way in which that "little group had been swallowed up in a larger public." Far too many people now read scientific books ("and expressed an opinion on them"), but many more didn't even bother with books, preferring to base their opinions on the "Scientific Jottings" columns found in every daily paper (exactly where the mutation theory made its US debut). As a young man, Linyard had regarded science as an "inaccessible goddess," but she now "offered her charms in the market-place" and as a result of this prostitution, was "not the same goddess, . . . but a pseudo-science masquerading in the garb of a real divinity" (6–7).

Linyard's obvious distaste for public science was reflected in his musings on its effects ("babies were fed and dandled according to the new psychology"). Such fashionable innovations irritated him, but his real hatred was reserved for popular scientific books, which "filled him with mingled rage and hilarity." They fed the public's desire for scientific knowledge with a debased, commercialized substitute, which prompted Linyard to produce *The Vital Thing*, "a skit on the 'popular' scientific book." He wrote it in the standard pseudo-religious, uplifting tone and piled on the fallacies, clichés, and false analogies until the result was so grotesque "that even the gross crowd would join in the laugh." (7). Despite being intended as a parody of popular evolutionary epics, his publisher persuaded him to market it as a serious book. Linyard agreed, convinced that the "elect would understand" his joke (even if "the crowd would not").

The book was energetically pushed using various ingenious strategies that echo twenty-first-century viral marketing ("Slips emblazoned with the question: *Have you read the 'Vital Thing'?* fell from the pages of popular novels and whitened the floors of crowded street-cars"). Linyard assumed that the eventual revelation that ordinary readers had been thoroughly bamboozled would convince everyone of the need to re-establish the appropriate boundaries between the scientific elite and the vulgar masses. He initially feared that his scientific colleagues would realize the book was a parody and spoil the joke, but his publisher reassured him that "even in this reverberating age the opinions of the laboratory do not easily reach the street." (18). He was right; Linyard's scientific colleagues did not read popular books. Meanwhile the press praised *The Vital Thing* so highly that it became a bestseller, taken seriously by its readers. The book was so expertly reviewed in one leading daily paper that Linyard himself could not help feeling that his ludicrous fallacies appeared "admirably as truths." (And these apparent truths were rapidly disseminated because

other periodicals simply copied or rehashed key details from the first review, rather than paying "an expert to 'do' the book afresh."[9]) Wharton's
fictitious author of a fictitious book is disturbed by that fact a fictitious
reviewer has mistaken his parody for scientific fact; her playful approach
to genre and her description of her period as the "reverberating age" captured how printed words became amplified and distorted as they were
reflected back and forth from one kind of publication to another (which,
as we shall see, regularly happened with real-world science).[10] Her story
illustrated how science became part of larger public conversations in the
new age of mass media, but most importantly it highlighted how difficult it
had become to decide who could participate in those conversations—and
who to take seriously.

Previous historians of this period's biology have generally shared Linyard's assumption that they knew what science was and where to look for
it: science is what scientists do—and its successes or failures are ultimately
judged by other scientists. Making those assumptions has reduced the
mutation theory to a footnote (at best), because the newly formed community of professional geneticists rapidly decided that de Vries had been
mistaken in his claims.[11] In standard accounts, mutation theory did little
more than lead some biologists astray, thus delaying the acceptance of
what are now regarded as better theories; the distinguished biologist and
historian Ernst Mayr went so far as to say that those scientists who "had
become imprinted with the de Vriesian mutation concept had to die out"
before modern conceptions could take hold.[12] Mayr is typical of those
historians who have judged the past in the light of what scientists currently
believe (and whose histories focus on the work of elite scientists). As a
result, biology's historians have failed to notice that de Vries's accelerated

9. Wharton 1904, 23; Ohler 2014, 116.

10. Wharton's friend Henry James had satirized aspects of the world being created
by these new media in *The Reverberator* (1888), which concerned a scandal-spreading
cheap newspaper of the same name. Wharton probably had this usage in mind when
she used the phrase.

11. See Cleland 1972; Endersby 2007, 128–69.

12. Mayr 1980 (1998). In the same volume, William Provine (1980b [1998], 55–56)
lists the "refutation" of the mutation theory as one of the major contributions genetics
made to the synthesis. Jan Sapp (2003, 145) also refers to the evidence from *Oenothera*
as something that "required refutation before the synthesis" could occur. And James
Schwartz (2008, 76) describes de Vries as having been led "astray" by *O. lamarckiana*.
Serious studies of mutation theory were pioneered by Allen (1969) and Bowler (1978),
and in recent years, several scholars have begun to reassess the mutation theory's
impact. See, for example, Kimmelman 1983; Theunissen 1998; Stamhuis, Meijer, and
Zevenhuizen 1999; Campos 2010; Stamhuis 2015.

evolution theory captured the imaginations of all kinds of people, from Edith Wharton to self-educated Utah socialists, from pioneering science fiction writers to theosophists. Mutation theory was central to the creation of experimental evolution and thus to many of the wider cultural shifts described in this book; its erasure from history has dramatically limited our understanding of biology's wider role in these decades and hence its long-term cultural impact.

The reception of de Vries's books (and the claims they contained) might usefully be compared with the fictitious careers of Linyard's *Vital Thing* or Dredge's *Arrival of the Fittest*, books that were pushed and puffed, reviewed and quoted (often by those who had not read them), which evaded their creators' intentions and took on lives of their own. (As when some misread *The Vital Thing* because they misunderstood its genre—and took the joke too seriously.) True to form, the few historians of science who have discussed mutation in detail have focused on de Vries's most significant scientific work, *Die Mutationstheorie* (1900, 1903), which described his many years of plant experiments, particularly those with *Oenothera lamarckiana*, a member of the evening-primrose genus. This detailed, 1,200-page book full of graphs and technical terms was arguably the definitive account of the theory (its forbidding bulk doubtless persuaded many who would never read it that there were lots of facts to back up de Vries's claims), but it could never have created a sensation. It was press reports, like the one in the *Youth's Companion*, that attracted unexpected and inappropriate readers, who spread the word and generated excitement by publicizing de Vries's claim that evolution was fast: new species appear "suddenly, without preparation or intermediate forms." And many interpreters promoted his view that natural selection played only a minor role in evolution, since "once formed, the new species are as a rule at once constant. No series of generations, no selection, no struggle for existence are needed."[13] Mutation explained the arrival of the fittest.

However, what created the most press interest was not de Vries's explanation of past evolutionary change but his theory's implications for the future. As numerous scholars have noted, forecasting and imagining are performative; they can make the anticipated future more, or less, likely to occur.[14] Koselleck argues that every prediction implies that the "previously

---

13. De Vries 1902b.
14. E.g., Kaushik Sunder Rajan, in *Biocapital: The Constitution of Postgenomic Life* (Durham, NC: Duke University Press, 2006), argues that speculative fictions of all kinds are part of a "promissory futuristic discourse" that underpins biotechnology start-ups. Quoted in Vint 2021, 9. See also N. Brown, Rappert, Webster, and Adam 2000.

existing space of experience is not sufficient for the determination of the horizon of expectation." Expectation and experience are dynamically related—changing one changes the other.[15] The claim that experiments would reveal the precise cause of mutation led to the prediction that new, artificial mutations must be possible, allowing new organisms to be built to meet human needs. To help such imaginary futures become a reality, the newly established Carnegie Institution of Washington (one of the first and wealthiest charitable foundations to support US science) established a new Station for Experimental Evolution (at its existing research station at Cold Spring Harbor, Long Island). At the new institution's opening ceremony—where de Vries was the guest of honor—the station's first director, Charles Benedict Davenport, described *Die Mutationstheorie* as "the most important work on evolution since Darwin's 'Origin of Species,' a work destined to be the foundation stone of the rising science of experimental evolution."[16] (Precisely the same comparison that had been made with Dredge's *Arrival of the Fittest*.) Thanks to de Vries's ideas, Davenport confidently predicted that biologists would eventually be able to engineer an "improvement of the human race."[17]

The scientists' initial excitement over mutation seems self-explanatory; it apparently resolved existing theoretical problems while suggesting lots of interesting new experiments. However, had the theory only interested scientists, it would be of little interest to the rest of us; its historical significance emerged from the fact that scientists and the public were learning about mutation's exciting implications at exactly the same time—and often from the same sources. The brevity and hyperbole of the earliest media reports were an invitation to interpretation. Experimental evolution was not developed and tested behind closed laboratory doors and then popularized; nor was it merely created and interpreted simultaneously—its interpreters were active participants in its creation. And this all happened in the full glare of unprecedented, early twentieth-century media attention, ensuring a much wider conversation about its significance and the futures it apparently foretold. The result was a culture of public science that largely ignored the supposed boundaries between the scientific elite and laypeople; everybody could—and did—join in the discussions about the kinds of futures that biology seemed to promise.

15. Koselleck 1985, 275–76. See also Hilgartner 2015, 34; Campos 2012.
16. Davenport, Jones, Billings, and De Vries 1905.
17. Carnegie Institution of Washington 1907, 92.

New Light

The *Youth's Companion* was typical of the ways in which industrialization transformed America and its periodicals. The magazine had been founded as a strictly Calvinist publication, with a suitably stern moral tone, crusading energetically against the various temptations that beset America's children in the age of rapid urbanization. However, as the numbers of children attending school grew, there was less concern about them roaming the city streets unsupervised. Moral crusades began to look a little old-fashioned and, perhaps under pressure from competing magazines, the *Companion* adopted a nonsectarian tone (while remaining Protestant), which it combined with an ever-more-businesslike approach to expanding its readership. After the Civil War, the paper used the new printing, publishing, and marketing technologies to build a circulation of over five hundred thousand; only the second magazine in America to achieve this feat.[18] (The *Ladies' Home Journal* had been the first—a reminder that these expanding readerships included many who had previously been excluded from public conversations.)

The *Companion's* success had been built on a distinctively American combination of religious fervor and hard-headed commercialism.[19] Alongside entertaining short stories with unobjectionable morals were adverts for patent hay-fever and toothache cures, improved fountain pens, brass-band instruments, and soaps. Similar ads appeared in the *Atlantic Monthly* and *Scribner's* as the railroads and cheaper postage created national readerships and a national advertising market. Popular magazines relied on adverts for the latest consumer items—such as sewing machines, typewriters, and pure, healthy, or novel foods—many of which embodied (or claimed to embody) some form of technological innovation. National magazines created a genuinely national American conversation, built directly on the production and sale of mass-produced commodities.[20] The *Companion's* readers were typical of those who, for a few cents a week, could enjoy a miscellaneous combination of short fiction, poems, improving maxims, and humorous anecdotes, combined with opportunities to join the new consumer culture—all infused with wholesome family values.

The *Youth's Companion* reported the mutation theory before any of the USA's more specialist, scientific journals, and it reached a much wider audience. Mutation's impact partly depended on the coincidence of it

18. Mott 1957, 16–17.
19. Mott 1957, 17–18; Ringel 2015, 151.
20. Mott 1957, 22–23, 17–20.

arriving in the midst of an explosive expansion in the number of column inches that needed filling. Overworked journalists and editors seized on any and every story that might catch the public's attention (and the advertiser's dollars). As a result, "New Light on the Origin of Species" reappeared in a range of US newspapers over the next few months, from the pro-Republican, African American paper the *Iowa State Bystander* to the Democrat-supporting *Chickasha Daily Express* (Oklahoma). Very similar stories appeared in at least thirty newspapers across the country.[21]

The implicit invitation to interpretation, which was crucial to mutation's success, becomes clearer when we look more closely at the *Companion*'s story. This was the complete text:

> **New Light on the Origin of Species.** Prof. Hugo de Vries, the well-known Dutch botanist and biologist, is credited with a "momentous discovery" concerning the origin of species among plants. Briefly stated, his observations indicate that new species appear suddenly by mutation, never as the outcome of progressive variation. He avers that he has been able, for the first time, to watch the formation and development of new species. A reviewer of his work in the English scientific journal, *Nature*, says: "The facts are so striking and convincing that an outsider, like the reviewer, cannot but feel that a new period in the theories of the origin of species and of evolution has been inaugurated."[22]

This brief paragraph exhibited several characteristics which came to typify the way mutation theory was reported, each of which added to its appeal. It used the language of mass journalism ("momentous" and "striking"),

21. The story appeared in *Cassville Republican* (MO), September 12, 1901: 3; *Chickasha Daily Express* (OK), September 5, 1901: 2; *Cook County Herald* (MN), October 5, 1901: 2; *Democrat* (Wichita, KS), August 31, 1901: 2; *Holbrook Argus* (AZ), November 16, 1901: 3; *Iowa State Bystander*, August 30, 1901: 3; *Leon Reporter* (IA), September 12, 1901: 9; *Lincoln Republican* (KS), September 5, 1901: 3; *St. Johns Herald* (AZ), December 7, 1901: 3; *Silver Messenger* (ID), December 3, 1901: 2. (See the Library of Congress's *Chronicling America* project database at https://chroniclingamerica.loc.gov/.) For circulation figures for *Youth's Companion*, see Patrick Cox, "The Youth's Companion," *H-PCAACA*, October 22, 2014, https://networks.h-net.org/node/13784/discussions/49795/2-youths-companion. A second, very similar, story titled "According to Nature" appeared in *Clarke County Democrat* (AL), October 24, 1901: 2; *Daily Republican* (PA), September 13, 1901: 2; *Indiana Weekly Messenger* (PA), September 18, 1901: 10; *Pittsburgh Press* (PA), August 4, 1901: 11. Other brief reports appeared in *Arkansas Democrat* (Little Rock, Pulaski, AK), July 31, 1901: 5; and *Sioux City Journal* (Sioux City, IA), March 7, 1901: 9. (Again, see *Chronicling America*.)

22. "New Light on the Origin of Species" 1901.

which would be increasingly used by scientists themselves.[23] The story emphasized de Vries's importance ("the well-known botanist and biologist"), which added to his credibility. Evolution was apparently fast ("new species appear suddenly"), thus standing conventional wisdom on its head. And the theory implied that Darwin had been wrong (evolution was never "the outcome of a progressive variation")—a topical claim in the light of the ongoing Darwinian debates, which would have been of particular interest to the magazine's predominantly Christian readership. And, finally, the story claimed that "for the first time" new species had been seen emerging; evolution was moving into the lab, the place where modern science manipulated nature to produce its wonders.

The "New Light" story offers another example of bricolage, as it took information from a specialized scientific publication (the British scientific journal *Nature*) and reinterpreted it for new audiences and purposes. Quoting *Nature* gave the story an authoritative source, and yet *Nature*'s contributor referred to themselves as "an outsider."[24] Mutation was so novel, still so undisciplined, that there were no "insiders" upon whom *Nature* could have called. Experts were even scarcer in the general media, which employed few (if any) scientific specialists. Experimental evolution's claims would have to be (at least partly) judged by the wider public. Their involvement was encouraged by the fact that the account was "briefly stated"; not only were readers not expected to digest 1,200 pages of technicalities, but the story effectively acknowledged the existence of gaps—there were further facts to be found and fresh implications to be teased out. Tasks that readers themselves might perhaps undertake.

The "New Light" story appeared in many other papers (possibly via a news service or syndication, or perhaps through simple plagiarism), at a time when newspapers were being increasingly used as evidence (rather than as reports of facts). As a result, newspaper stories about mutation became facts in their own right, which could potentially be cited and quoted.[25] A similar process occurred as longer, popular articles about mutation theory began to appear; each piece of journalism became a building

23. This style of writing was still comparatively novel at the time; see Chalaby 1998, 183–91.

24. Emphasis added. The *Nature* report appeared as an editorial under the title "Recent Scientific Work in Holland," on June 27, 1901, and was signed simply "J.P.K." The only other piece under these initials in *Nature* was a review of a German book on thermodynamics (1903). The author was almost certainly Dr. Johannes Petrus Kuenen (1866–1922), professor of physics at the University of Leyden. (See his obituary, *Nature* 110 [November 18, 1922]: 673–74.)

25. Matheson 2000, 559. See also Chalaby 1998.

block in a wider argument about biology and the future. Mutation stories reverberated through the press, and serial publication created a continuing conversation about the new theory and its potential impact (while reviews ensured that books were drawn into the same conversation). Just as the *Youth's Companion* made use of science to fill its pages and boost its circulation, many other publications did the same. And to meet the expectations of their audiences, some gave the new theory an aura of vaguely spiritual optimism. Like the fictitious Linyard's *Vital Thing* in Wharton's story, de Vries's idea was soon being interpreted in ways its author had never foreseen.

Genre-Bending

The timing of mutation theory's debut helped make it famous, thanks to the media hunger for novelties, but its arrival also coincided with the new century's optimistic mood. Despite the fears of degeneration that preoccupied many, particularly in Europe, these decades also saw a growing preoccupation with newness, evidenced by discussions of everything from the new journalism to the new woman.[26] While some Europeans were pining for past glories, Americans were more likely to be excited by the future. The young country's inventive, entrepreneurial spirit was epitomized by Thomas Alva Edison, known as the Wizard of Menlo Park, whose fame contributed to Americans beginning to dominate the world's patents—tangible evidence that the American century was dawning.[27] By 1900, the press was regularly reporting how breeders like Luther Burbank were extending the American spirit of innovation into the garden and farm (see chapters 3 and 4; figure 9.1). Widespread talk of "inventing" new plants was crucial to the context in which de Vries's discoveries were reported.

Just a few months after the first brief reports of mutation appeared in the US press, the successful monthly *Everybody's Magazine* devoted seven illustrated pages to a detailed account of the "seemingly magical" achievements of de Vries. Their article included a portrait of the "eminent" Dutchman, alongside photos of his experimental gardens and drawings of his plants, which illustrated de Vries's argument "against Darwin . . . that a new variety has a spontaneous birth, and not a gradual one."[28] According to the magazine, de Vries "has been experimenting, checking up Darwin,

26. Ledger and Luckhurst 2000, xiii–xiv.
27. Inkster 2003.
28. Lyle 1902, 599.

correcting him in instances, and thereby adding to the known code of nature's laws." (597). According to de Vries, evolution was no longer slow—new species appeared "by a leap or spurt that . . . is complete as soon as it comes into existence." As a result, he had "seen the originating of new species." (597). The author was Eugene Lyle, a journalist and pulp-fiction writer (mainly of Westerns) who does not appear to have written anything else on scientific topics.[29] His style emphasized the exciting possibilities—and dangers—of de Vries's achievement: "Creating a new race of plants, a race that has never lived before" entailed "molding at will the mysterious, terrible, and elusive forces of life."[30] Despite the hint of *Frankenstein*, the overall impression of de Vries was distinctly utopian; he had accelerated evolution, achieving "a terrific jump" to hasten nature and accomplish "within a few generations of a plant or animal what she [nature] might do only after unthinkable eras of time by natural selection, if at all." This article was one of the first to outline the biotopian possibilities of mutation, but it would not be the last; over the following decades, the theory would be regularly credited with the power to speed up and expand nature—creating new organisms that might otherwise never have occurred.

The *Everybody's* story juxtaposed poetic language ("a fairy romance in the flowery kingdom") with technical terms ("ligulate or strap-shaped florets") and referred to the theory by its German name, as *Mutationstheorie* (600). Readers would doubtless have had a sense of being privy to "proper" science, not an oversimplified version of it. When de Vries was quoted as saying that the "study of species and varieties is bringing about a momentous revolution in ideas," *Everybody's* readers doubtless felt they were witnesses to, perhaps even participants in, that revolution (601). The new ideas had practical import too; de Vries's new breed of clover, for example, was being tested by the Minnesota Agricultural College in the hope that it would prove better than ordinary clover (597–98). The ways in which de Vries and mutation were discussed in these early, popular articles—an eclectic mix of hyperbolic metaphor, scientific jargon, and practical payoff—rapidly came to typify the ways in which the new biology was being discussed.

When the "Plant Making" article appeared, *Everybody's Magazine* had an audited circulation of almost 160,000 a month; by contrast, the only detailed discussion of mutation that had appeared by this point in the US

29. Lyle was in Europe as a correspondent for *Everybody's* from 1902 to 1905 and visited de Vries; the article was clearly based on an interview.

30. Lyle 1902, 596.

specialist-science press was a three-page review in the *Botanical Gazette* (whose circulation would have been a few hundred, at most).[31] News of de Vries's work reached most people through general-interest articles, usually infused with an optimistic mood. It might be assumed that once scientists learned of de Vries's work, they would have turned to reputable scientific journals to discover more, but in reality the popular and scientific genres were almost indistinguishable. For example, in 1905 Winthrop John Van Leuven Osterhout, a professor of botany at University of California (figure 6.1), wrote a serious textbook aimed at school science teachers, *Experiments with Plants*, in which he recommended the article in *Everybody's* as further reading for those interested in mutation.[32] Osterhout's book, and textbooks more generally, will be discussed in chapter 6, but the use of an article from a 10-cent monthly as recommended reading for high school and university science classes is another form of bricolage—and of the boundary between two supposedly separate genres being breached. Osterhout evidently thought the article in *Everybody's* was accurate enough to recommend (and since he knew de Vries personally, was presumably confident that the Dutch professor shared his judgment). The magazine's writer had interviewed and quoted de Vries, so his article became a legitimate source of scientific knowledge. As a result, the biotopian tone of the *Everybody's* article would have acquired scientific status as it spread—via Osterhout's textbook—to students and their teachers, who thus learned to think about mutation in terms of its "seemingly magical" ability to achieve what nature might never have done.

The link between Osterhout's textbook and the article in *Everybody's* was not unusual; many mutation stories shared a common style and content, regardless of where they initially appeared. Nevertheless, as stories were reinterpreted for fresh audiences, their emphasis shifted according to each editors' sense of their market and readers. A useful example is a lengthy, rather technical, review of the present state of Darwinism, which was published in *Harper's Magazine* in February 1903. "Darwinism in the Light of Modern Criticism" was written by Thomas Hunt Morgan, then a biology professor at Bryn Mawr College in Pennsylvania, who was an early enthusiast for de Vries's theory (and would later become famous for his work with the fruit fly *Drosophila* at Columbia University).[33] Reflecting the magazine's well-educated market, Morgan addressed *Harper's* readers in long, sometimes convoluted sentences and used technical examples

31. Reed 1997, 112; Cowles 1902.
32. Osterhout 1905, 452–53.
33. See Kohler 1994.

from his own specialization, embryology, to explain some of the criticisms then facing Darwinism.[34] Nevertheless, Morgan attempted to make his article appealing to a general audience. He mentioned his own firsthand experience of visiting de Vries (as the *Everybody's* writer had done): "No one can see his experimental garden, as I have had the opportunity of doing, without being greatly impressed," because "on all sides . . . new species . . . have suddenly appeared." Based on these experiments, de Vries had concluded that "evolution may have come about without selection."[35] Nevertheless, Morgan's article acknowledged the persuasive force of Darwin's "brilliant idea," that natural selection mirrored the effects of artificial selection. This idea had had been vital in winning "general recognition" for the fact of evolution, so the *Harper's* audience were told that de Vries had built upon Darwin's achievements, not undermined them. By contrast, a successful Missouri paper appropriated the *Harper's* story to claim that "Darwin's theory of natural selection is not being strengthened, but rather the reverse, by later experiment and study" and described Morgan as "controverting Darwin."[36] The newspaper's staff—conscious, no doubt of their Midwestern audience's likely interests—picked the attack on Darwin as the point to emphasize.

Because mutation theory was widely interpreted as discrediting Darwin, it was regularly used (particularly in the USA) to attack apparently atheistic or materialistic readings of evolution.[37] This form of critique (which will be referred to for convenience as religious, even though some of its proponents would have rejected that term), characterized another of the earliest magazine articles to discuss mutation. In 1902, the Theosophical publication *New Century* described the "considerable claims" that had "been made for a new theory in evolution (biological), known as the mutation theory of DeVries."[38] The theory was accurately described

34. For *Harper's*, including circulation and audience, see Mott 1957, 43; Reed 1997, 52–53.

35. Morgan 1903a, 477–78.

36. "Are Darwin's Theories Sound?" *Mexico Weekly Ledger* (MO), January 22, 1903, http://chroniclingamerica.loc.gov/lccn/sn89067274/1903-01-22/ed-1/seq-4/. Although the newspaper's article appears to precede the magazine's, the *Weekly Ledger* gave its source as "Harper's Magazine for February."

37. For the US reception of Darwinism, see Glick 1988; Numbers 1998; Numbers and Stenhouse 2001; Gianquitto and Fisher 2014.

38. [Edge] 1902. The author was identified only as "HTE," but it seems likely that it was H. T. Edge, who wrote other articles on evolution in the theosophical press, e.g., H. T. Edge, "Studies in Evolution," *Theosophical Path*, May 1, 1916: 496. Theosophy was a late nineteenth-century movement that claimed ancient roots, whose primary tenet was that all existence (physical and spiritual) is a unity. Its practitioners

(for example, mutations enabled organisms "to overstep the species-line," whereas natural selection produced only minor changes that "speedily relapse"). However, *New Century*'s writer argued that de Vries had merely found *"what he was looking for"* (original emphasis) in *Oenothera*, because—like other scientists—he had searched nature for examples that appeared to support his preconceived hypothesis. This error was a fatal one in the opinion of theosophists, who believed that the evolution of all organisms, including humans, occurred through a continuous process of metempsychosis (transmigration of souls), of which only the material phases could be studied by science. As a result, scientists were unable to grasp the higher dimensions of reality, but "must continue to wander in darkness," bereft of the "keys offered by Theosophy."[39] The theosophical interpretation of mutation theory was an esoteric minority view yet exemplified the wider optimism that religious and secular interpretations of evolution often shared.

The founder of modern theosophy, Helena Petrovna Blavatsky, had attacked the spiritual blindness of Darwin and other evolutionists, yet she regularly used the term *evolved* to describe the ascent from lower to higher forms at the heart of her philosophy.[40] While Blavatsky's synthesis of Eastern and Western spiritual and religious traditions was unique, her positive, teleological interpretation of evolution clearly overlapped with the widely held assumptions of stadial evolution. Biological and spiritual evolution were commonly linked by many other varieties of the broad movement often called Higher, or New, Thought.[41] This loosely defined late nineteenth-century movement, predominantly based in the USA, emerged in reaction against an increasingly banal Protestantism (which seemed to reduce Christianity to a set of platitudes that guided believers toward worldly prosperity). New Thought was characterized

---

denied the existence of a personal God, personal immortality, and the validity of the Christian revelation. Theosophical ideas spread widely in the early decades of the twentieth century (see *The Concise Oxford Dictionary of the Christian Church*, 3rd ed., s.v. "theosophy," edited by E. A. Livingstone, published online 2014, https://www.oxfordreference.com/display/10.1093/acref/9780199659623.001.0001/acref-9780199659623-e-5746).

39. [Edge] 1902.

40. See, for example, Blavatsky 1877 (2012), 2, 8, 13, 14–15.

41. This movement comprised followers of the eighteenth-century Swedish scientist and mystic, Emmanuel Swedenborg, who claimed a "doctrine of correspondence" between the physical and spiritual worlds. Some of his nineteenth-century followers interpreted evolutionary theory as evidence of "the growing tendency to regard all things as connected, and related, and moving on in the paths of an immutable order" (Giles 1887).

by what William James called "an optimistic scheme of life, with both a speculative and a practical side," with diverse European and American roots.[42] Like Theosophy, New Thought used *evolution* as a synonym for *improvement*—physical, mental, and spiritual. A typical example was *New Reading of Evolution* (1907), written and self-published by Henry Clayton Thompson, who combined a miscellaneous range of sources into what he claimed was a new "Synthetic Philosophy of Individual and Social Life." While there is no evidence that Thompson's *New Reading of Evolution* sold well, it nevertheless offers a useful example of the way one ordinary reader attempted to make use of biology and evolution to find a sense of meaning in life. Among the ingredients incorporated into Thompson's bricolage was a quote from de Vries: "By natural selection species do not originate but perish." From this quote, Thompson (like many better-known writers) concluded that natural selection "really says very little about the origin of variations," which left the arrival of the fittest as an unsolved mystery.[43] Thompson's solution was that claim that "evolution is accomplished primarily by . . . Individual Intelligences coöperating with the Universal Intelligence with more or less harmony," which provided the foundation for his philosophy of individual growth and progress, reinforced by his firm conviction that the "general course of Evolution is upward and onward" (and would culminate in immortality, once humans had evolved to exist on "higher planes").[44]

Despite its title, Thompson's "New Reading" of evolution was distinctly familiar; as the *Washington Herald* noted, the book's "cheerful optimism is strongly suggestive of the new thought school." The *Herald*'s reviewer admitted Thompson's point (some scientific men believed natural selection might not account for the appearance of every living thing) but took exception to Thompson's anti-Darwinian conclusions. Not only had de Vries "witnessed the evening primrose" transforming "by mutation to another species," he had "convinced practically all biologists that new species may be formed by mutation," which the reviewer believed was sufficient to defend a purely materialistic theory of evolution.[45] Another reviewer

---

42. James 1902 (2011), 94–95, 91; Harley 1991; Lears 2009, 237–38.

43. Thompson 1907, 48, 50–51. Thompson gave the source of his quotation as *The Origin of Species by Mutation*, which was the title of an article by de Vries in *Science* (1902b). However, these words do not occur there, but can be found in Charles A. White's detailed report on de Vries for the Smithsonian Institution, in which he quoted sections of *Die Mutationstheorie* (presumably his own translations, as the book had not yet appeared in English). See C. A. White 1902a, 634.

44. Thompson 1907, 115–16.

45. Crane 1909.

sounded like Wharton's Professor Linyard, when they complained that the book's author appeared to have "no scientific or specialized training of any kind" and had presumed to philosophize on the basis of "having read a few books, mostly on biology and evolution."[46] Such attempts to restrict debate to the recognized elite were futile; science had become a part of wider public discussions. As unexpected readers and readings multiplied rapidly and unpredictably, neither reviewers nor scientists could control where the conversation would lead.

The connections between New Thought and popular evolutionary debates went deeper than isolated works by self-published autodidacts. The first English-language edition of *The Mutation Theory* was published in the USA by Open Court Publishing, which had been established to produce the *Open Court* magazine, dedicated to the "Religious Parliament Idea" (to unite all faiths against irreligion) and "to the work of conciliating religion with science."[47] The magazine and company published on typical New Thought topics, such as Eastern religions and culture, but also produced a "Religion of Science Library." This series included not only de Vries's works (including *Species and Varieties* and *Plant-Breeding*) but major biological texts by internationally renowned evolutionists (including August Weismann, George John Romanes, Theodor Eimer, Carl von Nägeli, and Edward Drinker Cope), as well as more obviously esoteric works such as Woods Hutchinson's *The Gospel According to Darwin* (which argued that "far from destroying or antagonizing the religious instinct, the spirit of worship, Darwinism broadens and quickens it").[48] De Vries wrote for *Open Court*, describing some of the "novelties destined for the immediate future" that Luther Burbank was creating, which might shed light "on the latest results of biological investigation," such as his own theory.[49] And it was the *Open Court*'s review of de Vries's work which first brought the phrase "arrival of the fittest" into the scientific debates around mutation (see p. 1).[50] The celebrated Scopes "Monkey Trial" has helped foster a simplistic "science vs. religion" narrative (see chapter 6), which makes the many links between evolution and various forms of religion seem

46. Helleberg 1908.

47. The magazine had emerged from the 1893 "World Parliament of Religions" (held in Chicago to coincide with the World's Columbian Exposition), whose main mover was a Swedenborgian, Charles Carroll Bonney, who hoped to unite all faiths against irreligion. See the Open Court archives, South Illinois University, Carbondale, IL, https://opensiuc.lib.siu.edu/ocj/aimsandscope.html.

48. Hutchinson 1898. The series was advertised in the back of Cope 1896 (1904).

49. De Vries 1906a, 645, 643.

50. Harris 1904.

surprising. However, the complex reality of early twentieth-century attitudes to evolution is better represented by William James's observation that the "idea of a universal evolution lends itself to a doctrine of general meliorism and progress," and was thus one that "fits the religious needs of the healthy-minded so well that it seems almost as if it might have been created for their use."[51] In reality, this interpretation *had* been largely created by its readers, who (like Henry Thompson) read a variety of sources, interpreted them, and then pulled together their own theories. For such readers, faster evolution—thanks to mutation—seemed to promise more-rapid spiritual progress.

Meanwhile, many Americans retained the belief that science and religion were necessarily opposed. As we have seen, those with conventional religious views tried to use de Vries and his theory to revitalize their long-standing critique of Darwinism, a trend that became particularly apparent when de Vries made his first visit to the USA. On September 21, 1904, de Vries addressed a large scientific audience at the International Arts and Science Congress (held in Missouri as part of the St. Louis World's Fair). Newspaper reports of his speech helped bring mutation to a much larger audience. The *St. Louis Republic* published an illustrated front-page report of the meeting that described de Vries as being at "the head of modern biologists" and told its readers he had "electrified" the congress by openly combating Darwin's theory and "advancing a theory of his own."[52] Under headlines such as "Darwin's Theories Assailed by Scientists' Congress" and "DECLARES DARWIN WRONG," the reporter explained that this was the first time de Vries's ideas had been placed before the general public. His detailed account of his "experiments and methods of research" had persuaded "many of those who heard him" that his ideas would "ultimately supplant the Darwinian theory." The *Republic*'s story acknowledged that de Vries did not "wholly discredit" Darwin but had argued that natural selection was "only a seine," a net that caught life's failures and not "a direct force of nature." Still, the story's main claim was obvious—de Vries challenged Darwin.

The claim that de Vries had undermined Darwin spread rapidly as the local paper's story was widely republished. For example, San Francisco's paper the *Call* ran it under the headline "Famed Botanist Disagrees with

51. James 1902 (2011), 94–95, 91.

52. Unless otherwise specified, all newspaper accounts are from the Library of Congress's *Chronicling America* project: "Darwin's Theories Assailed by Scientists' Congress," *St. Louis Republic*, September 22, 1904: 1–2, http://chroniclingamerica .loc.gov/lccn/sn84020274/1904-09-22/ed-1/seq-1/.

FIGURE 1.2. De Vries's claims were regularly interpreted as disproving Darwinism, often in strikingly imaginative ways.

Darwin's Theory," and the *San Francisco Chronicle* told its readers that "there is no evolution such as Darwin taught, in nature . . . the change from one species to another is immediate and abrupt. The law of nature, says de Vries, is not evolution but mutation."[53] The writer's view that Darwin's theory had been "flatly contradicted" was succinctly summarized in the article's opening sentence: "MAN didn't slowly evolute from the monkey after all." The same conclusion was reached by a writer in the *Nebraska Independent*, who argued that since de Vries had eliminated the "connecting links" in evolution, he "discards the monkey origin of man. Man appeared by a sudden 'mutation' and not by the billion year process of Darwinian evolution."[54] Ignoring what de Vries had actually said, several US newspaper writers and readers assumed that if "Darwin was wrong," the apes had

53. "Famed Botanist Disagrees with Darwin's Theory," *Call*, September 23, 1904: 1, http://chroniclingamerica.loc.gov/lccn/sn85066387/1904-09-23/ed-1/seq-1/; Wilson 1904.

54. "News of the Week," *Nebraska Independent* (Lincoln, NE), March 30, 1905, http://chroniclingamerica.loc.gov/lccn/sn88086144/1905-03-30/ed-1/seq-4/.

been pruned from our family tree. The media's almost insatiable demand for stories ensured that within a few months the original *St. Louis Republic* story had been widely republished—unchanged apart from its headline. It was headed "Goes Against the Theory of Darwin" by Minnesota's *Saint Paul Globe*, while Ohio's *Perrysburg Journal* was one of at least six papers that ran it under the headline "Attacks Darwin's Theory."[55] And Utah's weekly *Intermountain Catholic* flatly told its readers that de Vries "Does Not Believe in Evolution."[56] The paper had discussed de Vries's theory the previous year in a front-page story that expressed the long-standing Catholic position that scientific truths cannot conflict with religious ones, since both are God's work. Hence, any who claim to find such a conflict are apostates, like the "men of science and infidel philosophers [who] have directed their investigation and reasoning in their war against God and religion," such as Herbert Spencer and Darwin who denied the truth of creation. "But what of it?" the paper asked. "The latest discoveries show Darwin's theory to be out of date. A Dutch scientist, Hugo de Vries, who bears a national reputation as a naturalist, has, during this year, upset Darwin's theory of natural selection."[57]

The widespread debates over natural selection's precise role within evolution was utilized by some writers to interpret de Vries's criticisms of natural selection as an attack on evolutionary theory itself. For example, Eberhard Dennert's book *At the Deathbed of Darwinism* (see p. 18), used scientific controversies over the mechanism of evolution to claim that natural selection had been discredited.[58] The German edition of Dennert's book appeared in the same year as the second volume of de Vries's *Mutationstheorie* (1903), so Dennert had been unable to include de Vries in his

55. "Goes Against the Theory of Darwin," *Saint Paul Globe*, September 23, 1904: 4, http://chroniclingamerica.loc.gov/lccn/sn90059523/1904-09-23/ed-1/seq -4/. "Attack's Darwin's Theory" appeared in *Perrysburg Journal*, October 21, 1904: 7, http://chroniclingamerica.loc.gov/lccn/sn87076843/1904-10-21/ed-1/seq-7/; *Cameron County Press* (PA), October 27, 1904, http://chroniclingamerica.loc.gov/ lccn/sn83032040/1904-10-27/ed-1/seq-2/; *Topeka State Journal* (KS), September 22, 1904, http://chroniclingamerica.loc.gov/lccn/sn82016014/1904-09-22/ed -1/seq-5/; *Barbour County Index* (KS), October 26, 190, http://chroniclingamerica .loc.gov/lccn/sn82015080/1904-10-26/ed-1/seq-3/; *Starkville News* (MS), November 4, 1904, http://chroniclingamerica.loc.gov/lccn/sn87065612/1904-11-04/ed-1/ seq-1/; and *Kinsley Graphic* (KS), October 21, 1904: 7, https://www.newspapers.com.
56. *Intermountain Catholic* (Salt Lake City, UT), October 15, 1904: 3, http:// chroniclingamerica.loc.gov/lccn/sn93062856/1904-10-15/ed-1/seq-3/.
57. *Intermountain Catholic* (Salt Lake City, UT), August 15, 1903, http:// chroniclingamerica.loc.gov/lccn/sn93062856/1903-08-15/ed-1/seq-1/.
58. Dennert 1904, 28.

list of Darwin's critics, but when it was translated into English in 1904, the
Catholic priest Edwin Vincent O'Hara (later to become one of the USA's
more prominent bishops) wrote an introduction that updated Dennert's
evidence for "the decadence of the Darwinian theory" by adding de Vries
to the list of Darwin's critics. O'Hara asserted that the mutation theory
directly contradicted Darwinian natural selection, and quoted de Vries
as saying that the small variations Darwin had relied on "cannot overstep
the limits of the species . . . still less . . . lead to the production of new,
permanent characters"—and thus could not create new species.[59] O'Hara
renewed his attack on Darwinism the following year, when he again described the scientific controversies over evolution's mechanism to argue
that evolutionary theory as a whole was being abandoned by scientists.
He claimed de Vries was leading "the greatest revolt against Darwinism"
and listed an impressive array of European and American scientists who
had adopted the mutation theory (thus, paradoxically, contributing to the
fame of de Vries's idea). O'Hara quoted de Vries's view that species were
generally "invariable" (which perhaps hinted at the possibility that their
sudden transformations were indeed miraculous). With natural selection
dismissed as no more than "a speculative hypothesis," O'Hara concluded
that it was clear that science could not dispossess "God of his universe."[60]

O'Hara's anti-Darwinian arguments were another example of bricolage, and he demonstrated a detailed knowledge of contemporary biological debates in making it. To support his claim that scientists were rejecting
evolution, he listed the apparently growing number of American men of
science who had embraced de Vries's ideas (including Charles A. White,
of the Smithsonian Institution, and Orator F. Cook, at the US Department
of Agriculture's Bureau of Plant Industry, both of whom are discussed in
later chapters). In addition, prominent University of Chicago biologist
Jacques Loeb was quoted as saying that "the work of Mendel and de Vries,"
rather than that of Darwin, marked "the beginning of a real theory of
heredity and evolution."[61] Given all this skepticism, O'Hara concluded
that with "its foundation gone, what is the Darwinian world view but a
castle in the air?"[62] As evidence, he noted that Thomas Hunt Morgan
was a strong supporter of de Vries and quoted Morgan's *Evolution and Adaptation* (1903; see p. 208) to show that many American scientists believed

59. O'Hara 1904, 18.
60. O'Hara 1905, 722–23, 725, 728.
61. Loeb 1904, cited in O'Hara 1905, 725–28. For more on Loeb, see Reingold 1962;
Pauly 1987; Turney 1998, 67–72.
62. O'Hara 1905, 720.

that genuine evolutionary novelties could not be explained "by the purely mechanical agency of natural selection," which neither improved nor created new species.[63] O'Hara made his own use of scientific works to advance his religious agenda, and when his comments are compared with those of Morgan, the priest and the scientist sound remarkably similar. Morgan asserted that "we can profitably reject, as I believe, much of the theory of natural selection" because, as he explained, "Darwin's theory of natural selection is preeminently a theory of adaptation," which it explained more successfully than it could "the 'origin of species.'"[64]

The arrival-of-the-fittest problem helped dissolve the boundary between scientific and religious criticisms. The resultant confusion prompted some in the scientific community to intervene with supposedly more authoritative interpretations. The *Popular Science Monthly* (*PSM*) offered what it claimed was a more objective account of the St. Louis meeting, which described mutation theory as "demonstrating the fundamental thesis of Darwin" (that is, common descent) while "supplementing" his principles. Their writer concluded that "it would be premature" to judge which theory would ultimately prevail.[65] The *PSM* aimed to correct the errors of the popular press, which it noted "now publishes articles everywhere of a readable and light character on scientific topics, and no monthly magazine is complete without one or two such articles." This was exactly the point Wharton's Professor Linyard had noted, and he would doubtless have been pleased that the *PSM* aimed to set "a standard of accuracy and weight" that would separate real science from "the vagaries of the charlatan."[66] Nevertheless, the magazine's writer sounded rather like one of their more sensational competitors when they emphasized that de Vries had "been able to see with his own eyes the actual evolution of several new plant forms possessing the characters of true species" and praised his accomplishments "especially in the new experimental science of evolution."[67] A year later, the magazine published a detailed account of de Vries's work by distinguished Dutch zoologist Ambrosius Arnold Willem Hubrecht. He attacked "those who bear a grudge" against Darwinism and who "in their innermost heart" would discard both de Vries's ideas and "the whole theory of evolution."[68] However, even within the pages of the *PSM* the link between mutation theory

63. O'Hara 1904, 9–10; O'Hara 1905, 12.
64. Morgan 1903b, ix, 91.
65. Davis 1904, 25–27.
66. Cattell 1901, 511. According to Pandora 2009, 354, the *PSM* represented an important shift in approaches to publishing science.
67. Davis 1904, 25–27.
68. Hubrecht 1904.

and anti-evolutionary sentiment was not easy to break; the following year, the magazine ran an article about Burbank, which quoted the "Plant wizard" as saying that "the mutation theory of the origin of species seems like a step backward towards the special creation theory."[69] Even within the confines of one unambiguously pro-science periodical, the meaning of mutation remained unfixed.

Whether pro- or anti-evolution, newspaper and magazine reports brought de Vries's ideas to the attention of far more readers than any scientific journal could reach. Even anti-evolutionary attacks like O'Hara's contributed to a sharp increase in public interest in the theory; his version of bricolage entailed quoting scientists' own words, which ensured that the readers of publications like the *Catholic World* were as up to date with the latest biological theories as the readers of the *PSM*.[70] Perhaps the most important aspect of these debates was that whether scientists were correcting misapprehensions or promoting mutation's practical benefits, they were tacitly inviting nonscientists to join the debate. And readers of a magazine might well take successive articles by different authors as contributions to a single conversation, and of course they could combine what they learned from a variety of different publications to create their own, idiosyncratic version of the Darwinian debates.[71] Widely circulated news stories assumed that an esoteric biological theory was of sufficient public interest to be reported in a nonspecialist publication; writers, editors, and publishers assumed the public would *take* an interest in the new science, but the reporting also implied that the public *had* an interest—a stake in the conversation and its outcomes. These two versions of public interest made it possible for biology to function as public culture.[72]

Making Use of Mutation

After describing de Vries's announcement at the World's Fair, the *San Francisco Chronicle* reported that he had since spent the summer in California, "further demonstrating his theory of mutation." The key question was, "Can this mutation be controlled?" Its potential applications got even more coverage than anti-Darwinian attacks, as one newspaper

69. Burbank, quoted in Jordan 1905, 207. As Sharon Kingsland (1991, 498) notes, the article's author, David Starr Jordan, used Burbank's quotations in this article as a way of attacking de Vries and his theory.

70. O'Hara 1905, 722. The sources he quoted were White 1902a and Cook 1904b.

71. Tattersdill 2016, 9–10.

72. Broman 2012, 127–34, 23. For aspects of biology as public culture, see Squier 1994; Turney 1998; Nelkin and Lindee 2004; Thurtle 2007; Rouyan 2015.

after another described the promise of new, scientifically created plants. Although the results of de Vries's collaborations with the University of California were still being awaited, the investigators (who included Osterhout; see p. 46 and figure 6.1) were optimistic. Mutation's practical implications were emphasized by drawing readers' attention to the similarities that de Vries acknowledged existed between his goals and those of Burbank. However, Burbank relied on traditional methods to propagate his new plants, whereas, if mutation could be controlled, astonishing new plants could be produced more reliably and quickly.[73] The possibility of creating new kinds of plants *rapidly* was the second important reason for solving the puzzle of the "arrival of the fittest"—identifying the real origin of species appeared to have enormous commercial potential.

The year after the St. Louis World's Fair, the *New York Times*'s Sunday magazine reported "the first conclusive proof yet obtained" of artificially induced mutations in an article about de Vries's most enthusiastic American supporter, Daniel Trembly MacDougal.[74] MacDougal worked at the New York botanic gardens and had been injecting the ovaries of plants with "strong osmotic reagents" or "weak solutions of stimulating mineral salts" to identify which "agencies external to the cell may induce mutations" (technical details which doubtless gave readers the sense of being privy to the latest science). The injected ovaries produced new plants "notably different from the parent," which had proved "healthy, reached maturity, bloomed, produced seed, and are perpetuating themselves."[75] Work like MacDougal's fed the expectation that mutation theory would, literally, bear fruit. De Vries himself stressed this aspect of his theory; when he visited Washington, DC, for the first time in 1904, he was asked by a reporter for the *Washington Times* whether his results could "be applied to practical uses," to which de Vries answered, "Decidedly yes," and gave examples.[76]

As Garland Allen, Sharon Kingsland, and Helen Curry have all argued, the most tantalizing possibility offered by de Vries's theory was that it

73. Wilson 1904.

74. For more on MacDougal, see S. E. Kingsland 1991. MacDougal's achievements were also reported in the *New York Evening Post*, in a story that was reprinted in the *Washington Evening Star* ("Mutations of Plants," *Washington Evening Star*, December 26, 1905, http://chroniclingamerica.loc.gov/lccn/sn83045462/1905-12-26/ed-1/seq-7/).

75. Harding 1905.

76. "Striking New Theories of Man's Origin Propounded by a Holland Scientist," *Washington Times*, October 9, 1904, http://chroniclingamerica.loc.gov/lccn/sn84026749/1904-10-09/ed-1/seq-30/. See also Theunissen 1994b.

would eventually allow artificial mutations to be produced on demand, which might dramatically increase crop yields.[77] This claim became increasingly prominent in the US media when de Vries returned to America in 1912. The *Chicago Daily Tribune* told its readers that de Vries believed that without advances such as those he promised, "the world [would] face famine within a few centuries."[78] And the *Washington Post* headed its story "Fears World Famine," before explaining that de Vries (one of the world's "most eminent botanists") had "an entirely new and startling theory" that promised to create "entirely new forms of plant life" to feed the world's ever-growing population. Although his work was still at the theoretical stage, de Vries was quoted as saying that "if it should once become possible to bring plants to mutate at our will . . . there is no limit to the power we may finally hope to gain over nature."[79] Scientific control over nature was the ultimate biotopian dream, but—paradoxically—de Vries's acknowledgement that his goal was still some way off may have fueled popular interest in his ideas; an incomplete theory required more work, not least because media reports often implied that ordinary gardeners and farmers might play a part in fulfilling mutation's promise.

De Vries himself worked hard to ensure his ideas reached wider audiences by writing for a variety of nonspecialist magazines, including the New York *Independent*, the *PSM*, and *Harper's Monthly*.[80] In the popular *Scientific American*, de Vries defined the ultimate goal of mutation research as acquiring the ability "to take the whole guidance of [evolution] into our own hands."[81] And he struck a similar note when invited to give the convocation address at the University of Chicago in 1904, telling the crowd that "the application of new discoveries and new laws" would lead to the creation of new "industrial" plants.[82] Newspapers regularly focused on these ideas, as for example, when the *New York Times* reported de Vries's 1912 visit under the headline "Noted Holland Expert Tells How to Double Our Crops." The full-page story (complete with a large portrait) described him as being regarded "by fellow-experts as the world's greatest living botanist, hailed by them as the 'modern Darwin,'" and as "the man whose

77. Allen 1969; S. E. Kingsland 1991; Curry 2016, 26. See also Zevenhuizen 1998, 417.

78. "Backward America," *Chicago Daily Tribune*, December 6, 1912: 8, retrieved from ProQuest Historical Newspapers, document ID 392500951. The *Tribune's* story was reprinted in the *Arizona Republican*, January 31, 1913, http://chroniclingamerica .loc.gov/lccn/sn84020558/1913-01-31/ed-1/seq-4/.

79. De Vries 1905c, 688.

80. De Vries 1902a; 1903; 1905b.

81. Mott 1957, 306–7, 320.

82. De Vries 1904, 401; Mott 1957, 306–7, 320.

work may be the means of furnishing the world in generations to come with its daily bread." De Vries was, once again, compared to Burbank, but the article emphasized that the Californian breeder was merely creating the biggest plums and potatoes, while de Vries was "trying to discover the laws which govern such vagaries of nature."[83]

The *New York Times* told its readers that while Darwin had only examined the "survival of the fittest," de Vries was explaining the "origin of the fittest"—the appearance of "mutants," which marked the beginning of "a new species." Each new, mutant form was "of a pure type, never degenerating, and perhaps of double or triple the yield."[84] The practical value of understanding the mutant origin of species was obvious, and de Vries's emphasis on the need to preserve "pure" strains to prevent degeneration must have been particularly striking to American readers at a time when there were heated debates over immigration into America, which provide an important context for understanding America's excitement over experimental evolution.[85] Between 1903 and 1929 (exactly when interest in the mutation theory reached its peak), the USA passed seven different laws limiting immigration, invariably on racial criteria.[86] As these debates were going on, de Vries was arguing that the groups that scientists called species (Linnaean species, such as *Homo sapiens*) were in fact composed of many different forms and that the "real units are the elementary species," each defined by a distinctive mutation. It was only by identifying and modifying these pure, elementary species that major crop improvements could be made.[87]

As Europe's huddled masses poured in, some Americans wondered whether their country would be able to feed its rapidly growing population, hence the perceived need for rapid improvements to crop plants and animals. In 1902, Beverly Thomas Galloway, head of the US Department of Agriculture's Bureau of Plant Industry (recently established to apply botanical science to improving America's crops), told the American Association for the Advancement of Science that "as population increases," the demand for "plants better adapted to certain conditions and which can be produced at a minimum expense, will become greater and greater."[88] Galloway's fears might have been exaggerated, but they help explain the

83. "Noted Holland Expert Tells How to Double Our Crops" 1912.
84. "Noted Holland Expert Tells How to Double Our Crops" 1912.
85. Cohn 2017.
86. Tichenor 2002, 3.
87. De Vries 1905c, 11–12.
88. Galloway 1902, 51, 56–57. On Galloway, see Pauly 1996.

widespread attention paid to mutation; it seemed to promise precisely what America required—a rapid route to new and improved plants just when they were seemingly needed most. (The comparative neglect of mutation in Britain may be partly explained by the absence of mass immigration during these decades.) Thanks mainly to the press, de Vries and his theory became well-known in America in the years before World War I. He met Galloway during his 1912 visit to Washington and was even introduced to President Theodore Roosevelt.[89] As de Vries and mutation theory's fame spread, so too did the hope of creating new kinds of plants to feed the ever-growing numbers of Americans.

Biologists began to use purpose-built laboratories with increasingly specialized equipment in the late nineteenth century, and as the new century began, they found themselves in competition for new sources of funding. Governmental and industrial funding was still limited, but new charitable foundations emerged, notably the Carnegie Institution of Washington (f. 1902) and the Rockefeller Foundation (f. 1913).[90] As scientists became increasingly aware of the need for publicity to attract more money, many joined the broad public conversation about interpreting new biological ideas and developed a symbiotic relationship with journalists, editors, and publishers. The venues for this conversation included New York's *Independent* magazine, a broad weekly miscellany of general-interest news, comment, fiction, and poetry, which published four articles on mutation theory in 1902 alone (and a total of at least twenty-five articles and reviews that mentioned the topic in the period before World War I).[91] Among these was an original article by de Vries, "My Primrose Experiments" (September 1902), that was prefaced by a briefer piece by MacDougal, who contextualized de Vries's work within the Darwinian debates. According to MacDougal, these scientific disputes had come to resemble a rather sterile "debating society," characterized by partisan speculation. As a result, a new generation had arisen who relied on "results obtained by experimental methods." De Vries was a key leader of this new approach whose research had led to the most important contribution "since the time of Darwin."[92] This preface was followed by de Vries's article, which

89. De Vries's meeting with Roosevelt is mentioned in his account of his travels, *Naar Californië* (1905a). See Bavel 2000.

90. Abir-Am 1982; Kohler 1991.

91. Figures from ProQuest American Periodicals Series. Some of these articles are analyzed in later chapters.

92. MacDougal 1902, 2284. The *Independent* was the best-known of nondenominational religious papers; it had begun as a Congregationalist publication and never entirely forgot its roots (Mott 1957, 59, 288, 290, 301).

described how he had been able to see new species appear "by an abrupt leap or mutation" and offered the now-familiar speculations about its applications. He ended on an unmistakably biotopian note, with the observation that eventually "perhaps it will be possible for us to contribute something, little as it may be, to the evolution, of nature and to conduct it in courses that will be profitable to humanity."[93]

In contrast to the generally cautious (even dry) tone of de Vries's article, the *Independent* had previously published a piece headed "Jews and Primroses" that had described his ideas as "one of the most illuminating announcements of modern science." Their writer briefly explained the new theory and assumed other species must also be mutating. Perhaps, their writer speculated, the Jewish people might be experiencing a mutation period, which would explain how—despite being "the denizens of ghettos, cuffed by their rulers, poor beneath description, dwarfed in body and crushed in mind"—they nevertheless continually gave birth to so many exceptional figures (whether financiers, poets, or philanthropists—"Of all races in the world the Jewish seems able to produce the largest variety of type"). Someone like the British nineteenth-century prime minister Benjamin Disraeli must therefore be considered "a glorified Jewish Primrose," the *Independent* concluded, and perhaps other immigrant communities might also be mutating.[94]

The way the writer of "Jews and Primroses" moved rapidly from new discoveries about plants to speculations about human evolution might seem startling, but such links were relatively common as mutation became a regular topic (see chapter 7). For example, the *Popular Science Monthly* extended mutation's significance beyond plant-breeding when, in an article titled "Industrialism," it asserted that "as the biologist might say, the Industrial Age is a period of rapid mutation. . . . It is a day of hope and of optimism, such as the world has not hitherto known."[95] The claim that the twentieth century was characterized by rapid transformations was reinforced in a piece titled "Democracy, Nationalism and Imperialism," which argued that Japan's rapid modernization had "upset the theories of social evolution in about the same way as de Vries's discoveries and mutation theory modified the Darwinian hypothesis."[96] Such stories compressed

93. De Vries 1902a, 2285–86, 2287.
94. "Jews and Primroses" 1902; "Jews and Primroses," *Omaha Daily Bee*, June 29, 1902 17, https://chroniclingamerica.loc.gov/lccn/sn99021999/1902-06-29/ed-1/seq -17/. Many thanks to my former student Miles Bland for drawing this to my attention.
95. Slichter 1912, 363.
96. Beer 1907, 746.

time and space as they insisted the disruptive future was already here. Mutation theory promised human control over evolution—both biological and social—and because it changed both the tempo and mode of evolution, future changes would be both rapid and discontinuous, characterized by sharp breaks instead of smooth changes. (As we shall see in chapter 7, some interpreters argued that mutation provided a scientific warrant for revolutionary sociopolitical upheavals.)

Periodical publication created a sense of continuing conversation, as later articles could be read as responding to (or branching out unpredictably from) claims made in earlier ones. De Vries himself might not have read "Jews and Primroses," but some of the magazine's readers doubtless interpreted his later article as part of an ongoing discussion, partly about race and immigration. Serial publication further encouraged readers to interpret new sciences for themselves, to help judge their importance, and to participate in the dialogues through which twentieth-century science was created. Faced with so many articles making parallel arguments, many readers must have begun to think of the twentieth as the century of mutation.

Undisciplined Science

In 1910, the *Saturday Evening Post* was America's leading popular magazine, selling around 1.25 million copies a week—twice as many as most of its rivals.[97] In the highly competitive marketplace, the *Post* argued that magazines were like populist politicians—both must come up for reelection, "but for the periodical every Saturday is election day and every corner news-stand is a polling place. Its day of reckoning is always today."[98] Competition, fueled by the rapid growth in public education and literacy, forced publishers to diversify.[99] The 10-cent monthlies emulated the topicality of newspapers in order to distinguish themselves from the traditional highbrows, until all were heavily illustrated and topical.[100] In this context, a new scientific theory such as mutation could not fail to attract widespread press attention, particularly if the stories could be connected to readers'

97. Reed 1997, 106–8.
98. "The Popular Magazines" 1910.
99. The number of schoolchildren in the USA rose from seven to twenty million between 1870 and 1915, and adult literacy hit 90 percent among white Americans (Ringel 2015, 151; Pandora 2009, 354).
100. Frank Munsey, a leading publisher of cheap magazines, estimated that the new 10-cent magazines accounted for about 85 percent of US magazine sales by 1904 (Mott 1957, 6).

everyday concerns—whether those were immigration or increasing a farm's income. As the *Post's* writer argued, magazines should reflect public opinion, not expect to shape it: their stories must approximate "the experiences of the mass of men," including their "economic beliefs" and "political ideals."[101] Most popular magazines rarely covered science (and when they did, it was usually medicine, engineering, and astronomy rather than evolutionary biology); nevertheless, these sporadic stories reached many more readers than those in specialized scientific journals.[102] Those Americans who did hear about the mutation theory were far more likely to have learned of it from a newspaper or magazine than by any other means.

As we have seen, mass-market publishing provided the context for Wharton's "The Descent of Man," which—like many of her other works—reveals her deep interest in scientific matters.[103] As a rising and respected novelist, she seems utterly unlike the eclectic speculator Henry Clayton Thompson (see p. 49), yet both were avid readers of scientific books, particularly on biology and evolution, and their published writings are a record of how they interpreted what they read. They also shared an interest in evolution's potential religious implications. In Wharton's story, the publisher initially mistook *The Vital Thing's* genre and accepted its religious tone as heartfelt; he referred to it as Linyard's "apologia—your confession of faith."[104] The imaginary publisher's mistake was unsurprising given the widespread interest in evolution's spiritual implications; *The Vital Thing's* success was largely due to this optimistic, religious tone. This contrast had an exact counterpart in the real world: in the same month that Wharton's story appeared, *Harper's Monthly Magazine* reviewed Alfred R. Wallace's most recent book, *Man's Place in the Universe*, which struck the same note of evolutionary hope. A fictitious reviewer of *The Vital Thing* had praised its "ringing optimism" (which contrasted sharply to the scientist's more typical "whining chorus of decadent nihilism"), and *Harper's* praised Wallace for asserting that the "universe has been ascertained to have a mind, a heart, a soul, it resides here, and here alone, in the human race." Wallace's book typified a new spirit of optimism that "we may therefore call . . . the spirit of the twentieth century, and we may possibly discern in this the renewed light of the faith so long in eclipse."[105]

101. "The Popular Magazines" 1910.

102. Reed 1997, 252.

103. For more on Wharton's interest in biology, see Lewis 1975, 56; Hayes 1996; Saunders 2009.

104. Bender 1996, 320; Wharton 1904, 17.

105. Howells 1904, 640, 642.

Some readers probably found that same spirit in a work like Thompson's *New Reading of Evolution* (which could easily have served as the model for *The Vital Thing*), but wherever they found it, a large section of the public clearly had a taste for evolutionary optimism.

The confusion of genres in Wharton's story reflected the reality of the marketplace in which the mutation theory made its debut, where an ever-increasing variety of writings competed for the public's dimes. The *Saturday Evening Post*'s piece on popular 10-cent magazines emphasized their role in the process of interpretation between different audiences and genres of writing. According to their writer, the new magazines had "come to stay," because they fulfilled a need: "They occupy a place between the daily newspapers and the dear old ladies of the periodical press, who knit and embroider perfectly proper and perfectly elegant comforters and tidies of fiction and fact, pausing occasionally to lift a hand of shocked protest at the capers of the younger generation."[106] The image of the magazines mediating between the dailies and the rather highbrow "dear old ladies" epitomizes the ways in which an idea such as mutation reverberated back and forth between these different kinds of publication (figure 1.1). As we have seen, technical reports in specialized scientific journals like *Nature* could be reinterpreted as brief stories for the *Youth's Companion* or a newspaper. And in the same way, a cautious scientific essay by an established scientist such as Thomas Morgan could find itself filleted into a news brief in the *Mexico Weekly Ledger*. De Vries tried to adapt his message to different audiences, but in the process might find his work being unexpectedly applied to questions such as the status of Jews in American society.

However, the most important point about the diverse ways mutation was interpreted is that the movement was not one-way. As we have seen, an article in a 10-cent weekly like *Everybody's Magazine* made it onto university and high school reading lists because it had been referred to in a textbook. The audience for Osterhout's *Experiments with Plants* probably never came close to the 160,000 readers of *Everybody's*, but the claims in the original magazine article gained considerable scientific credibility once they appeared in his book. Numerous other examples will follow in later chapters: interpretations transformed general-interest writing into more specialized kinds, and—to some extent—transformed journalism into scientific fact in the process. As they reverberated, articles might gain credibility as they shed readers, or gain readers while shedding some of their scientific gravitas. Not only was mutation subject to the judgment

106. "The Popular Magazines" 1910.

of the general public, it was being interpreted by and for nonspecialist audiences at exactly the same time as the scientific community were first hearing of it—there was no delay while the new theory "diffused" from an elite to a popular audience.

Most of the media coverage discussed in this chapter reflects the widespread enthusiasm for mutation theory in the USA. By contrast, the most obvious feature of the British press coverage is how little of it there was.[107] Although the original report that prompted the "new light" story in the USA had appeared in the British magazine *Nature*, the *Daily Telegraph* seems to have been the only daily that noted it.[108] The British monthlies and reviews gave de Vries's work a little more coverage, but it was mostly confined to the more expensive and intellectual magazines, whose writers and editors tended to offer their readers technical details rather than sensationalism. For example, the *Nineteenth Century and After* (as the journal became in 1900) featured a regular "Science News" column by the exiled Russian anarchist Pyotr [Peter] Alexeyevich Kropotkin.[109] He discussed the mutation theory, but it was neither the first nor the most prominent item in his twenty-one-page roundup. Kropotkin situated de Vries within a "young branch of the science of evolution" that was sometimes called "experimental" or "physiological" morphology (the term *experimental evolution* was also less common in Britain). Kropotkin concluded that de Vries's "facts are so significant, and yet so new, that their bearing upon the theory of evolution cannot yet be appreciated in full."[110] This open conclusion (combined with the length of the article, the mass of rather technical detail, and the lengthy discussion of other evolutionary ideas and theories) diminished the impact of de Vries's work and provided a striking contrast with most US coverage.

107. For example, the *Times* mentioned de Vries's work in passing on only a couple of occasions prior to World War I, each brief and in the context of reporting a serious scientific lecture. See "Royal Horticultural Society," *Times*, September 29, 1909: 20; "Professor Bateson on Genetics," *Times*, May 15, 1912: 4.

108. The *Telegraph* briefly mentioned de Vries's conclusion that "new species appear suddenly by mutation, and never as the outcome of progressive variation," but concluded that his ideas "seem to confirm Darwin's theory." No source was given but the final words (in quotation marks in the newspaper) were from the same *Nature* report cited in "New Light." See "Science Notes," *Daily Telegraph*, July 1, 1901: 4. Another similarly brief report summarized an account of de Vries's work in the *Revue Scientifique*: "Evolution by Explosion," *Edinburgh Evening News*, March 26, 1902: 2.

109. Kropotkin earned his living as a journalist, writing mostly about science, after his arrival in Britain in 1886. He made his own, distinctive contribution to evolutionary theory, a 1902 book titled *Mutual Aid: A Factor of Evolution* (see chapter 7).

110. P. Kropotkin 1901, 430.

While many US writers delighted in discrediting Darwin, the British press was more deferential, not least because of the living presence of Alfred Russel Wallace, co-discoverer of natural selection and a staunch defender of Darwin's legacy. The grand old man of British Darwinism was firmly opposed to mutation theory (which he dismissed as "a mountain of theory reared upon . . . an almost infinitesimal basis of fact").[111] And Wallace, like many writers in British magazines, also tended to conflate de Vries's work with Mendel's, which further reduced the mutation theory's impact.[112] Since de Vries was one of those who rediscovered Mendel's long-neglected work, the confusion of the two men's work was perhaps inevitable, but they were even more commonly conflated in Britain because of William Bateson's influence. Bateson was one of the first prominent British biologists to immerse himself in the new laboratory-based research that spread rapidly in the USA (where he did some of his studies). His book *Materials for the Study of Variation* (1894) marked a departure from both the British style of natural history (exemplified by Darwin and Wallace) and the gradualist conclusions that emerged from it. Partly because of his doubts about the efficacy of natural selection (the arrival-of-the-fittest problem), Bateson emphasized discontinuities in nature (as opposed to slow, gradual change); when he first read Mendel's newly rediscovered work, Bateson became convinced that it explained the phenomena he had observed.[113] He became an early and enthusiastic promoter of Mendelism, so much so that many British writers assumed Bateson had rediscovered Mendel's work.[114]

Because Bateson's earlier work on discontinuous variation was well-known, some British writers portrayed de Vries as following in Bateson's footsteps. The upmarket *Quarterly Review* offered a fairly standard British view when it criticized those who had adopted the mutation theory for prizing novelty above everything: as a result "we constantly read popular accounts of the 'new species' which de Vries has experimentally produced" (this little sneer at "popular accounts" doubtless went down well with the *Quarterly*'s elite readership). In reality, their writer asserted, "there is not the slightest evidence that the Dutch professor has produced any new

111. Wallace 1908. See also Bowler 1978, 57.

112. A similar elision of the two men's work was also widespread in the specialized scientific literature; see Stoltzfus and Cable 2014.

113. Radick 2023, 92–121.

114. For examples where Bateson was given either priority or equal credit with de Vries, see Lock 1906, 17; Montgomery 1906, 133–34; Lankester 1907, 125; Drinkwater 1910, 2; Geddes and Thomson 1911, 120–23.

species."[115] A similar view was apparent in the prestigious British literary miscellany the *Monthly Review*, who invited well-known plant breeder Arthur J. Bliss (whose "Bliss Irises" are still on sale today) to explain "hybridisation and plant breeding" to their readers in 1906. Bliss characterized Mendel's work as the culmination of a long tradition of European scientific breeding and noted that its rediscovery coincided with de Vries announcing "the results of his experiments" and "especially of those comparatively large and sudden variations, or breaks, to which he has given the name of 'Mutations.'" In Bliss's view, de Vries's results should be seen as "confirming and amplifying views which Bateson had been urging for some years previous in England"—implying the Englishman might claim priority in this area. Bliss noted that the third international plant-breeding conference was about to take place in London, at which "important results" based on "the lines of research opened up by the discoveries of Mendel and De Vries will no doubt be communicated."[116] Mendel and de Vries were effectively being given equal billing with Bateson (who was to coin the term *genetics* at the conference); the *Monthly*'s readers probably came away with the impression that all three were utilizing essentially the same theory to advance plant-breeding.

Bliss's article closed with a suggestion that "the gardens of some future generation, with the black Tulip and the blue Rose and many another flower that to-day is but an ideal, will witness to the skill and patience and enthusiasm of the plant-breeder."[117] However, it was not primarily new scientific theories that would produce these extraordinary plants, but "the artist—the breeder himself," whose achievements depended on practical gardening skill, patience, taste, judgment, and "an almost prophetic insight into the properties of his plants." In Bliss's view, "above all he must have imagination, and not be afraid to use it—anybody can raise plants, but it is the man with imagination who will raise the flowers of the future."[118] His confidence in the achievements of traditional plant-breeding methods implied a certain skepticism about newfangled notions, including de Vries's claims. And it is suggestive that Bliss's predicted "flowers of the future" were ornamental novelties (in keeping with his claim that breeders were artists), not new crops. The most obvious contrast between Britain and the USA was the lack of mass immigration into the country. Indeed, in the period prior to World War I, the British were more likely to

115. Poulton 1909, 16.
116. Bliss 1906, 86.
117. Bliss 1906, 102.
118. Bliss 1906, 96–97.

worry about the way the demands of Empire seemed to be draining their manpower (the gendered term being entirely appropriate in this context), particularly at a time when their native population appeared to be degenerating. Feeding an ever-larger number of hungry mouths was not a British concern at this time (particularly in their colonies, where repeated famines were routinely ignored). Hence, one of the most attractive features of the mutation theory to Americans—the promise of rapidly increasing agricultural yields—was of considerably less interest in Britain.

The contrasts between Britain and the USA are analyzed further in later chapters, but the earliest reports of mutation theory suggest some obvious ones. Mutation theory was an upstart science, incomplete and demanding more interpretation. As Katherine Pandora has shown, part of America's republican ethos was that scientific knowledge should be "held in common," not restricted to elites. Some Americans believed that the compressed journalistic language of the mass-circulation publications embodied republican virtue; brief, accessible texts were more democratic and better suited to keeping busy people up to date with rapidly changing knowledge.[119] By contrast, British readers were encouraged to defer to the expert opinions that trickled down from the secluded bastions of the ancient universities. And the British media were, on the whole, rather more conservative and deferential (particularly toward both Darwin and his living defenders) than their boisterous American counterparts.

In the century's early decades, experimental evolution was not yet (and might never be) a recognized scientific discipline; it was an undisciplined science that promised an undisciplined future, one in which many supposedly established boundaries would be violated. Its upstart identity was embodied in claims to being new, exciting, rapid, and revolutionary—all of which found echoes in America whose citizens embraced similar adjectives for themselves (and were sometimes contemptuous of those who "knew their place" and were content to stay there). The mutation theory's glittering American career brought biotopian ideas to their widest audience, not least because US horticulture and American utopianism had already prepared the ground in which speculative heredity would take root.

119. Pandora 2009, 353.

2

Remaking Nature

In 1907, several US newspapers announced that utopia was being founded on Long Island, just outside New York: "The dream of Bacon, who saw in the New Atlantis gardens a land devoted to the modification of animals and plants at man's will, is being realized by the Carnegie Institution at its new 'Station for Experimental Evolution.'" The story emphasized humanity's new power by adding such headlines as "Creation of Species," or "Man in his new role of inventing creator" thanks to "EXPERIMENTAL EVO-LUTION."[1] Various newspapers repeated it, and the best-selling *Saturday Evening Post* had previously reported the Station's work in similar terms. They noted that sports had once been considered random accidents, but recently "a scientist of high reputation named De Vries succeeded in 'inventing,' if the phrase be permissible, seven entirely new kinds of primroses by such 'mutations,' as he calls them." Adding that, along with Mendel's law, the study of mutations would prove "of incalculable value to us." There was, the writer concluded, "no telling what wonderful discoveries" Davenport and his team might make in future.[2] The reverberating age transformed the language in which biology was discussed, as scientists and journalists began to describe the future in similar terms, transforming the prosaic experimental growing plot into a utopian garden.

---

1. Watkins 1907b. The story was reprinted in several other papers, including "Creation of Species," *New-York Daily Tribune*, February 24, 1907: 2; *Times-Democrat* (New Orleans, LA), February 24, 1907: 35; and "Man as Creator," *St. Louis Globe-Democrat*, February 24, 1907: 63; "Man as Creator, Wonders of New Station for Experimental Evolution," *Illustrated Weekly Magazine, Los Angeles Times*, February 24, 1907: 11. A briefer version, which included the reference to the "Dream of Bacon," appeared in the *Hutchinson News* (KS), December 6, 1907: 7, reprinted from the *Chicago Tribune*. See also Campos 2017, 165; Rouyan 2017, 120, 303.

2. Bache 1905, 13.

FIGURE 2.1. The biotopian possibilities of experimental evolution were vividly captured by this artist's description of a human being as "some raw material" for biologists' imaginations to work on.

Davenport, the station's first director, first made the eye-catching claim that humanity had assumed the role of creator of species at the laboratory's opening, when he had also credited de Vries with founding the new science (see p. 40).[3] The new station would try to both understand and to "control" evolution's laws, which would not only make organisms "to meet our requirements of beauty, food, materials and power," but eventually enable "an improvement of the human race." This startling claim was highlighted by the newspaper's subheading—"HOW TO PRODUCE SUPERMAN"—under which it was explained that Davenport and his colleagues were currently working on "inventing to order beetles with certain spots, [and] guinea pigs with four toes." These and similar projects appeared (from a "superficial glance") "to be useless and silly," but their writer quoted Davenport's claim that "when we know the law we may control the process." The journalist speculated that once scientists knew how to make four-toed guinea pigs, "will not the same law enable science

3. Davenport, Jones, Billings, and De Vries 1905.

to breed, at will, . . . human geniuses, with lofty morals, acute senses and blood highly resistant to the bacilli of disease?"[4] The scientific remaking of human nature seemed almost inevitable.

The much-reprinted story quoted above was written by John Elfreth Watkins Jr., who wrote on a wide variety of topics including occasional science articles that specialized in bold predictions about the future, often with utopian themes ("**Strawberries as Large as Apples** will be eaten by our great-great-grandchildren for their Christmas dinners a hundred years hence").[5] A busy man, he perhaps forgot to mention that he had not interviewed Davenport to learn about "the dream of Bacon," but had lifted the quote from the Carnegie Institution's 1904 yearbook (Davenport had made the claim in his speech at the station's opening). The linguistic shift underway at this time is epitomized by the fact that the language of a formal opening address by a professional scientific researcher was sufficiently journalistic in tone that it did not look out of place in a daily paper.[6] Watkins quoted Davenport several times in his article, and phrases such as "the principles of evolution will show the way to an improvement of the human race" sound like the kind of slightly exaggerated claim that a scientist might say in an interview, but these quotes were also taken directly from a formal annual report.[7] Davenport had chosen to address his patrons and the station's board of directors in a style that allowed his words to be seamlessly inserted into a newspaper report. The new language reflected a change within science itself; Davenport was neither the first nor the last twentieth-century scientist who needed to impress his funding body with the importance of his team's research. The sober, serious scientists and the freelance creators of futuristic hype were creating a common language, built on the prospect of creating a scientific utopia.

The excitement with which the press greeted experimental evolution's pronouncements may surprise historians, since, as Philip J. Pauly showed, nineteenth-century America had already witnessed a "horticultural transformation" that substantially changed the country's landscape. Analyzing those earlier efforts, and the utopian dreams they aroused, reveals continuities that illuminate the specifically American elements of twentieth-century visions. Nevertheless, sharp contrasts also emerge, which further clarify what was new about biotopianism. As Pauly showed, nineteenth-century American farmers and horticulturalists engaged in large-scale

4. Watkins 1907b.
5. Watkins 1900, original emphasis. See also Watkins 1906; 1907a.
6. Carnegie Institution of Washington 1905, 33.
7. Carnegie Institution of Washington 1907, 91–92.

collaborative efforts that reworked many aspects of the landscape. They created fresh combinations of plants that were imagined as either re-creating the original, supposedly natural landscape, or deliberately im-proving upon it. Many newly arrived Americans saw the largely treeless prairies as an empty wilderness (and often blamed the supposedly savage indigenous people for having created it). Efforts to rectify the problem included widespread tree-planting to reforest the plains. Arbor Day (1872) encouraged community tree-planting efforts, usually involving children, which created a new American form of spring festival; and the federal Timber Culture Act (1873) paid farmers to plant and preserve trees.[8] The scale of American tree-planting efforts resulted in large areas of prairie being covered in a novel mix of American trees from across the conti-nent, intermingled with selected imports from all over the world.[9] These achievements certainly laid the groundwork for the hope that the "dream of Bacon" might be realized, but the mutation theory catalyzed a new biological mood that shifted the expectations of what human gardening might achieve. As Pauly emphasized, earlier horticulturalists generally as-sumed they were restoring or repairing a natural landscape, putting it back to the way it "ought" to be.[10] By contrast with this essentially conservative vision, de Vries and Mendel offered new tools for modifying nature. And the biotopians also overturned the previously unexamined assumption that nature offered its own guidance. Despite some obvious continuities, the resultant utopian dreams took on radically new forms.

Conservation (as the very term suggests) took it for granted that the past—history, tradition, and heredity—determined the present. One re-sult was a nostalgia for earlier landscapes, apparent in pioneering mid-century works on conservation, such as George P. Marsh's *Man and Nature*, which was shaped by the author's mourning for the trees that had defined the landscape of his childhood.[11] Conservationists often saw appropriate cultivation techniques as a way of restoring fallen nature, a view that was partly shaped by the prevailing nineteenth-century under-standing of heredity as the past's legacy (see p. 29). Conservationists and eugenicists often shared a commitment to ancestral heredity, and hence they imagined the human stock being scientifically managed like a breed-ing herd of animals. American eugenicist Madison Grant epitomized the link when he argued that the large wild animals of North America were a

8. Pauly 2007, 80–83, 92–93.
9. Pauly 2007, 96–98.
10. Pauly 2007, 191.
11. Pauly 2007, 82–83.

valuable natural resource that needed to be protected from civilization's advance, just as the country's white people (the "Nordic" stock, as he called them) needed to be conserved from the hordes of rapidly reproducing, racially inferior peoples who were flooding into the country.[12] Grant's vision—like those of many eugenicists—was based on a supposedly natural white superiority—which was inherited via a shared bloodline that had to be protected from contamination. Although the biotopians were not noticeably more progressive on racial issues, they were generally uninterested in preserving the past or protecting existing plants or animals—they looked to the future and imagined creating entirely new organisms.

Despite their differences, many biotopians shared the earlier horticulturalists' fascination with the practical democratic work of gardening. As Leo Marx has argued, America produced a specific form of pastoral literature that celebrated a human-made landscape, one that combined the natural beauties of the supposedly unspoiled new world with the results of farming and gardening. Early European accounts of America described it as Edenic in its perfection (one seventeenth-century writer admitted his description would seem "to be but Utopian, and legendarie fables" to those who had not seen the New World for themselves).[13] However, by the late eighteenth century, writers like Thomas Jefferson were more focused on the work of cultivation that was needed in order to redeem the wastes and wilderness. These ideas combined to depict America as both unspoiled wilderness and well-tilled farmland, creating a palimpsest of Golden Age myths, frontier fables, and old husbandmen's tales that underlay what Marx called the "middle landscape." America's rapidly growing cities embodied the relentless activity of history, busy modernizing and urbanizing, while the countryside remained almost outside history, immune to its disruptive influence (figure 6.2). Keeping these irreconcilable images in mind allowed Americans to think of their land as a second Eden without thinking about its artificial qualities (particularly the work—and the often-genocidal violence—that its creation had required).[14]

The middle landscape was epitomized in the horticulturalists' attempts to create a distinctively American "Prairie" style of gardening (which, for example, emphasized indigenous plants, yet excluded those that were toxic to introduced livestock) produced a carefully edited artificial/natural

12. Allen 2013.

13. *A True Declaration of the Estate of the Colonie in Virginia* ... (London, 1610), quoted in Marx 1964 (2000), 34.

14. Marx 1964 (2000), 3–29, 108–38. See also Fender 1992, 58–60; Ellingson 2001.

landscape.[15] For the most part, American gardeners and farmers rejected the wilderness as firmly as they rejected cities—there was little trace of wilderness preservation in mid-nineteenth-century American utopias, which often imagined a future park city, built on massive ecological engineering (including the eradication of unwelcome species).[16]

One of the leaders of the horticultural reimagining of America was Liberty Hyde Bailey, dean of the College of Agriculture at Cornell. An enormously popular writer (whose sixty-five books sold almost a million copies in his lifetime), he became a nationally known figure when he chaired President Theodore Roosevelt's 1907 Country Life Commission.[17] Bailey's publications ranged from technical scientific subjects to practical gardening, and included philosophical musings on democracy and nature, such as *The Holy Earth* (1915), which have led to him being hailed as a pioneer of modern environmental thinking.[18] Bailey was firmly committed to the Jeffersonian ideal of the independent yeoman farmer as the very bedrock of American society, an ideal that had—among other things—led to the founding of the US Department of Agriculture. Efforts to preserve the yeomen led, paradoxically, to large-scale government funding of American agriculture, including the Hatch Act (1887), the second Morrill Act (1890), and the Smith-Lever Act (1914). The result was a network of land-grant colleges, experimental research stations, and rural extension courses, all intended to improve agriculture and rural life. The Hatch Act had brought Bailey to Cornell, where he would play a key role in expanding and standardizing efforts to produce what he called an "intelligent class" of "awakened" farmers who could build "a new day" in the countryside.[19] Bailey's influential and widely read ideas about nature, farming, and scientific agriculture show some of the ways in which the bucolic myth of the American farmer was modernized as it was transmitted to the twentieth century, while retaining a long-standing fascination with Edenic wilderness and the normative role of nature. Partly as a result of these contradictory elements, America's biological futures would be imagined as developing along distinctly different lines to those that emerged a little later in Britain (see chapter 5).

15. Pauly 2007, 165–94.
16. Burt 1981, 175.
17. Jack 2008a, 1–3, 17, 30.
18. Jack 2008a, 5–6.
19. Sorber 2018, 137.

Nature Is the Norm

Bailey's writings and speeches on the practical and spiritual benefits of closeness to nature caught the attention of many of his fellow Americans. Progressives in particular agreed that working the land was both healthy and morally uplifting. In recent years, his work and ideas have attracted considerable (somewhat overdue) historical attention, so no attempt will be made here to do justice to their full range.[20] I will focus instead on the paradoxical nature of Bailey's commitment to both tradition and modernity. As Pauly argued, Bailey believed strongly in rural progress through scientific farming, and he embraced evolutionism as its scientific basis (he argued that farmers helped their crops to evolve into better ones).[21] Yet his faith in evolution and science was combined with an equally firm commitment to the normative value of nature; analyzing this combination further clarifies what separated biotopianism from earlier, superficially similar attitudes and practices (such as mass tree planting).

Bailey's early writings focused on the practicalities of growing (for example, *Field Notes on Apple Culture* [1886] and *The Pruning Manual* [1898]), interspersed with more scientific works such as *The Survival of the Unlike* (1896), which argued that cultivated plants provided evidence of evolution while demonstrating that there was no contradiction between evolution and belief in God (who alone could explain the ultimate origin of the universe).[22] As Daniel Rinn has argued, Bailey saw the fence around a garden as an arbitrary barrier between nature and culture, practice and theory, and between the wisdom of gardeners and that of scientists; he sought to pull down the fence and get those on both sides of it to recognize and value each other's expertise.[23] Bailey collaborated with Cornell's Anna Botsford Comstock to promote the nature-study movement, which, as Sally Kohlstedt has shown, offered children practical, object-based nature study in order to bring some of the experiences of rural life into schoolrooms. Both Bailey and Comstock were expert teachers and writers, whose work inspired a generation of teachers to follow their example.[24] However, Bailey did not regard nature-study as primarily a way of recruiting young people to the cause of science, but more as a way

20. Morgan and Peters 2006; Jack 2008b; Armitage 2009, 89–104; Kohlstedt 2010, 77–110; Rinn 2018; Bailey, Stempien, and Linstrom 2019.
21. Pauly 2000, 117.
22. Banks 1994, 9.
23. Rinn 2018, 121. See also Bailey 2019.
24. Kohlstedt 2010, 85–100.

of bringing them into "sympathy" with nature. "Nature-Study is not primarily a natural history subject," he wrote, "it is primarily a pedagogical idea. . . . Nature study is not a science. It is not knowledge. It is not facts. It is spirit. It is concerned with the child's outlook on the world."[25] Moral education was promoted above scientific (which led to some criticism from Bailey's scientific contemporaries and contributed to the subject's gradual decline around World War I). Bailey's distinction between the ethical and scientific aspects of education maps onto the divide in his thinking between the need to conserve rural values while acknowledging the demands of modernity.

Bailey's interest in moral education was apparent when he published *The Outlook to Nature* (1905). As the prestigious Chicago-based literary magazine the *Dial* commented, Bailey had previously been "a guide to practical people in the affairs of garden and field," but the new book announced him "as philosopher and friend to all who love the out-door world."[26] The *Outlook* was based on lectures Bailey had given in Boston and began with a description of a typical US city scene, with people pouring through the streets "as if moved by some relentless machinery," a world in which "no one stopped and no one seemed to care." History's disruptive drive for change had infected twentieth-century America with a hunger for "the new, the strange, and the eccentric."[27] Those who shared his anxieties over the moral implications of an increasingly urban lifestyle would doubtless have agreed when he argued that Americans needed to be refreshed with the spontaneous, healthy spirit of the countryside, which he termed "the outlook to nature." In his view, the "race" depended upon the "farming occupation, and therefore that we need to conserve this occupation in order to recruit and reinforce the native strength of our civilization." (69–70).

Bailey clearly shared Jefferson's view that the farmers were the bedrock upon which the republic had to be built, but he was not opposed to either progress or cities.[28] Indeed, as Leo Marx argued, the key to Jefferson's thinking was his belief that America had resolved the problem of choosing between raw nature and overcivilized culture. America *was* the best of both worlds, neither primitive Arcadia nor an industrialized utopia, but a middle landscape of classical shepherds dressed, as Marx put it, in American

25. Bailey, *The Nature-Study Idea* (1903), 5, quoted in Kohlstedt 2010, 87.

26. "For All Those Who Love the Outdoor World" 1905. For science and the *Dial*, see Setz 2018.

27. Bailey 1905, 1–2, 5–6. All subsequent page references are to this edition.

28. Jack 2008a, 25.

homespun. Jefferson had realized that this ideal might never be fully realized but believed it would nevertheless serve to guide the country away
from the old world's extremes.[29] The majority of non-indigenous Americans saw cultivation as central to their very claim to the land, which made
farmers the quintessential Americans, the pioneers who had carved civilization out of the wilderness. From J. Hector St. John Crèvecoeur (*Letters
from an American Farmer* [1782]) onward, the question, "Who is an American?" was frequently met with the answer, "A farmer."[30] Brutish nature
was not the best environment; it was improved—cultivated—nature that
Crèvecoeur admired. America's middle landscape was neither too raw nor
too refined; that was what produced the prized qualities of Americans.[31]

The legacy of these earlier ideals was apparent in Bailey's writing, as
for example when he celebrated rural life while extolling industrialization
as "the glory of our time" and "a new epoch" for civilization. He called
for new forms of literature and poetry to do justice to this new age, commenting that the "flights of science and of truth are, after all, greater than
the flights of fancy." He denied that nature-lovers espoused a "serene and
weak utopianism," arguing instead that only hard work ensured progress.
Moreover, Bailey asserted that one could neither do this work nor really
love nature without scientific and technical knowledge (28, 37–38).[32] His
lecture on the "school of the future" stressed the need for all Americans
to receive a more technical education, in which science and business skills
would be as important as poetry. Bailey denied that rural America was
backward or reactionary, praising the "really marvelous development of
machinery" and of general improvements in rural organization and education (80). Despite sharing with many of his contemporaries a suspicion
of urbanization (the city "breeds or attracts most of the crime," 82–83),
he nevertheless argued that the US needed its cities as much as it needed
its farmers. Anyone who was going "to lead an effective and reposeful life
must be in sympathy also with artificial environments, [with] factories
and streets," because the "repose of the nature-lover and the assiduous
exertion of the man of affairs are complementary, not antithetical, states
of mind" (7–8).

The modern city celebrated remorseless change for its own sake, but
while Bailey accepted the necessity of progress, he fought to prevent
slower, rural alternatives from being abandoned. His "outlook to nature"

29. Marx 1964 (2000), 121–29.
30. Crèvecoeur 1782 (1904).
31. Marx 1964 (2000), 112.
32. He made similar claims in many articles; see, e.g., Bailey 1901b.

aimed to ameliorate and enhance the complex, artificial urban life—not
abandon it. His progressive agenda for America (which was echoed in that
of his admirer Roosevelt) included bringing many of the city's advantages
to the country; Bailey called for modernizing rural schools and churches
and for introducing traveling libraries, itinerant lecturers, improved roads,
rural electrification, telephones, trolleybuses, and a new literature that
would celebrate farming without affectation or caricature (134). To get
that new literature and the other benefits of modern life to country people,
he argued that the country was "much in need of a parcels post" (125).[33]
Despite his love of simple country life, Bailey was no Luddite (he did
pioneering work on growing plants under electric lights).[34] And given
his love of progress, he could sound like a biotopian at times, particularly
in his enthusiasm for the possibilities of human power over nature. He
argued, for example, that a "new art profession is just now rising" from
the traditions of landscape gardening and landscape architecture. This
profession would eventually give people the power to remake the natural
landscape according to human standards; even the "regulation of the scenery of mountains is not too large for its grasp. It will be one of the great
art efforts of the future" (88–89).

Bailey also shared his biotopian contemporaries' fascination with scientific plant-breeding. In an article titled "Making New Plants" for the widely
read *Collier's* magazine, Bailey noted that (mostly thanks to the publicity
surrounding Luther Burbank) there was intense excitement around the
claim that understanding the laws that governed the plant kingdom would
enable people to make entirely new, more useful plants. In 1902 he told
the Society for Plant Morphology and Physiology about de Vries's (then
very new) mutation theory, arguing that it was persuasive because de Vries
had carried his ideas "into the realm of actual experimental investigation."
Bailey doubted whether all species originated by mutation, but accepted
that some probably had, a fact that would "have a profound and abiding
influence on our evolution philosophies." He used the same address to
discuss the equally novel Mendelian theory, emphasizing its basis in modern scientific methods (including Mendel's use of statistically significant
numbers of plants). Thanks to both de Vries and Mendel, whose work he
presented as tightly connected (a common interpretation at the time),
Bailey argued that future scientists, breeders, and agriculturalists would
all look to "experiments in actually growing the things under conditions

---

33. The US Parcel Post was one of the tangible results of Roosevelt's Country Life
Commission, which Bailey chaired (Armitage 2009, 103–4).

34. For the electric lighting experiments, see Banks 1994, 8–9.

of control." And he concluded that the "experimental method has finally been completely launched and set under way."[35]

As more Americans went to college and studied the new biology, Bailey forecast that they would create new kinds of farms and gardens, and he argued that the "most satisfactory garden is not the one that is most perfect in the eyes of a gardener, but the one that has the most meaning"—for its human maker.[36] Such visions of a more artificial future make Bailey sound distinctly biotopian, yet he constantly stressed the moral superiority of the simple, rural life. Despite having shown that Americans were preeminently creatures who shaped and were shaped by artificial environments—whether gardens, farms, or cities—Bailey still insisted that the "natural environment is the more important, because it is the condition of our existence." For him, nature came first, in every sense—we evolved out of it and were shaped by its laws, and so nature provided the ultimate guide to how we ought to live (49). His view was summarized in the claim that "nature is our governing condition and is beyond the power of man to modify or to correct" (despite his forecast of a future art of designing mountains). If humans wanted to know how best to live, he believed, they needed only to "look upward and outward to nature" (7). *The Outlook to Nature* asserted that Americans should model themselves on the plainspoken "country-man" (one of whom Bailey claimed to be; he came from a Midwestern farming family). He wanted to "emphasize the things that are free," to escape the commercial bustle of America's twentieth-century cities and help his listeners escape modernity's "contention and noise." As he explained, "I preach the plain and frugal living of plain people" (4, 134). In certain moods, Bailey sounded like a rural revivalist, preaching a back-to-the-land ideal in a style that was reminiscent of those who preached a return to the old-time religion of their forefathers. In practice, Bailey was all too aware of the disasters likely to befall the untrained city dweller who simply bought a farm in the hopes of becoming a son of the soil overnight. A belief in ancestral heredity is perhaps apparent in his argument that keeping good farmers on the land was more important than creating new ones; farmers, he seemed to imply, were born, not made.[37]

Nevertheless, Bailey's faith in an earlier, simpler, better America was unmistakable when he urged his countrymen to be younger at heart,

35. Bailey 1903b, 445, 454.
36. Bailey 1910.
37. Jack 2008a, 17.

spontaneous, and natural. "Therefore," he wrote, "I preach the things that we ourselves did not make":

> the sky in rain and sun; the bird on its nest and the nest on its bough; the rough bark of trees; the frost on bare thin twigs; the mouse skittering to its burrow; the insect seeking its crevice; the smell of the ground; the sweet wind; the leaf that clings to its twig or that falls when its work is done. (10–11)

This was an appealing vision, and, unsurprisingly, Bailey is rightly hailed as a pioneering conservationist. He coined the word *biocentric* ("life-centered") to capture his philosophy, which, as Kevin Armitage has argued, blended evolutionary naturalism with farmer's rights and a commitment to democracy. Bailey has also been claimed as a founder of "planetary agrarianism," which attempts to craft a philosophically coherent alternative to the consumer-focused form of sustainable agriculture (which critics claim is all-too-easily co-opted by agribusiness).[38]

However, for Bailey biocentrism was essentially a religious conception; the creation came first, and humans were part of it, along with everything else on the planet. During the early years of the twentieth century, only John Muir offered a more radical ecological philosophy, but Muir's view was what would now be called "ecocentric," placing nature firmly and squarely at its heart.[39] Bailey, by contrast, placed more emphasis on the vital role people had to play in managing and preserving nature. Rinn sees Bailey as offering an alternative to the polarizing, ecocentric/anthropocentric binary that often characterizes human attitudes to the natural environment.[40] By contrast, I argue that Bailey's "biocentric" philosophy was full of contradictions that illuminate its *anti*-biotopian nature.

In explaining the spiritual and practical benefits of the "outlook to nature," Bailey wrote: "If nature is the norm, then the necessity for correcting and amending the abuses that accompany civilization becomes baldly apparent" (7–8). The "if" in his claim was redundant; he clearly believed nature *was* the norm and that the natural world provided the ultimate guide to what was good. As he would later write, "One does not act rightly toward one's fellows if one does not know how to act rightly toward the earth."[41] He turned to nature for guidance as readily and

38. Morgan and Peters 2006.
39. Armitage 2009, 105–7.
40. Rinn 2018.
41. Bailey 1916, 3; Rinn 2018, 129.

as confidently as the French Enlightenment *philosophes*, who averred that nature was the "source of good laws, of useful arts, of the sweetest pleasures and of happiness."[42] For many eighteenth-century thinkers, America epitomized the ideal state of nature, land and people untouched (and thus uncorrupted) by civilization. Robert Beverley, in his *History and Present State of Virginia* (1705), even insisted that the new world was so bountiful that it seemed untouched by original sin (its people seem "not to have been concern'd in the first Curse, Of getting their Bread by the Sweat of their Brows").[43]

The claim that humanity could be happy by simply following nature's laws was a secularized natural theology, a naturalized ethics built on the loosely defined claim that whatever was natural was morally good. As Darwin's ideas became more widely accepted, various thinkers developed this notion into what was sometimes called evolutionary ethics; whatever behavior aided evolution (invariably interpreted to mean "progress") was ethically correct. Some developed such ideas into the claim that—since competition was clearly a law of nature—war, imperialism, and capitalism were ethically desirable. Meanwhile, their opponents contended that cooperation between members of the same species aided their survival, from which they concluded that socialist or anarchist views had received evolution's ethical imprimatur (see chapter 7). Lorraine Daston has argued that naturalized ethics arose from a specific late eighteenth-century understanding of nature. As various enlightened thinkers critiqued the natural theologian's claim that God and Nature were coterminous, some began to ask whether nature alone might do the work once entrusted to God—of providing universal, invariant moral laws.[44] Daston cites a *philosophe* who claimed "Do you want to be happy? Listen to nature"—a claim clearly echoed by Bailey's demand that his readers "look upward and outward to nature."[45] The French Republic's *Déclaration des droits de l'homme et du citoyen* (Declaration of the Rights of Man and the Citizen, 1789), attributed all "public ills and corruption of governments" to the ignorance or neglect of "the *natural*, sacred, and inalienable rights of man" (emphasis added). That combination of "natural" and "sacred" as antidotes to civilization's ills was one Bailey would have embraced happily,

42. Antoine-Clair Thibaudeau 1795, quoted in Spary 2000, 10. See also "Science and the Enlightenment," in Outram 2019, 108–22.

43. Marx 1964 (2000), 75–77.

44. For a brief overview of the naturalistic fallacy and its historical use, see Daston 2014.

45. Platon Blanchard, *Catéchisme de la nature* (Nature's catechism, 1794), in Daston 2014, 583.

despite his political conservatism (a reminder that nature's authority could be—and still is—deployed to support reactionary arguments as readily as revolutionary ones).[46] Of course, many nineteenth-century thinkers rejected nature's guidance. Well before Darwin published the *Origin*, Alfred Lord Tennyson was already expressing his doubts about nature's goodness, afraid that nature "cared for nothing" and was hostile to all human hopes and values.[47] As evolutionary thinking took hold, the idea that nature might offer consolation (as even Darwin sometimes hoped) was a faith many found hard to cling to.[48] Nevertheless, Bailey was a thoroughgoing evolutionist who nevertheless retained a faith in nature's guiding goodness. As he argued in *The Holy Earth* (1916), the "final test of fitness in nature is adaptation, not power. Adaptation and adjustment mean peace, not war."[49] And, as Armitage notes, Bailey's favorite example of this process was farming: a farmer needed "to learn how to adapt one's work to nature," to adjust to the local climate and soil. For Bailey, the goal was not to conquer nature (because "nature is the norm"), but "to live in right relation with his natural conditions," which he argued was "one of the first lessons that a wise farmer or any other wise man learns" (11). For Bailey, God and Nature were not at strife, because both were equally outside us, aspects of "the things that we ourselves did not make," and thus equally available to provide moral guidance.

Yet despite his affinities with earlier thinkers, Bailey was still a twentieth-century scientist, which resulted in numerous contradictions in *The Holy Earth*. He quoted a letter from an anonymous American farmer that captured the bucolic virtues that Bailey saw as inseparable from life on the land:

> Too many people confound farming, with that sordid, selfish, money-getting game, called "business," whereas, the farmer's position is administrative, being in a way a dispenser of the "Mysteries of God," for they are mysteries. Every apple is a mystery, and every potato is a mystery, and every ear of corn is a mystery, and every pound of butter is a mystery, and when a "farmer" is not able to understand these things he is out of place. (37)

---

46. Daston 2014, 579–80; Williams 1980, 78–79. See also Daston and Vidal 2004 (especially the editor's introduction).

47. Tennyson 1850, 80.

48. Endersby 2009.

49. Bailey 1916, 83. All subsequent page references are to this edition.

Bailey's unnamed farmer claimed to be a guardian of the earth, and Bailey concurred, asserting that the farmer "does not even own his land," but "is the agent or the representative of society to guard and to subdue the surface of the earth; and he is the agent of the divinity that made it." There are many traces of modern ecological thinking in Bailey's writing, yet his choice of the Biblical "subdue" to describe humanity's relationship to the rest of nature leads him inevitably into the contradictions that the Judeo-Christian tradition bequeathed to Western notions about nature.

Bailey averred that "a man cannot be a good farmer unless he is a religious man," because only then would he recognize his obligation to increase the land's productivity, so that both his crops and fellow citizens could be fruitful and multiply (32). Dominion over the earth had to be exercised morally ("The morals of land management is [*sic*] more important than the economics of land management"). Tenants who farmed responsibly, increasing the fertility of their land through careful management, should be enabled to become its owners, because it "is hardly to be expected . . . that tenant occupancy will give the man as close moral contact with the earth and its materials as will ownership" (52). He was happy (writing before 1917) to describe America's farmers as practicing "a communism that is dissociated from propaganda and programs." Such men (and they are always men in Bailey's writing) recognized that "there is no inalienable right in the ownership of the surface of the earth"—it was given by God to all humanity, for all time; he condemned "the awful sin of partitioning the earth by force." Yet he was adamantly opposed to state ownership of the land because he shared with many of his fellow Americans a suspicion of potentially tyrannical government. Yet nor was he simply anti-government, arguing that it was "part of our duty to the race to provide liberally at public expense for the special education of the man on the land." The government should also protect farmers "from exploitation and from unessential commercial pressure," so that they should continue their sacred duty of caring for and improving the land (33–35).

Bailey's political conservatism and traditional Christian values (he described himself as committed to "the conservation of native values") help explain why his enthusiasm for scientific novelties was invariably tempered.[50] For example, when he discussed "Making new plants" in *Collier's*, he noted that the possibilities he described might seem to challenge to "the old idea—which some of us have not yet really outgrown—that all the kinds of plants and animals were fashioned at first as they now are by the Creator." But for Bailey there was no contradiction; "the handiwork

50. Bailey, quoted in Jack 2008a, 29–30.

of the Creator" was evident in the laws of nature. The discovery of evolution in no way challenged Bailey's faith; he simply concluded that "Man is a part of the evolution record . . . [a] partaker in the process, not a passive looker-on."[51] And humanity's ability to manipulate plants was simply proof of divine foresight; living things were "really very mobile and plastic," which allowed them to respond "to the changing conditions of the earth," including those wrought by humans.[52]

For Bailey, science allowed better cooperation with nature, not a conquest of it.[53] In 1903 he described the newly rediscovered Mendelian theory for the widely read New York magazine the *Independent* and gave its readers considerable technical detail. Yet, while accepting that Mendelism was important, Bailey commented that the "wildest prophecies" had been made about its impact, from which he demurred. The new theory would allow more precise and rigorous experiment, clarifying and reinforcing good breeding practice (such as "breed for one thing at a time," which Bailey translated into Mendelian terms as breeding for "unit characters"), but such practices would not—he believed—revolutionize practical plant-breeding, which would continue to seek adaptation and adjustment between people and plants.[54] Despite his interest in modern methods, Bailey never fully embraced the future-oriented speculative heredity that Mendel and de Vries helped launch.

Rinn interprets Bailey as synthesizing rival traditions, but he more often seems caught on the horns of a series of dilemmas. The image of the garden was at the heart of these contradictions (a claim that might be made for the whole Judeo-Christian tradition). Bailey firmly believed that every city home should have one, to keep busy city dwellers in contact with the soil and the rural values it contained. In *The Outlook to Nature*, Bailey described how "I step from the house, and at once I am released. I am in a new realm. This realm has just been created, and created for me. I give myself over to the blue vault of the sky; or if it rain, to firsthand relationship with the elements." (100) Bailey's writings regularly extolled the moral virtues of nature, such as the cleansing power of the "quick smell of the soil," yet he is describing a suburban garden, not the countryside (which in any case was, for Bailey, always epitomized by a farm, not a wilderness) (100–103).

51. Bailey 1905, 287; Rinn 2018, 137.

52. Bailey 1910.

53. This view was common among early American settlers, that they were fulfilling nature's/God's purposes by developing natural resources; see Nye 2003.

54. Bailey 1903a, 182.

If gardens were to bring rustic virtues to the city, suburbanites should avoid what Bailey disparaged as the flashy "look-at-me kind" of gardens: he disparaged "the American yard" that resembled "the literature of the period, in being striking, curious, or wonderful" and recommended English gardens instead, "as secluded and as personal and individual as a library or a study," a place to enjoy privately, rather than to show off to others (98–99). Yet an English garden is no more "natural" than an American one; every human garden is as artificial as a human city, and at times Bailey seemed to teeter on the brink of realizing this fact. He described gardening in ambiguous terms:

> I am part of the drama; I break the earth; I destroy this plant and that, as if I were the arbiter of life and death. I sow the seed. I see the tender things come up and I feel as if I had created something new and fine, that had not been seen on the earth before; and I have a new joy as deep and as intangible as the joy of religion. (103)

The gardener is Adam, subduing the earth, holding dominion over it, "arbiter of life and death," as he condemns the unworthy weed to the bonfire while he struggles to emulate God and create "something new and fine." Bailey ultimately believed that nature had been "created for me," and thus (because it was one of "the things that we ourselves did not make") could be a source of moral guidance. And yet he seems willfully oblivious to the artificiality of the supposed nature (farm or garden) he celebrates. Bailey's conservative combination of traditional, American, religious values made him unwilling to accept the full moral responsibility that comes with "breaking" the earth and choosing which of its plants should live or die.

Bailey's contradictory views were typical of much nineteenth-century American thought, which expressed a nostalgia for a wilderness that had long since been tamed. The scenery of the period's utopias is not wilderness but gardens or parks, and Donald C. Burt describes utopia's "well-manicured landscape" as the epitome of Marx's Middle Landscape—neither fully primitive nor entirely urbanized. Another ingredient that contributed to this contradictory vision of artificial nature was what David Nye has called "second creation" narratives, which naturalized the technologies used by European settlers, thus making their roles in the transformation (or, as their critics would argue, destruction) of the American landscape seem predestined.[55] The contradictions of managed nature are evident in the magnificent highway systems that frequently crisscross

55. Nye 2003, 6.

the landscapes of late nineteenth-century utopias; these systems carve
their way through the landscape but usually have flowerbeds and trees
planted between their carriageways. These suburban utopias relied primarily on physical (rather than biological) technologies, which would tame
and manage nature, space, and time by giving city dwellers easy access to
leisure in carefully preserved and managed pseudo-wilderness.[56]

Despite an emotional commitment to unspoiled nature, nineteenth-
century American visions invariably offered a thoroughly domesticated,
human-managed landscape. Where wilderness survived, it was closed off
in a reservation or preserve, a site for leisure and contemplation, before
returning to the bustle of reality. These characteristic themes and images
can all be found in Milan C. Edson's utopian novel *Solaris Farm: A Story
of the Twentieth Century* (1900), which embodied many of nineteenth-
century horticulture's ideals.[57] Firmly rooted in the Jeffersonian tradition
(it was dedicated to "the sons and daughters of the farms of the Republic"), the book described how a young man, Fillmore Flagg, is able to fulfil
his dream of reforming farming by falling in love with a beautiful young
heiress, Fern Fenwick, who wishes to put her vast wealth to work on just
such a project. Together the alliterative couple produce a network of cooperative farms that bring an end to rural isolation, debt, and backbreaking
toil, by ending the power of the "land monopoly" and making the "latest
and best machinery" affordable for every farmer.

Fern is a committed spiritualist and medium who is able to summon
her father (who has not died but undergone "a happy transition to a higher
life"). Speaking via "a large, peculiarly shaped trumpet of aluminum," he
dictates "the ethics of planetary evolution" to Fern and Fillmore, along
with a mind-numbingly detailed plan for the new cooperative farm movement. Death has not shaken her father's commitment to the onward and
upward vision typical of new thought (nor has it done much for his literary style); history, he avers, is "a continuous upward movement . . . from
the mineral to the vegetable; from the vegetable to the animal; from the
animal to man," who represents "the apex of progress in the constantly
ascending spiral of the evolution of life" (33, 71).

Edson made no mention of Bailey in his book, but *Solaris Farm* shared
the same progressive visions of American evolution and horticulture. The
book argued that the ideal conditions for people to develop in were to be
found "in the rural districts, far from the turmoil and strife, the smoke
and poisonous gases of the great city" (while acknowledging that rural

56. Burt 1981, 179, 182–83.
57. Edson 1900.

life needed to be improved with "healthful amusements, books, and . . . educational facilities" [79–80, 76–77]). Women would benefit from the same scientific education as men and enjoy communal childcare (which included an innovative electric-powered circular train that served as a perpetual perambulator and merry-go-round for up to three hundred children of all ages). Despite its celebration of traditional, rural values, Edson's vision was neither reactionary nor anti-technology.

Nevertheless, *Solaris Farm* was deeply committed to the nineteenth-century vision of ancestral heredity (the utopia was made possible by Fern's inheritance of her father's money and her mother's abilities as a medium). The book imagined a future in which women would be economically independent and educated but would nevertheless be expected to play their traditional and supposedly natural role in society. However, instead of being "mothers of a feeble race of puny children" ("born as the very dregs" of a "tainted heritage"), future women would produce "a race of perfect children" as the farms sought to "unite stirpiculture . . . with agriculture," by breeding only the best (174, 80–81).[58] The emphasis on the past was equally apparent in the conservationist ideas that emerged in the book: large scale tree-planting would restore "the fertility and productiveness of the valleys and plains," improve the climate, and would realize Bailey's vision of a new art by transforming the entire Rocky Mountains range into the "Pride-of-the-World-Park," a "matchless reservation" devoted to the "preservation of [the country's] great variety of natural curiosities, and of American Game" (81–83, 211).

Original sin played no part in Edson's vision, not because it had been eliminated by science but because it never existed in the first place (to assume "a dominant principle of evil . . . denies all progress, all moral reform, every noble aspiration, every good deed, all evolution, all science and all reason" [384]). Dead Dad even urges his daughter and son-in-law to recruit the rich to their "New Crusade" to spread the cooperative ideal, reminding them that "the hearts of nine-tenths of the wealthy, are good and true. Their natural promptings are to do right; to use their riches for the advancement of science, and for the cause of humanity" (428). *Solaris Farm* condemned the "old barbaric law of the survival of the fittest," but neither Darwin's nor Spencer's name occurred anywhere in the book. Nature was neither hostile nor indifferent to human values but instead offered a perfect guide to a good life. Cooperation was "life-promoting and

58. The term *stirpiculture* was coined by J. H. Noyes at the utopian Oneida Community to describe a version of positive eugenics inspired by his readings of Darwin and Galton (McGee 1891; Richards 2004).

poverty-banishing, . . . just and right, because it is the nearest in harmony with natural law." Fern's late father insists that "the object and purpose of this planet is the evolution of human beings, their continued growth and development, until the state of perfection for the entire race is reached" (72–73). No eighteenth-century *philosophe* ever expressed the naturalistic fallacy more clearly.

Solaris Farm is a deservedly forgotten book (its clichéd and banal style is excruciating). Nevertheless, it is evidence of how widespread the ideals of the nineteenth-century horticultural transformation had become. Although the novel was set in the near future, the actual early decades of the twentieth century did not see the triumph of the earlier outlook to nature. Despite the fact that biotopianism was indebted to some aspects of these earlier stories and attitudes (not least because the garden remained the central image for utopia), the focus began to shift after 1900 as the new biology created an increasing fascination with the possibilities of engineering living things.

Bailey was a self-described conservative who saw the countryside and its traditional values as a brake upon the speeding cities. Despite some similarities, his commitment to America's rural values and myths kept him from fully sharing the biotopian mood. For example, his response to state and national campaigns to eradicate invasive plants was that "weeds are beyond the reach of the sheriff."[59] Nature's laws were above human ones, so if weeds were fitter than the crops, the weeds would survive. However, numerous late nineteenth-century American utopias had already begun to imagine futures in which supposedly unnecessary species had been destroyed.[60] As we shall see, their twentieth-century successors were biotopian sheriffs, like H. G. Wells (chapters 5 and 6) and Charlotte Perkins Gilman (chapter 7), who enthusiastically imagined futures from which all weeds and pests had been eliminated.

In many regards, Bailey resembled the celebrated Californian horticulturalist Luther Burbank (Kohlstedt argues that Burbank was the only person in early twentieth-century US horticulture who was better-known than Bailey).[61] Both were fundamentally conservative in their outlook yet fascinated by new scientific ideas, including the mutation theory. In key respects, both inherited the complex American relationship to nature that

59. Liberty Hyde Bailey, "Coxey's Army and the Russian Thistle: An Essay on the Philosophy of Weediness," in Bailey, *The Survival of the Unlike*, 4th ed. (New York: Macmillan, 1901): 200–201; quoted in Pauly 2007 (135).

60. Burt 1981, 178–79.

61. Kohlstedt 2010, 85.

is embodied in Marx's concept of a natural/artificial "middle landscape." Bailey remained convinced that nature was the norm, while Burbank's claim to have created new plants led to his becoming caught up in an almost cultlike wave of publicity. As the next two chapters show, Burbank began his career as part of Bailey's world of nineteenth-century horticulture but was ultimately reinvented by his fans and transformed into an embodiment of American biotopianism.

3

Paradoxical Futures

As the USA's media market became increasingly crowded, new magazines struggled to stand out, yet fresh titles kept appearing. There were over six thousand different periodicals in print in the USA at the start of the twentieth century (and almost as many had disappeared over the previous twenty years).[1] When the *World's Work* joined the throng in 1900, its publisher, Walter Hines Page, believed his unusual mix of experience (as a newspaper man, editor of the *Atlantic Monthly*, and book publisher) would make his new monthly distinctive.[2] It claimed to provide "A History of Our Time" by offering more depth and detail than a newspaper, but each issue began with the newsy topicality of the "March of Events." The September 1901 column included a typical foretaste of the new century through a range of stories emphasizing technological progress: the country's telephone and telegraph networks were expanding; American engineers had just built the world's largest railway viaduct; and the first turbine-powered passenger steamship had been launched. The future was also bearing down hard on the recently colonized peoples of the Philippines and Puerto Rico, where the USA was engaged in "the best work now in hand anywhere in the world to build up a backward people," who were facing a "new era" and being prepared for "civilization" (figure 3.1). Some US commentators had justified the Spanish-American war on evolutionary grounds, asserting that the supposedly more advanced Anglo-Saxons had not merely a right but a duty to take control of countries whose inhabitants would, in the opinion of future president Roosevelt, require "the slow growth of centuries" before they had could match the "immense

---

1. Mott 1957, 11.
2. Mott 1957, 11, 53. A UK edition was also published by William Heinemann, which carried a different subtitle: "An Illustrated Magazine of National Efficiency and Social Progress."

FIGURE 3.1. Uncle Sam preparing to transform some of the USA's newly colonized "backward people" into productive workers. Luther Burbank's work with plants aimed at a similar transformation.

reserve fund of strength, common sense, and morality," which their colonizers claimed to possess.[3]

Meanwhile, back in the USA there had "been a larger crop than usual of the customary hot-weather labor troubles" (particularly in Chicago, where "a large force of carpenters struck because they were not allowed to have as much lemonade as they wanted"). Meanwhile, the steel workers had tried to stop "the largest combination of capital in the world" from controlling "its business affairs," but the magazine's editors were delighted to report the unions' defeat, thus ensuring that America's "industrial supremacy" would "be maintained in these days of world competition." Despite setbacks, America's industry flourished and its population rose, meaning more food would be needed, but rural America was readying itself for the challenge. Just as Liberty Hyde Bailey had hoped (see previous chapter), farmers were adopting "the improvements of modern science," which included "improved farm machinery and implements," while the rural postal and telephone networks put the countryside into "closer communication with the cities." As the population expanded, Americans would need more

3. Quoted in Gossett 1997, 328–29.

land, but the *World's Work* reported that President McKinley had recently opened the Oklahoma territory belonging to four native American nations to "white settlers." In order to avoid the "unseemly" scenes of mad races and claim-jumping that had accompanied every previous "great land opening," Congress had purchased almost five million acres, to be distributed by lot for "the public benefit." The "public" did not, of course, include the indigenous owners (the Kiowa, Comanche, Apache, and Wichita peoples), who claimed they had been threatened and tricked into signing the treaty documents. However, the magazine assured readers that the "Assistant Attorney General has vouched for the legality of the plan employed," and native Americans would not be able to block the government's plan.[4]

The *World's Work* provided one vision of what twentieth-century America would be: expansionist, racist, politically conservative, but technologically advanced—endlessly modernizing and transforming itself. Production lines and skyscrapers became emblematic of the American century, but the countryside was also undergoing technological transformation as plants and animals were modernized and reinvented to become more efficient. The same issue carried one of the earliest detailed articles about the Californian plant breeder Luther Burbank, the "plant wizard" as the press loved to call him, who argued that when "intimately observed over a long period, Mother Nature is revealed as a spendthrift and almost careless old lady, wasting and throwing her treasures about."[5] However, Burbank believed the plant breeder had the power of "mending nature," and described his own work as "adopting the most promising individual" plant (from among "a race of vile, neglected orphan weeds" with "hoodlum tendencies") and then "gradually lifting it by breeding and education to a higher sphere."[6] Unkempt nature was nasty, dirty, and—worst of all—inefficient, but Burbank was confident that humans (particularly male ones) knew better. Science would do for plants what US imperialism claimed to be doing to the people of Puerto Rico and the Philippines—civilize them by turning them into modern productive workers. As we shall see, many of Burbank's plant-improvement projects aimed to make plants more efficient, productive, and rational than nature herself could make them. Burbank himself was a creation of both America's media market and its rural mythology. And as his many admirers promoted and interpreted him, he came to personify a distinctively

4. "The March of Events" 1901, 1127–28. See Hagan 1985.
5. Burbank 1926, 365.
6. Burbank 1901, 5–6.

American vision of a future based on the commercial possibilities of biological invention.

For early twentieth-century Americans, the idea that Burbank was a technocrat—even a horticultural imperialist—would have been deeply shocking; the press celebrated him as an almost saintly, hardworking man of the soil, deeply in touch with nature. As Katherine Pandora has shown, he became the world's most famous plant breeder during these decades because he seemed to exemplify a series of distinctly American virtues: a simple, self-taught man; who served the public good by enriching his fellow Americans (but not himself); and although he invented new crops for the new century, he remained directly connected to traditional, American rural values (he even lived behind a white picket fence).[7] Yet, despite exemplifying all these bucolic values, Burbank was a creation of the twentieth century: a product (literally) of the newly reverberating media age. Pandora has argued that Burbank's popularity (and the media coverage that promoted it) forces historians of science to look beyond the conventionally defined scientific world to what she calls "science in the vernacular"—the rich variety of ways in which various publics domesticated science and its implications.[8] Building on her insights, this chapter explores how Americans would come to see Burbank as epitomizing a particular version of the promises of experimental evolution. Exploring the paraliterature that surrounded him, from scientific books to commercial plant catalogues, reveals how his various interpreters transformed him into a strange chimera of scientist, gardener, saint, and businessman—a set of contradictions that came to characterize the way Americans imagined the biological future.

Thanks to the media, Burbank became a living trademark, one whose fame depended on the rapidly expanding press market (Burbank himself acknowledged that the "printing presses are eating up our forests more rapidly than they grow"—and suggested he might need to invent a new "paper tree" to meet the demand.)[9] With over a hundred pages a month to fill, the *World's Work* was one of many new titles devouring America's woodlands; Burbank's inventions fed the demand for news stories just as his new crops promised to feed America's ever-growing population. Burbank was a master publicist who, despite his reputation for being a simple man of the people, was very much at home in the new media world and the ever-expanding advertising market. He actively promoted

7. Pandora 2001. Bailey described the picket fence in his article.
8. Pandora 2001, 487.
9. Burbank 1923, 68.

his achievements in a series of catalogues with the eye-catching title *New Creations in Fruit and Flowers*. The *World's Work* accepted him at his own estimation when they called him "A Maker of New Fruits and Flowers," who had gone beyond improving nature's gifts to actually inventing new species. The article was by Liberty Hyde Bailey, who also blended rural conservatism and the latest horticultural science, a mixture that became typical of the rapidly growing Burbank legend.[10] Bailey's article encapsulated the contradictory elements that came to characterize both Burbank and American biotopianism, and for almost thirty years, a torrent of similar stories would bring these paradoxical qualities to their largest-ever readership.

The first contradiction involved Burbank's commercial success. Like most articles on Burbank, Bailey's stressed the millions of dollars his products had made for their growers. Yet while Burbank could have become rich by selling his novelties, his enthusiasm for plants led him to make numerous experiments "for the mere zest of it," and hence "he does not make money." Burbank argued that a successful plant breeder would have "no time to make money," and Hugo de Vries celebrated "Burbank's ideals" after meeting him, adding that since he had no interest in "accumulating money," he could work for "the general welfare of his fellow beings."[11] Given Burbank's idealistic indifference to profit, Bailey argued that "some philanthropist" would "render a good service to mankind" if he were to relieve Burbank of all financial pressures and let him devote himself to research (as chapter 4 shows, "some philanthropist," in the shape of Andrew Carnegie, would indeed come forward—with rather mixed results). The image of Burbank as an altruistic servant of humanity was as prominent as reports of the commercial value of his new plants.

Burbank was also incongruously presented as both democratic and elitist. Bailey, like many other writers, described him as possessing almost saintly personal qualities: he was never boastful (but was "inclined not to talk of his work"); he was as "retiring and mild-mannered" as the "simple, vine-covered cottage" he lived in (you talk to him, Bailey reported, "and before you know it you love him"). He was a man of the people, whose science was so accessible and democratic that anyone could take up plant-breeding. And yet—like most of those who wrote about Burbank—Bailey also emphasized his unique perception, amazing memory, refined technical skills, and "remarkably acute judgment," which gave him rare, almost occult, insight (1209–10). It was these qualities that prompted many,

10. Bailey 1901a. See also Bailey 1901b.
11. Burbank 1904, 38. See also Hansen 1905, 32; De Vries 1905d, 333.

including Bailey, to argue that Burbank's work should be publicly funded. At times, this supposed man of the people seemed like an elite of one.

Burbank was also depicted as both deeply scientific and as an untaught genius. Bailey stressed that Burbank had learned "technical botanical matters" from impeccable sources (Darwin's *Variation of Animals and Plants under Domestication* was his "chief inspiration," and he possessed "much of the spirit of the great master"). The Californian's results had also been checked "with every scientific precaution," and so Bailey insisted that "Luther Burbank is not a wizard," but a scientist. Nevertheless, Bailey still described him as possessing a rare "gift of prophecy," a man who worked by intuition, never wrote anything down, and conducted experiments "as naturally as he would dig a new lily from the fields." Burbank "has not studied the books" (despite Bailey having listed several titles) and because he "has not been taught . . . he is free" (1213). The claim of untaught genius sat rather awkwardly alongside Burbank's supposed scientific credibility.

Perhaps the most paradoxical aspect of Burbank's public image was his attitude to his plants. Bailey's article claimed that "Burbank loves all plants," an idea that would be reiterated endlessly in later articles. At times Burbank was feminized, portrayed as if he were a loving mother whose children were plants, yet the same writers would describe Burbank's method as making the plant "break" through crossing—"Thereby are their customary characters upset" (1212). The vision of Burbank "breaking" and "upsetting" plants suggested a rather tough kind of "love," and numerous writers dwelt on the way Burbank destroyed failed experiments with huge bonfires that consumed thousands of plants. Alongside the endless celebrations of Burbank's gentle warmth and care, there are subtle yet unmistakable hints of an unsentimental ruthlessness.

The picture that emerged from Bailey's article, which reverberated through many of those that followed, was utterly contradictory: Burbank was an altruistic huckster, an elitist democrat, and an unnatural parent—who was both gentle (feminine) and violent (masculine). And, as chapter 4 explores, he was also presented as a scientific mystic. What linked these contradictory qualities together was Burbank's relationship to nature. He became famous just as Americans were proudly realizing they were the world's leading industrial, urbanized nation, whose food was increasingly produced by large-scale industrialized agriculture. Burbank—the radical, biological inventor—was part of that modernization. As Bailey commented, "He knows no cross that he may not attempt," and the results of Burbank's "daring" were startling; he was endlessly correcting, improving upon, or challenging the work of that "spendthrift and almost careless old lady," Mother Nature. Burbank was endlessly credited

No. 107. **Luther Burbank's** White Blackberry, Iceberg. Its creation required the sacrifice of many thousand plants.

FIGURE 3.2. Luther Burbank could be either a nurturing or harsh parent to his plants. This souvenir postcard, sold at his Experimental Grounds, celebrated his willingness to sacrifice thousands of his plants if he deemed it necessary.

with not merely improving plants but making brand-new ones—notably the Primus berry (a cross between a dewberry and Siberian raspberry)—while the media's focus on his traditional, rural values meant he also represented an incoherent hope of getting "back to nature" (albeit the "middle-landscape" version of nature, more apparent in farms and gardens than wilderness).[12] As we saw in the previous chapter, Bailey straddled many of the same contradictions but cannot be called a biotopian because for him nature remained the ultimate guide to what was good, wholesome, and American. Despite important similarities, Burbank was pushed across the line because his admirers insisted on the *un*naturalness of his work.

Burbank's contradictions were also quintessentially American, not least that he was both a Jeffersonian farmer and a twentieth-century American inventor.[13] He was often called "the Wizard of Santa Rosa" (as Edison was "the Wizard of Menlo Park"), and being the Edison of the garden meant he was tasked with embodying both America's rural mythology *and* its urban reality—nature and culture (figure 9.1). To understand what biology was doing to America's sense of its own future in these decades, we need to analyze each of the paradoxes Burbank personified.

Speculative Futures

While journalists presented Burbank as a rather otherworldly, child-like figure, they also stressed his hardheaded practicality. The attention he received in 1905 (when he received a large grant from the Carnegie Institution of Washington; see below) resulted in thousands of visitors descending on Burbank's home; large notices had to be posted stating "POSITIVELY NO VISITORS ALLOWED." The following year a notice was printed "By the Friends and Relatives of Luther Burbank," which explained that Burbank was "one of the busiest men in the world." It claimed that "Mr. Burbank has nothing for sale," because he was simply "an originator" of new plants.[14] This notice was one of many attempts to persuade people that the extraordinary claims made about his achievements were more than just a sales pitch, yet (in typically contradictory fashion) the flyer included a long list of the companies from whom Burbank's "new creations in plant life" could be purchased.

Burbank's inventions ranged from new flowers (such as the Shasta Daisy) and improved fruits (including stoneless plums and his oxymoronic

12. J. S. Smith 2009, 8.
13. Marx 1964 (2000).
14. Friends and Relatives of Luther Burbank 1906.

white blackberries) to supposedly new species (such as the Primus berry). However, his most famous creation was probably the spineless cactus, the result of many years breeding cacti of the genus *Opuntia* (commonly known as prickly pears) to try and create one that was completely devoid of thorns. Burbank believed nature's energy was being "wasted on the spicules and thorns," so he tried to teach Mother Nature better habits: "by taking away from her" the need to grow spines he had "left her free to put all her energy upon producing food."[15] Burbank predicted that his new cactus would become an important source of forage for cattle, allowing the world's semi-desert regions to be grazed, and he claimed he had also made its fruit more nourishing for people. The *Los Angeles Times* told its readers that "Luther Burbank, Plant Creator . . . Believes He Can Reclaim All of Great Southwest's Arid Vastness." The paper quoted him as saying, "The cactus will be the most important plant on earth for arid regions."[16] Meanwhile, *Sunset* magazine reported that Burbank considered the cactus as "one of his greatest successes, because it means the reclamation of the desert."[17]

William S. Harwood, one of the most energetic of Burbank's boosters, described the cactus in the popular *Century Magazine* (one of the world's most widely read monthlies, selling around 125,000 copies a month).[18] Under the typically hyperbolic title "A Wonder-Worker of Science," Harwood offered an "authoritative account of Luther Burbank's unique work in creating new forms of plant life," which included the cactus that would "convert the desert into a garden."[19] The new thornless plant had also been adapted to grow in colder climates and to produce "a delightful, nutritious food for man and beast." As a result, Burbank believed the cactus alone would "afford food for twice the people now upon the earth" (658). And Harwood stressed that Burbank's new invention was only "one of the hundreds upon hundreds of works he is engaged upon for the welfare of man" (not, of course, for personal profit). The benevolent cactus maker brought to Harwood's mind (as it would to many other commentators) the Biblical prophecy that the "the desert shall rejoice, and blossom as

15. Burbank, quoted in Harwood 1905c, 661.
16. "Wizard's Wisdom" 1907.
17. "The Sage of Santa Rosa" 1905, 542.
18. For *Century Magazine*, see Mott 1957, 1–4, 8–9, 43; LaFollette 1990, 33; Reed 1997, 105; Cairns 1907–1921 (2000).
19. Harwood 1905c.

the rose" (Isaiah 35:1). Burbank's cactus promised that humanity need not wait for paradise but could create it themselves.[20]

Harwood interviewed Burbank for his articles (which were so successful that they became a book, *New Creations in Plant Life* [1905]) and quoted him as saying that it was a "great effort on the part of the [cactus] to produce all these spines." By breeding a thornless variety, Burbank ensured that "the plant will at once become more docile and pliable, and can be easily led into almost any useful occupation in which plants are employed," including the colonization of deserts (662). Even the most apparently delinquent plant could be civilized and put to work to serve humanity. In an earlier article, he had anthropomorphized the plant as "Mr. Cactus," who insisted on wearing "all those tacks, pins and needles." He hoped to persuade Mr. Cactus that "you would look much better if you would drop those ugly thorns," and he loved to demonstrate his triumph by rubbing his face with a cactus paddle to demonstrate its smoothness.[21] (Burbank can be seen doing this in *A Visit with Luther Burbank*, a 1922 film which was seen by around five million people when shown in over four thousand theatres.[22]) Even Mother Nature's prickliest creations could be tamed by the breeder's skill.

Burbank often claimed that "nature will do her part always. She never lies; she never deceives"; yet he also claimed that "often, in the sight of man and from his standpoint, she fails."[23] Nature and her uglier creations needed human help, but fortunately, scientifically trained men were willing to come to her aid, ensuring that, in Harwood's words, "the new cactus begins a new era in its family, an era of unexampled prosperity, and the era of good will and not enmity to man" (156). The same language of modernization, efficiency, and improved productivity was used to describe both plant-breeding and America's efforts to redeem the supposedly flawed human natures of the Filipinos and Puerto Ricans (see p. 91).

The spineless cactus became a much-used example of the futuristic, economic value of Burbank's wizardry. The widely circulated *Success*

20. The same prophecy was cited in Howard 1906 (454). As Krishan Kumar (1987, 13–14) has argued, Isaiah embodies the Jewish prophetic tradition that the New Jerusalem will be here on earth, not in the hereafter (Judaism never took much interest in eschatology—God's promise was the land of Israel on earth).

21. Burbank 1901, 5–6.

22. Bailey 1906, 243–45. See also *A Visit with Luther Burbank*, Ford Education Weekly film no. 34 (Detroit, MI: Ford Educational Library, 1917, released 1922), https://search.alexanderstreet.com/view/work/bibliographic_entity%7Cvideo_work%7C2463124.

23. Quoted in Harwood 1905c, 661.

magazine (which aimed to present the esoteric ideas of New Thought [see p. 48] in a mass-circulation illustrated monthly) published a lengthy article on Burbank that hailed the spineless cactus as "the greatest thing he has ever done." The earlier claims about its impact received further inflation, with *Success* claiming that "His Edible Cactus May provide food for Four Times the world's population."[24] Even the supposedly sober *Scientific American* made similar claims when it described how America's deserts "may be clothed in the spineless cactus at no late day," in which case the plant's "value would be incalculable."[25] Scientific endorsements of the cactus's practical value could also be found in numerous biology textbooks, where it was regularly described as one of Burbank's most important economic breakthroughs.[26] And de Vries himself discussed the cactus, repeating Harwood's claim that it "may, in time, double the population of the earth."[27]

At times these visions of blooming deserts read like science fiction—alien worlds being terraformed with the help of tame, anthropomorphized plants to support human colonists. The vast profits that were supposedly on offer to those who grew Burbank's creations illustrate Michelle Murphy's observation that "speculation" is a tool for both imagining and managing potentially risky futures. Applied to financial risks, speculation attempts to profit from the future, while science fiction—often called "speculative fiction"—ponders the wider implications of possible futures—Burbank's work epitomized both meanings.[28] However, in practice the cactus revealed the considerable gap between speculation and reality.

Burbank had long adopted a strategy of selling the rights to each of his new varieties and leaving a commercial company to handle its sales. Predictably, a new one was set up to market the spineless cactus (along with some of his other plants): the Luther Burbank Company (f. 1912), one of several that bore his name at various times, which had apparently promised Burbank $30,000 for the rights to his cactus (although he would later sue them, claiming that most of this amount was never received).[29] His ambiguous relationships with these companies allowed him to emerge unscathed and blameless if they failed (as they invariably did).

24. Clark 1905, 456–58.

25. Brown 1905, 220. See also Jordan 1909, 59–64.

26. For example, Coulter 1914, 160–63; Gruenberg 1919, 452–53. See also Smith 2010, 45.

27. De Vries 1907b, 230.

28. Murphy 2015. See also Uncertain Commons 2013.

29. Smith 2010, 43–44.

The Luther Burbank Company exemplified the way in which Burbank himself had become a product, a signifier of speculative futures. Their catalogues carried an image of Burbank, with a facsimile of his signature and the words "This Seal guarantees a genuine Luther Burbank Production." It would supposedly protect unsuspecting buyers from the

> unscrupulous dealers who, taking advantage of the name "Burbank" hoist [*sic*] on the public green carnations, hardy bananas, half wild, thorny cactus for Burbank thornless ones, blue roses, seedless watermelons, cigars, soap, real estate, magazine articles, obtaining money or positions under false statements of having been in his employ, and a thousand other similar schemes.[30]

Obviously the company wanted to preserve Burbank's image as someone aloof from commercial matters (not least because that image added value to their trademark). The catalogue tried to exploit his fame while absolving him of mercenary motives by including a preface by Burbank, claiming nobody could be "a successful creator of new forms of plant life, and a successful merchant" in one lifetime. Devoting his time to "selling" would leave him none in which to create "new forms and improved varieties"—or to share them with "the world"—so "a corporation has been formed which will manage, market, and carry on exclusively the business of selling the various new forms of plant life" that he had created.[31]

In reality, Burbank's spineless cacti proved a major disappointment. When grown from seed, they soon reverted to the original thorny form, and it proved too expensive to keep propagating them from cuttings. Moreover, as both Bailey and de Vries had pointed out, cacti have thorns for a reason—to stop hungry desert animals eating them.[32] Cacti could not be grown as fodder unless they were fenced and protected. And while cacti do, of course, grow in deserts, they grow very slowly except when it rains; producing sufficient fodder for herds of cattle (and perhaps fruit for humans to eat too), would require expensive irrigation. Ignoring all these issues, various speculators (led the Luther Burbank Company) promoted the cacti, and their investors—lured by Burbank's fame and reputation for working wonders—lost their money. Burbank's reputation suffered major damage from which it never fully recovered.[33] Nevertheless,

---

30. *Luther Burbank's Spineless Cactus* 1912, 2.
31. Luther Burbank, in *Luther Burbank's Spineless Cactus* 1912, 1.
32. Bailey 1906, 243; De Vries 1906a, 645.
33. See Smith 2010 for a full account of the cactus and its failure.

his paradoxical image as both disinterested and businesslike, an altruistic huckster, helped spread the characteristic tropes of the biotopian approach across numerous genres of writing, particularly in the USA. In the process, such writing created a set of persistent expectations about the power of biology to enrich the world.

Every Man His Own Burbank

Numerous writers utilized the appealing paradox of Burbank as an elitist democrat, stressing his ordinariness and urging readers to learn from him; by copying his simple techniques anyone could participate in the future he promised.[34] Burbank's fans made use of the claim that everyone could be their own Burbank to promote a domesticated, participatory, experimental evolution. Backyard evolution was epitomized by a little booklet called *Give the Boy His Chance*, whose cover depicted the eponymous "boy" as a latter-day Tom Sawyer, clad in overalls and straw hat, with a rake over his shoulder, evidently expected to become a man like Burbank, in touch with nature (figure 8.3). The route to such self-improvement was not contemplation of wilderness or reading nature poems but conducting experiments. The pamphlet's double-page color centerfold showed Burbank's seed boxes and was captioned, "Here are more than two hundred different experiments. Space occupied 20 × 30 feet. Couldn't you spare this much space?"[35] The writer gave no details of the experiments "the boy" should be doing, nor why he should be doing them; the mere effort of planting a garden, calling it "experimental" and invoking "Burbank" would apparently imbue it with almost magical properties.

Give the Boy His Chance was produced by the Luther Burbank Society (LBS), which was effectively the marketing arm of the Luther Burbank Press, one of numerous companies established to publish the full details of Burbank's methods. Their prospectus claimed that their supposedly forthcoming books would not merely be "a source of entertainment and instruction to young and old," but would also assist in "inbreeding a lasting love for nature" (13–14). Plant experimentation was presented as equally improving to plants and readers, and the pamphlet began by asking "How shall the boy begin?"; it then responded (catechism-like), "By working with the plants themselves, by learning to understand Nature and to love her responsiveness" (15). To get "the boy" started, all his parents needed to do was provide a plot of ground, however modest; city dwellers as well

34. Pandora 2001; J. S. Smith 2009, 9.
35. *Give the Boy His Chance* 1913, 2, 15–16.

as farm boys could be transformed into junior Burbanks. (Among other things, the booklet promoted an ideal of American masculinity; girls were mentioned just once in a later pamphlet—as appreciative consumers of the improved flowers "the boy" would produce.) Once given his plot, a boy would learn "to create entirely new fruits and food plants," or even to "perfect plants which yield entirely new substances for manufacture—new chemical elements which have their definite bearing on lowering the cost of living" (16). Comparing new plants to new chemical elements, particularly at a time when there was an intense public fascination with radium, would have helped persuade these would-be backyard Burbanks that they were participating in the latest science.[36]

The LBS produced a whole series of these pamphlets, each of which implied that by taking up experimental gardening, readers were joining an important movement. Several bore a notice that announced them as "FREE PUBLICATIONS" (rather like religious tracts) and added that "those who have a genuine interest in the improvement of existing plants and the creation of new forms of plant life, whether for pleasure or for profit, are invited to apply for the free, color-illustrated monographs of the Society."[37] The pamphlets offered a homely, accessible, and morally improving version of experimental evolution. For example, *Start the Boy Right* stressed how easy it was to participate in experimental biology ("any boy could do the same thing") and hinted at its moral benefits ("every boy should be encouraged" to "see everything in nature that is good," and he should then be encouraged to preserve "that which is good and destroy that which is not good" while perpetuating "the good by multiplying it").[38] Burbank's junior male followers were encouraged to emulate the master's assumption they knew better than Mother Nature, whose works should be judged by purely human standards. Similar themes were also apparent in *How Nature Makes Plants to Our Order*, which emphasized human control over nature and concluded with a quote from Burbank: "Does this mean—do you ask—anything for the human race? Yes. In the hands of the plant breeder rests the story of the future. Upon him, more, perhaps, than upon any other agency depends the progress which succeeding generations are to achieve."[39] Inspired by such claims, young readers might come to see their backyards as places where speculative futures could be brought into existence.

36. Campos 2015. See also Curry 2016, 21–44.
37. Burbank 1913, 32.
38. *Start the Boy Right* 1914, 14, 26.
39. Burbank 1913, 32.

The Luther Burbank Press and Society were the latest of a series of companies that attempted to publish Burbank's work and cash in on his enigmatic status as an elitist democrat (his wisdom was so rare as to be worth paying for, yet it would permit everyone to participate in his work). These assumptions had been apparent in Harwood's *New Creations in Plant Life* (1905), which largely launched the Burbankian boom; it described his methods in detail, assuming that both amateurs and professionals would want to emulate him, since Burbank claimed that "any man who takes up plant-breeding with patience and intelligent interest" would benefit humanity.[40] Doubtless inspired by the success of Harwood's book, various publishers tried to produce more comprehensive accounts from which ordinary gardeners could learn.[41] (The troubled history of publishing Burbank can be glimpsed in the fact that the earliest attempt was made by Dugal Cree, of Minneapolis, in 1907; Cree Publishing Company was taken over by Oscar Binner of Chicago, circa 1911, and renamed Cree-Binner; it briefly became Binner-Wells Co. and was finally called "Oscar E. Binner Co.: Luther Burbank's Publishers." The books were eventually produced in 1914 by an entirely different company—which went bankrupt a year later.)[42]

Despite the bankruptcies, mergers, and other failures that marked this history, the paraliterature these various companies produced offers evidence of the key meanings that Burbank's interpreters managed to attach to him. For example, Binner's company produced several brochures to advertise their forthcoming books, such as *Luther Burbank: How His Discoveries Are to Be Put into Practical Use* (1911), which repeated all the standard stories with a folksy charm apparently intended to appeal to a wide audience (Burbank "has added to farm incomes a grand total in the neighborhood of six hundred million dollars," which was more than the entire estimated earnings of Standard Oil since its founding, yet was too "engrossed in his experiments" to profit from them. Meanwhile, "my friend, who had eaten Burbank potatoes all of his life . . . asked me what this man had done that was practical!").[43] Binner's brochures focused on practicality, claiming that "the scientific part of Mr. Burbank's work, important and interesting as it is, is not what the farmer needs" (10). Burbank's words

40. Harwood 1905b, 24.

41. The local papers reported that Rev William Mayes Martin of Minneapolis had arrived in Santa Rosa to begin writing the books for Cree ("Set of Books on Burbank Work" 1907).

42. Binner 1911; Howard 1945, 388–401; Glass 1980, 138–39; Dreyer 1985, 174–89.

43. Binner 1911, 3.

were to be published in five easily affordable volumes, supplemented by an unfeasibly large range of additional publications (including "several thousand separate handbooks, ranging in size from a tiny bulletin-pamphlet to two-to-three-hundred-page volumes, each dealing in the fullest possible measure with some one specific subject—some for general sale—others to be used as school-texts"). The marketing effort was to be bolstered by a series of pamphlets ("in process of publication for free distribution"). Some were to be written by Binner himself, but most were further examples of bricolage—reprinting magazine articles by both scientific writers and journalists (11).[44] The publicity materials included various celebrity endorsements: the governor of Nebraska wrote, "I should prize this book as among the choicest books in my library and will be delighted to get it"; while the governor of Washington State believed that the publication of Burbank's "methods and discoveries in popular form should prove an enduring blessing to mankind." Another Binner pamphlet, *Luther Burbank's Bounties from Nature to Man* (1911), promised would-be investors (in large, bold type) that both the Mikado of Japan and the King of Italy were keen to make Burbank's work available to their respective peoples, and so **"The popular edition will therefore have a field of about fifty million prospective purchasers to draw upon."**[45]

The utopian possibilities of Burbank's work were repeatedly stressed: the books would be "of such vast commercial and social importance as has never been told by man in any age." Burbank would succeed where God himself had apparently failed: "The Almighty made at one time all the earth that He will make, there will be no second crop, but population increases with each generation and the demand for food is just as certain to grow as the quantity of earth is to remain stationary" (11). One plant catalogue quoted Burbank as saying his "great work" with the cacti was like "the discovery of a new continent" and would bring "a new agricultural era for whole continents like Australia and Africa."[46] According to Burbank, the "very existence of the human race in its present state of civilization" depended on improving plants.[47]

And yet, despite Binner's tireless efforts, the promised books never appeared under his imprint. The work in progress was sold to a new "Luther Burbank Press," with the LBS being established to aid with marketing.

---

44. I have not been able to locate a single copy of any of the announced pamphlets, so it is more than likely that they never appeared.

45. Binner 1911, 35. See also: Howard 1945, 391–94.

46. Quoted in *Luther Burbank's Spineless Cactus* 1912 (5).

47. Binner 1911, 3, 39, 14, 21–22, 18.

The new venture's main instigator was Henry Smith Williams, a medical doctor and journalist who wrote over a hundred books on a vast range of topics, including several on science.[48] The LBS claimed that Burbank had rejected previous publishers' offers because they "were not content to wait till the work was done *as Luther Burbank would have it done.*"[49] Other attempts failed because they had had tried to limit "the field of the writings to pure science," but Burbank had held out until he could offer "the clear, practical exposition of every-day methods" which more scientific publishers "would have made secondary to theories and science" (15). By contrast, the LBS volumes would be useful anywhere "in Illinois—or Abyssinia," and would help everyone from "the woman who has but a kitchen garden" to "the man with a thousand-acre tract" (10–12, 5). Predictably, the strongest selling point for the LBS was that those who bought the books would learn Burbank's methods and profit from them ("even the layman may understand and apply and profit"). And yet, equally predictably, Burbank's supposed altruism was stressed to help sell the books (he had apparently "retired from all business . . . in order that he may devote himself to giving the whole result of his life work to the world" [8, 11–12, 10]).

The Luther Burbank Press and Society printed thousands of copies of their brochures, yet, like their various predecessors, both went bankrupt around 1915. A few years later, the New York firm P. F. Collier and Son republished the Burbank books, condensed into eight volumes and retitled *Luther Burbank: How Plants Are Trained to Work for Man* (1921).[50] Fresh promotional efforts included a booklet called *Half-Hour Experiments with Plants* (1922); almost fifteen years after Cree's first abortive attempt to bring Burbank to the world, Colliers were still promising "facts about plants stranger than fiction" as well as "expert guidance in every detail of plant culture; interpretations of the laws and principles of nature," thanks to Burbank's "priceless secrets."[51] The Colliers pamphlet also continued the long-established tradition of encouraging readers to conduct their own homegrown experiments, such as telling them "How to Burbank Your Geraniums." And, of course, the now-familiar utopian claims were repeated: "Nature has time without limit," but humans could not wait "for better and still better food and must take a hand in hastening and directing plant

48. Glass 1980, 142. According to the publisher's preface for Williams's book *Drug Addicts Are Human Beings* (1938), it was his 119th book; see https://archive.org/details/DrugAddictsAreHumanBeingsTheStoryOfOurBillion-dollarDrugRacketHow/page/n1/mode/2up.

49. Luther Burbank Society n.d. (c. 1913), 14.

50. Howard 1945, 391–94.

51. Burbank 1922.

improvement"; nature was sluggish, but Burbank showed how to increase evolution's tempo. Even worse, nature was wasteful, so readers were told how to save "plants from their own extravagances" by persuading them to fruit earlier or mature faster. And, most importantly, the experiments described were ones that "you may actually put into practice in your garden" (1).

The participatory culture that surrounded Burbank was exemplified by the way the Colliers booklet used quotes from newspapers to demonstrate the success amateurs had enjoyed by "Burbanking Flowers into New Form" or the creation of "Miracles in Garden of City's Own 'Luther Burbank.'" Above the quotes was the bold title, "What Others Have Done Working in the Manner of Burbank"—to which was added "Why Not You?" (1–4, 21–24). The publishers could easily have added many more examples from the newspapers of the day; among those they omitted was a story from the *New York Herald* titled "How to Do What Burbank Does," which illustrated the simple, cheap equipment needed to obtain wonderful results. Readers were told that "a suburban garden" could become "a small experimental station for the making of new flowers." The piece criticized those who mystified Burbank's work, insisting that experimental plant-breeding was accessible to all.[52] Among those who read this story was seventeen-year-old Hermann J. Muller (see chapter 5), who cut it out and kept it—and went on to become one of America's most important pioneering geneticists.[53] Perhaps the idea that anyone could do what Burbank did may have inspired his career? There is no way of knowing, but the article's presence in his personal archive is another example of the way ideas about the exciting possibilities of new biologies circulated back and forth between publishers, readers, writers, and scientists. The name "Burbank" (and the thousands of pages written about or supposedly by him) persuaded a vast array of ordinary people that they could—and would—participate directly in creating speculative biological futures.

Unnatural Parent

Although the interpretations of Burbank as an elitist democrat or an altruistic huckster helped make him famous, the sense that he was in some ways an unnatural parent was the most paradoxical (and thus, I would argue, the most biotopian) aspect of his public image. Despite being childless

52. Barron 1907.
53. Carlson 1981, 23. My thanks to the staff of the Lily Library, University of Indiana, who helped me locate the original clipping.

himself, he evinced a great love of children (which some ascribed to him being as "sweet, straightforward, and as unspoiled as a child").[54] He often compared plants to children and children to plants, and appeared to possess an instinctive love for both—yet was famously willing to burn tens of thousands of the plants he had raised and cared for in order to create a single new type. The media commonly portrayed Burbank as loving, gentle, and nurturing—a stereotypical mother—but the same writers stressed that he could also be a stern disciplinarian—a tough, even vengeful, father. Burbank seemed almost to be both male and female. A few writers suggested there was something "queer" about his creations, in that they were odd, peculiar, and somewhat suspicious (the word already carried a pejorative sense of referring to the supposedly unnatural; see p. 114). As we shall see in chapter 5, some British biotopians were to celebrate the idea of perverting nature; such arguments were seldom heard in the USA at this time, but popular writings about Burbank occasionally suggested that he challenged nature's normative value.

Burbank was presented as the antithesis of the intimidating, expert scientist. A key feature of his apparently domesticated science was his instinctive love of children, which became more prominent after he published an article on childcare titled "The Training of the Human Plant" (1906), which was expanded into a successful book the following year.[55] Burbank argued that children were like plants (we should "cultivate them as we cultivate plants, . . . make them the very best they are capable of becoming"). And in words that were to be much-quoted, he argued that no child "should see the inside of a school-house" before they were at least ten—as long as they were being "reared in the only place that is truly fit to bring up a boy or a plant—the country, the small town or the country, the nearer to nature the better" (129). His book was widely reviewed, and the *Scientific American* reflected a common opinion when they described Burbank as a "loving apostle of childhood, with ideas just as radical and just as sensible concerning children as concerning plants."[56] Like Bailey and many other earlier thinkers, Burbank argued that rural values brought out the best in Americans (he admitted that less fortunate city children might have to go to school early to keep them safe). The right environment was crucial, and Burbank convinced many that successful nurture could overcome the deficiencies of a child's or a plant's inherent nature.

54. Quoted in Harwood 1905b, 368. Republished in Jordan 1909, 80–81. See also Pandora 2001, 484.

55. Burbank 1906, 129; Burbank 1907. All page references are to the magazine article.

56. Berry 1908, 261.

Burbank's love for both children and plants added human warmth to magazine stories about plant-breeding. Garrett P. Serviss was an experienced journalist, specializing in scientific topics (he had a BSc from Cornell), and was also a pioneer science fiction writer.[57] He was one of many who took advantage of the opportunity Burbank presented by writing two successful articles about him for the best-selling *Cosmopolitan* magazine in 1905.[58] Any readers who had also read Harwood's articles (which had appeared in *Cosmopolitan*'s direct competitor *Century* a couple of months earlier) would have found those by Serviss eerily familiar—many of his details appear to have been lifted directly from the earlier pieces. Nevertheless, Serviss possessed a knack for dramatizing the contradictory aspects of Burbank's role as a plant parent, describing how the Californian saw potential new plants as "a myriad of dim eager faces, hidden behind nature's draperies—starved, neglected children for whom there is no room and no hope." Sadly, their mother (nature) had a "multitude of pressing duties" which left her with "no time, no thought and no place for them." These were potentially new species, yet when "occasionally, one peeps forth with momentary boldness" it was "only to be rudely thrust back" (164–65). Serviss made similar points in a later article: "nature usually frowns upon departure from her customary lines" and "stamps out independence." A conservative and inflexible mother, "she has little mercy for the nonconformist among her children." By contrast, when Burbank spotted a potential new flower variety, he "would be the friend of this friendless child of nature," rescue it and so provide "the chance that nature denied it."[59] According to Serviss, Burbank was a better mother than Mother Nature herself.

Harwood had described Burbank's character in terms of the stereotypically feminine virtues of "gentleness" and "tenderness," as someone who believed that "trees, plants, flowers" all need "loving care and respond to it as far as is in their power," but plants' responses paled alongside those of children (315). In similar vein, Edward J. Wickson, who wrote several successful articles on Burbank, described his character as combining masculine characters ("masterful traits") with conventionally feminine ones ("a spirit of exquisite tenderness").[60] And the Luther Burbank Company's official catalogue described how Burbank had created his spineless

57. For Serviss, see Fisher 1929.

58. Serviss 1905b; 1905c. For *Cosmopolitan*, see Mott 1957, 6; Reed 1997, 50–51; LaFollette 1990. For Serviss, see Fisher 1929; Searles 1947.

59. Serviss 1908. For the *Chautauquan*, see Mott 1957, 54.

60. Wickson 1902, 12.

cactus by selecting just seven plants from a hundred thousand that seemed promising; these he watched over "as carefully as a mother her nursing babe" (while the other 99,973 would have been destroyed).[61] The regular emphasis on Burbank's maternal qualities made the way he sometimes treated his "plant children" almost shocking. Wickson noted that crossing plants could be compared to shuffling the cards in a pack, but felt this analogy was "too gentle" to describe Burbank's practice; it would be better compared to "the use of dynamite," since Burbank violently "explodes" what "has been" in order to release "that which is to be."[62] Serviss also argued that hybridization "shattered a plant," thus creating "a myriad of variant forms" and expanding the horizon of horticultural expectation. Burbank then examined the results, which he found "as different as the faces in a crowd." It was only after careful scrutiny that the inscrutable Plant Wizard "picks out a few, a very few—sometimes but a single one from among thousands—and decrees that these only shall live," while the rest perished (64–65). The University of California's Vernon Kellogg also emphasized Burbank's concern for "the few tenderly cared for little potted plants" in his garden, while noting that he mimicked "Darwin and Spencer's struggle and survival in nature" by creating "great bonfires of scores of thousands of uprooted others, the unfit."[63] Picking a few lucky faces from crowds of thousands recalls America's selective immigration policies, and if Burbank's ruthlessness is seen as a form of plant eugenics, it offered dark forebodings of where such policies were to lead. De Vries told readers that while Burbank was breeding a new kind of blackberry, "he picked out the best from 60,000 specimens, . . . dug up the rest and burned them."[64] A souvenir postcard (figure 3.2) sold at Burbank's home celebrated this achievement and explained that the white blackberry's creation had required "the sacrifice of many thousand plants." Burbank played the role of natural selection itself, exhibiting the harsh and uncaring qualities that led some to conclude that nature was indeed devoid of moral guidance (see chapter 5).

Burbank's bonfires implied that some plants had natures upon which no amount of nurture could ever stick; only his firm methods would eventually achieve what nature could not. According to *Scientific American*, Burbank proved that "plants could be made to respond to a dominant will"; every aspect of the plant "might be controlled or altered" to produce

61. *Luther Burbank's Spineless Cactus* 1912, 12.
62. Wickson 1905, 14.
63. Kellogg 1909, 88–90.
64. De Vries 1905d, 340.

new types, "never dreamed of or imagined."[65] Serviss made the same claim: that Burbank proved there was no aspect of a plant that was "beyond the reach of human interference," so nature could be "made to follow the dictates of man's wishes" (69). As a result, he was giving humanity new plants "*such as have never been known before*" (68, original emphasis), some so different they could "only be described as new creations," while a few even "bridge the supposed impassable chasms between species, and between genera." (67). Serviss stressed the modernity and unnaturalness of Burbank's work when he observed that his "experiments prove that the plant-world is plastic to human touch, and that we may shape it at our will" (63). And the similarities between Burbank's work and the goals of experimental evolution was explicit in the assertion that "*man* can produce species and do it in a dozen summers!" (68, original emphasis). The common language of dominance aligned Burbank with conventional masculine values, long associated with science's mastery of feminized nature (a topic that later chapters will return to).

Burbank achieved more than unaided nature because he supposedly knew better than Mother Nature. When the governor of California, George Cooper Pardee, spoke at a banquet in Burbank's honor, he called him "a genius capable of playing tricks with Nature" who had "set the seal of his disapproval upon much that to him and us seems wrong in Nature's handiwork."[66] And Harwood noted how Burbank's skills helped him in "supplementing Nature where necessary, tenderly outwitting her, if needs be" (233–34). Such claims offer a sharp contrast to Burbank's assertion that "all scientists" had to learn to listen "patiently, quietly, and reverently to the lessons, one by one, which Mother Nature has to teach."[67]

In addition to repeatedly celebrating Burbank's ability to lead (or force) nature into new pathways, the press occasionally hinted that there was something unnatural, perhaps even monstrous, about his creations. In 1907, the *Los Angeles Times* previewed the city's Pacific Lands and Products Exposition under the headline "Plant Freaks to Be Shown," adding "Wizard Burbank Will Exhibit Some Queer Ones." The paper told its readers that "the greatest single attraction" would surely be Burbank's recent work; one of the organizers had "solemnly stated" that Burbank would exhibit "a tree, part palm and part oak, on which will be growing simultaneously oranges, apples, bread fruit, mangelwurtzels [*sic*], watermelons and sweet potatoes." The secretary of the Santa Rosa chamber of commerce

65. Brown 1905, 220.
66. California State Board of Trade 1905, 16.
67. Burbank 1895, 59.

admitted that the bizarre forecast of a palm/oak tree was a slight exaggeration but added that "plant freaks almost as inconceivable would be on exhibit." The journalist noted that the "Burbankisms" on display would include his celebrated spineless cactus, whose possibilities "are of course staggering—most things Burbankesque are." Since it could be eaten by both cows and people, visitors to the exhibition would be able taste the future—both candy and cake made from Burbank's cacti would be on sale. Perhaps the "thrifty housewife" of the future would avoid expensive meat and "hie blithely out onto the desert and lay in a supply of cacti."[68] While Europe had scarcely begun daydreaming about a biological utopia, America seemed ready to bring it to market.

Despite celebrating Burbank, articles like that in the *LA Times* frequently border on parody, and comic examples of "Burbanking" nature became routine examples of impossible claims (for example, somebody claiming they would "Burbankize a breed of roosters that lay three eggs a day").[69] A correspondent in the US agricultural weekly the *Country Gentleman* commented that even one of the "great daily papers" had repeated a supposed news story about a man in California who had crossed a guinea hen with a parrot to produce a talking chicken and "hoped to become the poultry Burbank of the Pacific Coast."[70] And the *Atlantic Monthly*'s 1908 review of recent biology books (see p. 19) commented that, thanks to experimental evolution, "a benevolent and all-powerful despot backed by a scientific commission could 'Burbank' the soberness of Jew or Chinaman into the most drunken of races, and make the saloon as innocuous as the public library," a lighthearted example that nevertheless demonstrated the potentially dystopian aspects of applying Burbank's methods (at least for saloonkeepers).[71]

The boundaries between journalistic exaggeration and full-blown fiction became distinctly porous around Burbank; roosters that laid eggs might seem no more unlikely than cacti without spines; the "Burbankesque" promised some strange but enticing possibilities. At the same time, joking about "freaks" and "Queer Ones" might also hint at a slight anxiety about various kinds of perversity; the disapproving implications of the word "unnatural" occasionally haunted the writing around Burbank. (It is intriguing, for example, that the *OED* records one of the earliest American

68. "Plant Freaks to Be Shown: Wizard Burbank Will Exhibit Some Queer Ones" 1911.
69. Meagher 1919, 204.
70. Massey 1907.
71. Brewster 1908, 123.

appearances of "queer," meaning homosexual, just a few years later—in the *LA Times* where this report on Burbank appeared.)[72] When Harwood described "How Luther Burbank Creates New Flowers" for the readers of the *Ladies' Home Journal,* he admitted that Burbank's experiments sometimes produced "strange forms . . . queer efforts of Nature to accomplish something, she seemed not to know what."[73] He had made the same point a couple of years earlier, telling *Century Magazine*'s readers that Burbank's experiments sometimes "produce a whole series of monstrosities, the most strange and grotesque plants that ever took root in the soil of the earth. Some of these plants are hideous, and all such are put to death."[74] (And if imitation is indeed the highest form of flattery, Serviss flattered Harwood mightily when he acknowledged that Burbank's experiments sometimes went awry, so that he accidentally created "vegetable monsters, which ought not to live!" [164].) The hint of horror engendered by Burbank's "queer" "plant freaks" may explain why Serviss tried to absolve Burbank of blame for creating these monsters ("useless," "repellent," and "unfit to live"); they emerged, he explained, "from the deep of the past," and "nature's past, like that of a human life, is not made up entirely of beautiful and desirable elements." It was nature, not Burbank who was to blame because "she has had her tragedies and her sins and the memory of these can never be eradicated" (415). Luckily for nature, the world-famous Californian breeder was ready and able to "put to death" her more shameful progeny.

While most magazinists distanced Burbank from anything monstrous, others suggested that his achievements were an affront to God (as was his use of the term *creation* to describe his work). Harwood claimed that Burbank had once been "forced" to listen to a sermon that had denounced him "as a foe to God and man, one who was interrupting the well-ordered course of plant life, destroying forces and functions long established and sacred, reducing the vegetable life of the world to a condition at once unnatural and abnormal" (357). (Harwood, characteristically, gave no details of the incident, but it circulated widely enough for the West Coast

72. *Oxford English Dictionary,* s.v. "queer": "He said that the Ninety-six Club was the best; that it was composed of the 'queer' people . . . He said that the members sometimes spent hundreds of dollars on silk gowns, hosiery, etc. . . . At these 'drags' the 'queer' people have a good time," citing the *Los Angeles Times,* November 19, 1914: ii. The word was used as a noun in Britain slightly earlier. "Queer" was apparently in widespread use in the US armed forces in the period immediately after World War I (Canaday 2009, 67, 81).

73. Harwood 1907, 11.

74. Harwood 1905c, 669.

Congregationalist magazine the *Pacific* to hope that if any minister "ever did exactly this," they would know better than to attack the saintly Burbank and not be "so rash in the future.")[75] Hints of religious discomfort about the unnaturalness of Burbank's work may also have arisen because of the moralizing language with which his work was regularly described. As noted above (see p. 93), Burbank claimed plant-breeding could redeem promising plants from "a race of vile, neglected orphan weeds with settled hoodlum tendencies," but cautioned that such improvements required aggressive interventions in the downtrodden plant's life.[76] Harwood quoted these words in his *New Creations in Plant Life*, and added that, just as it might require a shock to persuade a human to forsake a life of crime, a plant might need an "overpowering shock" if it was to "irrevocably break with the past. As in the case of the man, so with the flower" (167–68). And those "hoodlum" plants who failed to reform were not given a second chance but were "gathered in large bonfires and burned" (39). The *Open Court* also claimed that some plants showed "outlaw tendencies" (a product of "the blood of atavism") and that if they showed "no definite or hopeful perturbations, there is a massacre."[77] And Charles Shinn (see p. 173) described how Burbank examined "millions upon millions" of seedlings before "choosing, destroying," thus making him "their very god incarnate."[78] Burbank was depicted as both creator and vengeful Old Testament deity—as much a relentlessly strict father as a loving mother.

The utopian possibilities of experimental evolution were always entangled with the darker, oppressive side of managing or controlling nature. Burbank's remorseless destruction of his failed offspring could only be justified if the result was improved plants that were better than their "hoodlum" forebears. Monstrous errors threw doubt on the whole redemptive mission, as did any hint that what Burbank did was unnatural. In his autobiography, Burbank claimed that "a magazine writer who should have known better" once published a story titled "Burbank versus Nature," which he claimed enraged him because "there never was a man lived on this planet who had more respect for Nature."[79] And yet, Burbank himself had described Mother Nature as a profligate whose works were disfigured

---

75. It appeared in several local US papers, e.g., *Covina Argus* (CA), May 27, 1907: 7; *Holbrook Argus* (AZ), May 27, 1907: 2; and in both Britain and Canada (Howard 1906; Graham 1905). Burbank himself repeated it in his autobiography (Burbank and Hall 1927, 77–78). See also "The Pacific Coast in Brief" 1905.

76. Burbank 1901, 5.

77. Woodbury 1910, 307.

78. Shinn 1901, 10.

79. Burbank and Hall 1927, 84. I have been unable to locate an article with this title.

by such things as ugly, wasteful spines, and who sometimes failed to meet human needs and expectations. And there were even occasional hints in Burbank's writing that odd or queer plants were just what he was looking for: when breeding "we destroy much that is unfit," yet he was "constantly on the lookout for what has been called the abnormal." Such monstrosities might be the starting point for a new variety; as he asked, "How many plants are there in the world to-day that were not in one sense once abnormalities?"[80] Burbank regularly contradicted his own claim that nobody "had more respect for Nature."

Burbank asserted that the environment shaped a plant over its evolutionary history, so the "plant's lifelong stubbornness" had to be "broken" to force "a complete and powerful change in its life." The contrast between ancestral and speculative heredity was apparent in his desire to overcome a species' ingrained habits. His willingness to "break" stubborn plants and reshape them sat uncomfortably alongside his repeated analogies between plants and children. When it came to plants, "the hope of all progress" rested on "rigid selection of the best and as rigid an exclusion of the poorest," which inevitably raised the question of how those children that Burbank and others called "abnormal" were to be treated: "Shall we," he asked, "as some have advocated, even from Spartan days, hold that the weaklings should be destroyed?" His answer was an unambiguous "No"; they needed to be loved and nurtured, and he was confident that "the cultivation of abnormal children" would transform them "into normal ones."[81] The basis of this claim was that children's natures were far more "sensitive and pliable" than those of plants, and so they could be improved with much gentler means ("the influence of light and air, of sunshine and abundant, well-balanced food," combined with "music and laughter"). By using such means, "we may teach them as we teach the plants to be sturdy and self-reliant."[82] Despite the supposedly close plant/human analogy, it seemed children were to be treated to Burbank's gentle, feminine side, while plants suffered the harsh, masculine one.

The Training of the Human Plant advocated a version of eugenics, but it was as idiosyncratic as most of Burbank's other opinions. As will be explored in more detail in the book's conclusion, mainstream eugenics assumed the supremacy of nature over nurture, but Burbank was one of several who used his understanding of current biology to reverse this assumption. Like many of his contemporaries, he believed in Lamarckian

80. Burbank 1907, 50–53.
81. Burbank 1907, 50–51. See Pandora 2001, 512.
82. Burbank 1906, 137.

inheritance—the claim that acquired structures and behaviors could become heritable (see p. 12). He contended that "heredity is the sum of all past environment," which implied that improving an environment would improve the organisms that developed within it—whether plants or people.[83] America would gain better children if they were raised in more natural, healthier surroundings, a claim which led to numerous editorials about the desirability (or not) of "Burbanking children." Many women, particularly progressive educators, used this aspect of Burbank's thinking to oppose standardized education and rote learning.[84] Burbank's *Training of the Human Plant* was probably the most popular book to make one of experimental evolution's key arguments—that what was true of plants was true of people. And the fact that a childless plant breeder came to be regarded as a potential expert on childcare is evidence of the new biology's imaginative power. Although the unnaturalness of his work was less explicit than it would become in the work on the British biotopians, his fame was nevertheless fueled in part by the appealing paradox of Burbank's slightly queer relationship with Mother Nature.

83. Burbank 1907, 81–82; 1921, 116. The phrase was also regularly quoted in articles about Burbank, e.g., Harwood 1905d, 836; De Vries 1905d, 338; Wickson 1905, 10; De Vries 1907a, 676; Jordan 1909, 16; Williams 1915, 268. For Burbank's Lamarckianism, see Gould 1996, 290–91; J. S. Smith 2009, 139–40; Crow 2001, 1391; Stansfield 2006, 95.

84. Pandora 2001, 511–12, 515. She also argues persuasively that this aspect of his work helps explain his lack of credibility in the eyes of some male scientists.

4

Hybrid Futures

Was Burbank a scientist? Because experimental evolution was still a new, comparatively ill-defined field, it was not obvious who should be expected to answer this apparently straightforward question. The question was further complicated by the fact that Burbank's reputation included a hint of the spiritual, which appealed to those Americans who yearned for values that transcended materialism (in every sense). Fans of New Thought interpreted evolution as a grand narrative of metaphysical progress. Others found moral values in the American landscape. And, of course, those who campaigned against Darwinian evolution were, in part, protesting against the way science seemed to rob their lives of meaning.[1] Media representations of Burbank often tried to appeal to the values of these varied audiences, which resulted in the Californian grower being presented as a kind of scientific mystic, supposedly unifying two incompatible worlds. This representation was another of the paradoxical aspects of Burbank's public image that contributed to his extraordinary fame. The popular American biotopianism embodied in Burbank was rooted in science, yet somehow transcended it, which added fresh strands to the language that surrounded biology.

Rather appropriately, the idea that Burbank was both scientist and mystic first emerged from a commercialized celebration of natural beauty—*Sunset* magazine, which had been founded in the late nineteenth century by the Southern Pacific railway company to promote travel to California. Along with its competitor *Overland*, the magazine played a key role in creating the iconography of natural California: big trees, big mountains, and vast fields of golden wildflowers.[2] In November 1901,

---

1. Livingstone, Hart, and Noll 1998; Numbers 1998; Numbers and Stenhouse 2001; Lindberg and Numbers 2003; Clark 2008.
2. Shapiro 2019.

Sunset offered its readers stories about vacations on Lake Tahoe and griz-
zled cowboys, together with the first of a series about another supposedly
natural Californian icon, Luther Burbank. These articles were where the
eye-catching claim that Burbank had actually created a new species—the
Primus berry—first appeared, so when the mutation theory's claims about
new species began to appear in the American media, Burbank's achievements were regularly mentioned (which, as we shall see, brought him into
an awkward relationship with de Vries).[3] Articles like *Sunset*'s reinforced
both Burbank's scientific credibility and the more otherworldly aspects
of his thought, a mixture that would become a source of tension between
him and the scientific community that he aspired to join.

Sunset published three articles on Burbank that were republished as a
single pamphlet, which at first glance placed Burbank in an unambiguously
scientific context because they were written by Edward J. Wickson, dean
of the College of Agriculture at the University of California and director
of their Experiment Stations.[4] Wickson claimed Burbank was well-read
in biology, and similar claims were made by his colleague, Winthrop Osterhout, the university's professor of botany (see p. 46), who was quoted
as saying that Burbank had become "widely known to scientists" because
"he resembles Darwin" in his ability "to penetrate behind the facts to the
laws." Supported by this testimony, *Sunset*'s articles confidently asserted
that Burbank's "achievement in science . . . will forever link his name with
those whom the world counts greatest in the interpretation of Nature."[5]

3. As with so many Burbank-related claims, this one was copied from writer to
writer, invariably without acknowledgement. *Century Magazine* claimed that the Primus berry "disproves the dictum of scientists that new species cannot be produced
by man" (Harwood 1905c, 666). And a few months later *Cosmopolitan* repeated that
it was "the first fixed species . . . ever artificially produced" (Serviss 1905c, 68). The
British evangelical magazine the *Quiver* claimed that the berry "gave the first shock
to the theorists," who said new species couldn't be created (Howard 1906, 455). The
New York Sun called it "the first recorded fixed species produced by man" ("Wizard of
Horticulture" 1911). And the book *Luther Burbank: His Life and Work* described it as
"the very first new species of plant ever produced under conscious human direction"
(Williams 1915, 84–85). See also Largent 2000, 70.

4. Wickson had originally been "Professor of Agricultural Practice," and he had
also worked as a journalist and editor of the *Pacific Rural Press*. Much of his work
entailed bridging the gap between the farming community and those who wanted to
see the University of California become a major academic research institution (Smith
2012, 35–52).

5. The articles were "Man," *Sunset* 8, no. 2 (December 1901): 56–68; "Methods,"
Sunset 8, no. 4 (February 1902): 145–56; "Achievements" [fruit], *Sunset* 8, no. 6 (April
1902): 277–85; and "Achievements" [flowers], *Sunset* 9, no. 2 (June 1902): 101–12. All

The articles and their claims reverberated widely: the popular US monthly *Current Literature* quoted Wickson to demonstrate that Burbank had helped elevate "horticulture toward the lofty plane of biology"; and, de Vries himself told readers of the New York–based magazine the *Independent*, he had prepared for his first visit to Burbank using Wickson's articles (which were "as valuable for scientific as for practical purposes").[6] Nevertheless, Wickson's writing also seemed to cast doubt on Burbank's scientific status: visitors to his home would find he had "no library, no laboratory, no case of medals and certificates" and "indulges in no display of instruments and accessories." The implication that Burbank's genius was innate, not acquired, meant his biological imagination was not restrained by orthodox learning (he had "cast aside the elaborate armament of his scientific brethren lest it should impede his movements"). Burbank was presented as a practical man, seeking results not theories, but also as something more than a scientist. Wickson described him as a "gifted seer" and a "prophet," one who entered "fields . . . the scientific worker would have ignored," and who worked "*above* the pathway of the contemporaneous scientists" (emphasis added). It was thus unsurprising that the scientists "should fail to recognize him for a time" (9–11, 29, 15).[7] He quoted Osterhout's claim that Burbank was a "true scientist," but not because he had spent long years grinding away at careful experiments; Burbank "seems to discover great laws by a flash of genius, such is the swiftness of his intuition" (30). Yet this image of untutored genius was somewhat undermined by such claims as the one that Burbank's gardens were a "a laboratory for the study of variation" (despite the fact that he supposedly had no laboratory), and that he possessed "patience akin to Darwin's." Wickson was convinced that Burbank's scientific credentials meant his achievements could only "be judged . . . by his peers, men of science" (29), yet by writing this sentence in *Sunset* magazine, he was inviting a much broader audience into the conversation.

As noted previously, Burbank described his approach with the claim that scientists had to learn to listen "to the lessons . . . Mother Nature has to teach"; a humble attitude was vital, since "she conveys her truths only to those who are passive and receptive," truths that had to be followed

---

citations are to the booklet, whose text was unchanged from the articles: Wickson 1902, 5, 9, 13, 29–30.

6. "Burbank, the Horticultural Wizard" 1905; De Vries 1906b, 1135.

7. Wickson (the son of a Presbyterian minister) seems to hint that Burbank resembled Christ on the Road to Emmaus. See also Smith 2012, 35–36.

"wherever they may lead."[8] Wickson, like many of Burbank's admirers, interpreted this claim as meaning that a purely materialistic view of nature was inadequate (18), so it is somewhat surprising to find the same quote in a paper titled "Scientific Aspects of Luther Burbank's Work," by a young academic scientist, Vernon Kellogg (who wrote *Darwinism To-Day*, which Edith Wharton read; see p. 20).[9] Kellogg was among the first American biologists to adopt the newly rediscovered ideas of Mendel, and he saw in Burbank a chance to demonstrate both the progressive values and practical usefulness of modern biology.[10] He had joined the newly founded Stanford University in the 1890s, where he co-taught a popular course on evolution with Stanford's president, David Starr Jordan (also a strong supporter of Burbank).[11] Jordan would later write that Burbank was a botanist "in the highest, the original meaning of the word," whose "special field is that of plant genetics; here he is artist as well as scientist."[12] Jordan was far from unusual in depicting Burbank as a kind of hybrid (between art and science, in this case). And Jordan would later quote Burbank (in the *Popular Science Monthly*) as claiming that a "dead material universe moved by outside forces is in itself highly improbable," hinting that even scientists needed to attend to its higher, spiritual dimensions.[13]

Burbank's comment about the implausibility of materialism was retained when the article was republished (alongside Kellogg's) as *The Scientific Aspects of Luther Burbank's Work* (1909), in which Jordan tried to manage the incongruity of Burbank's more mystical language: "Whether we accept this or not, whether or not indeed we can conceive what it means," Jordan wrote, such unorthodox ideas were essential to "the progress of knowledge" since they might suggest "many new avenues of experimentation."[14] The authors carefully avoided offending Burbank, since they were trying to enlist him in their campaign to increase public support for science.[15] They even arranged for Burbank to be appointed

8. Burbank 1895, 59.
9. Kellogg 1909, 111.
10. Largent 2000, 4–12, 62–63.
11. Jordan had been important in lobbying Carnegie to get Burbank his grant, convinced that his work could provide raw data to support and enhance experimental evolution. See, for example, in a letter from David Starr Jordan to Andrew Carnegie (March 2, 1904), Jordan asked CIW to give Burbank a grant of "perhaps $10,000 a year." He made very similar points in a letter to Walcott, written on the same date: Burbank 1904–1905.
12. Jordan 1921, 29–30.
13. Jordan 1905, 225.
14. Jordan 1909, 77.
15. Largent 2000, 67–68.

"Special Lecturer in Evolution" at Stanford, at $300 per annum (he only gave two lectures a year, and the unorthodox nature of his own education meant these "lectures" consisted of Kellogg or Jordan questioning Burbank in front of a classroom of students).[16] *Scientific Aspects* was part of the campaign to recruit Burbank, but that project required interpretation—translating him into more conventionally scientific contexts and language. *Scientific Aspects* was another example of bricolage, making use of the paradoxical public image that the media had created by adapting it to their specific needs; in its introduction, Jordan argued that to call Burbank a "wizard" injured his scientific reputation and trivialized his achievements, yet he also insisted that "Burbank's ways are Nature's ways" (echoing Burbank's own views).[17]

Burbank's popularity made him potentially valuable, but that same popularity also made it impossible to separate his scientific achievements from his more esoteric claims. The image of Burbank as a scientific mystic was central to his fame, not least because it was such a useful paradox for journalists to hang their stories on: Harwood quoted Burbank as saying that the "universe is not half dead, but all alive," which inspired Harwood to describe Burbank's work as "teaching a plant to change its mode of life."[18] Such claims led one British religious magazine to describe Burbank as "an explorer into the infinite" whose work helped unite science and religion.[19] Meanwhile, the US Theosophical magazine *New Century Path* argued that "the vegetable world seems to recognize in [Burbank] the sympathetic nature which can call out its latent forces," supposedly proving that science glimpsed only a fraction of the larger truths of spiritual evolution.[20] The spiritualized evolution talk that grew up around Burbank echoed Liberty Hyde Bailey's claims about the need to "look upward and outward to nature" (see p. 79). For early twentieth-century Americans, dismayed by the ever-increasing incursions of urbanization, industrialization, and technology into their lives, Burbank offered a future that preserved much of what they valued about their imagined, rural past.

However, the hybrid future that Bailey and Burbank promised, which recognized the need for progress while conserving valued traditions, depended as much on science as on nostalgia. Mass production, mass

16. Largent 2000, 69.

17. Jordan 1909, x.

18. Harwood 1905d, 824, 837. Katherine Pandora 2001, 505–6 has argued that Burbank's vitalism was a key aspect of his popular appeal, while also being the least acceptable to the scientific experts.

19. Howard 1906, 451.

20. Malpas 1904.

marketing, and the mass media persuaded most Americans that there could be no going back to earlier days, so it was essential that those who wrote about Burbank emphasized his scientific status just as firmly as they did his spirituality. Harwood's "authoritative account" of Burbank's work began by quoting de Vries, who had told a gathering of scientists that Burbank was "a great and unique genius" whose creations surpassed California's native plants. Claiming that a desire to see Burbank and his gardens had been his chief motive for visiting America, he added that Burbank's achievements could only have been made by "one possessing genius of a high order."[21] And Harwood told his readers that more recently, Burbank had been invited to address an audience of "five hundred professors and post-graduate students," a remarkable spectacle because a few years earlier the more unnatural aspects of his work had had supposedly led to him being "denounced by scientific men as little less than a charlatan, a producer of spectacular effects, a seeker for the uncanny and abnormal, an enemy to all true scientific progress, a misleading, though powerful, prophet of a new order of things that could never come to pass" (656–57).[22] Harwood's summary blended the unnatural parent with the scientific mystic, transforming Burbank into a hybrid more surprising than any of the plants he created.

The paradox of Burbank's scientific-mystic status was dramatically highlighted in 1905 when he was granted his clearest scientific recognition yet, in the shape of a huge grant from the Carnegie Institution of Washington (CIW), widely reported to be $100,000 over ten years.[23] The press was entranced by the spectacle of the supposedly once-scorned scientific mystic being not only respected but funded by the scientific community, and their coverage of Burbank reached new heights of hyperbole. Among its founding goals, CIW had been tasked with discovering "the exceptional man . . . whenever and wherever found, inside or outside the

21. De Vries, quoted by Howard (1906, 454), who says the speech was given at a dinner in San Francisco in July 1904 but offers no specific source. The same passage occurred in Gregory 1911 (218), which also gave no source for the quotation.

22. Harwood, characteristically, gave no sources for these claims.

23. Harwood 1905c, 658. The grant was for $10,000 a year but its full term was not specified. Carnegie's president, R. S. Woodward, wrote to Burbank's close friend Judge Samuel F. Leib, complaining about press reports that the grant was to be for ten years, stressed that "no such agreement exists." The $10,000 per annum was only available "at the pleasure of our Board of Trustees; so that we are quite at liberty to cut off his appropriation whenever it may appear desirable to do so." See R. S. Woodward to Judge Samuel F. Leib, August 4, 1908, San Francisco, California, in Burbank 1908. Woodward made the same point in other correspondence and the grant was canceled after five years.

schools," and to provide the support that would allow such a man to do the work only he could accomplish.[24] The institution had been founded by the Scottish-born steel millionaire Andrew Carnegie in 1901 to address what he perceived as a lack of original, basic research in the USA (at a time when the country had no real government body for national science policy).[25] Carnegie was proud of being a self-made man and insisted that funding should not be restricted to well-established researchers (there was an explicit policy of judging recipients on their individual merits, not their institutional affiliations).[26] CIW's archives include numerous examples of people who wrote in, apparently spontaneously, to urge that Burbank was exactly the kind of "exceptional man" the institution had been founded to support, and several sent copies of Wickson's *Sunset* articles to describe the "remarkable work" he was doing.[27] Once again, journalism became scientific evidence which, together with letters from prominent Californians and Carnegie's direct personal intervention, eventually led to Burbank being granted the enormous sum of $10,000 a year (for a period widely but erroneously reported as ten years).[28] It was the second-largest grant that CIW had made up to that point (and the largest to an individual). Both press and public were fascinated, prompting Wickson to write a further article for *Sunset*, which highlighted the paradox that "the greatest establishment for the promotion of original research in the world" was funding Burbank—who had never "had a day's scientific training."[29]

Soon after the grant was awarded, the California State Board of Trade organized a banquet in Burbank's honor at which numerous dignitaries praised his contributions to both the state's economy and science. Jordan asserted that when scientists like de Vries met Burbank, "they knew

24. Carnegie Institution of Washington 1902, xiii.

25. Reingold 1979, 314.

26. Kohler 1991, 15–16; Reingold 1979, 320.

27. Examples include Louis Blankenhorn to Secretary, CIW, May 14, 1902; Thomas L. Gulick to Daniel Coit Gilman, March 18, 1902; both in Burbank 1902–1903; and W. W. Morrow to C. D. Walcott, February 14, 1904, in Burbank 1904–1905.

28. J. S. Smith 2009, 179–81. Glass (1980, 134) believed there was a "verbal commitment" from someone at CIW to continue the grant for ten years, but he offers no evidence apart from Burbank's indignation when the grant ended. The archives of CIW contain no evidence about the term of the grant; see below for full details. For more details, including Carnegie's intervention, see Reingold 1979, 318–22.

29. Wickson 1905. Much of this article was republished by other papers, e.g., as "Luther Burbank's Hypothesis," *Imperial Press* (CA), May 13, 1905: 8, accessed via the Library of Congress's *Chronicling America* project, http://chroniclingamerica.loc .gov/lccn/sn92070143/1905-05-13/ed-1/seq-8/.

TO THE CARNEGIE
$10,000,000
PENSION FUND
FOR
COLLEGE PROFESSORS

Life.]   [New York.

Mr. Carnegie's Gift to the Schoolmasters.
" The Rush Begins."

FIGURE 4.1. The scale of the Carnegie endowment created widespread excitement, especially among academics. Some of Luther Burbank's admirers recommended him as exactly the kind of "exceptional man" that Andrew Carnegie wished to fund.

him for a man of science."[30] And Judge William Morrow (one of those who had lobbied CIW) asserted that "Mr. Carnegie believes that man is destined to become an absolute ruler in the kingdom of Nature," and that the institution's funding of Burbank would help achieve this clearly biotopian goal. Morrow evoked *Gulliver's Travels* when he asked: If (as the king of Brobdingnag asserted) a benefactor of mankind was one who made two ears of corn grow where only one grew before, "what shall we say of Burbank, who makes tons of new varieties of vegetation to grow where none grew before?"[31] Burbank had joined Carnegie's crusade to see that "man's dominion will be extended," and Morrow claimed that "Burbank, who is an old soldier in this army, will return with much spoil, and additional honors."[32] CIW's grant helped make Burbank part of the biotopian campaign; from blooming deserts to massive profits, Burbank's speculative future was on the march.

Carnegie and Burbank

Harwood had first reported the CIW grant in the second of his *Century Magazine* articles (and the additional publicity probably persuaded his publisher to develop them into a book), which explained that the grant

30. California State Board of Trade 1905, 25–26.

31. According to Swift (1726 [2002], 113), such a benefactor would "do more essential service to his country, than the whole race of politicians put together."

32. W. W. Morrow, "The Carnegie Institution, and What It Has Done for California" (read by Arthur R. Briggs; Morrow was unable to attend), in California State Board of Trade 1905, 31.

would be used to send "trained experts who are in close touch and sympathy with Mr. Burbank," to work alongside him to help organize the "important data" he had accumulated.[33] Visitors often commented on Burbank's haphazard record-keeping, noting that he relied on his memory and intuitive sense of which plants were worth studying. His scorn for the plodding work of scientific bookkeeping was all part of his image as an eccentric genius, but CIW's trustees were concerned that it would lead to valuable information being lost. Several years earlier, Bailey had expressed concern that no scientific record was being made of Burbank's "rich experimental results" (which was why "some philanthropist" should step in and fund him; see p. 95).[34] As the work of translating Burbank into appropriately scientific terms got underway, his paradoxical qualities became a source of tension. The process exhibited some of the problems science would face in the reverberating age, as magazine and newspaper writers reported every step (along with unsubstantiated and exaggerated claims), and Burbank himself used the media to manipulate the Carnegie Institution's managers.

In his official report for 1905, CIW's president, Robert Woodward, explained that preserving Burbank's data had been a key motivation for their grant. He also noted that although Burbank's work was "well known in a *popular* way" (emphasis added), his achievements had "yet to be interpreted to men of science as well as to the interested public," and so the institution had joined the long list of would-be interpreters whose diverse understandings of both Burbank and science were competing for the public's attention. Given the noted breeder's unorthodox methods and reputation, CIW decided that a reliable account of his work would require appointing a qualified scientist, who would prepare a "scientific account of the ways, means, methods, and results of Mr. Burbank's work."[35] This account was expected to take about five years, but it never appeared, and CIW's archives provided detailed evidence of the often-difficult relationships between Burbank and those who tried to work with him.

CIW had first considered interpreting Burbank a couple of years earlier. Its archives contain an application for $50,000 to facilitate "Investigations in the Evolution of Plants" by establishing a research laboratory and gardens "near Santa Rosa, California." Setting up a research station from scratch would have been prohibitively expensive, so the proposal

33. Harwood 1905b, 285–86.
34. Glass 1980, 134–35.
35. R. S. Woodward, "Report of the President," in Carnegie Institution of Washington 1905 (22).

recommended taking advantage of Burbank's "experimental gardens" (which supposedly contained over a million varieties). He was not only considered "the greatest genius who ever worked in this field" but had also "expressed a willingness to cooperate in such an undertaking." The applicants commented that Burbank was too busy with practical work to become "a writer on evolutionary subjects"; nevertheless, his carefully collected facts would be invaluable to more theoretically minded biologists.[36] The proposal effectively suggested a division of labor that had its roots in standard accounts of Baconian science: practical workers like Burbank supplied facts, while scientifically trained experts would supply the theory and become the interpreters of nature.[37]

The 1903 application was signed by several members of the "Evolution Committee of the Botanical Society of Washington," all of whom also worked for the US Department of Agriculture (USDA).[38] Its chair was David Fairchild, a USDA plant collector and a leading figure in the Botanical Society.[39] The committee's justification for the proposed research was that although Darwin's doctrine of evolution was "now generally accepted," the precise causes of evolutionary change "remain[ed] unexplained." Specifically, the persistent difficulty of accounting for the origin of evolutionary novelties (the arrival-of-the-fittest-problem) had led many biologists to abandon natural selection (at least as "the active cause or principle of evolution"). The proposed research station aimed to end the Darwinian debates by studying an alternative hypothesis: that change resulted from a "normal evolutionary motion independent of external causes." The planned research was intended to establish the true a law of "organic motion."[40]

For those in the know (a very small group), the references to an "evolutionary" or "organic motion" showed that the new station was intended to investigate Orator F. Cook's "kinetic theory of evolution." Cook, one

36. Fairchild 1903, 2–4. Fairchild was one of those who supplied Burbank with the first spineless cacti samples he used in his breeding experiments (*Luther Burbank's Spineless Cactus* 1912, 4).

37. Yeo 1985 (2001).

38. As Philip Pauly has shown, Washington was home to an ambitious, tight-knit group of young biologists with connections to the USDA. Their de facto leader was Beverly Galloway, who promoted scientific crop-breeding—including the mutation theory—as the solution to feeding the USA's rapidly expanding population (see chapter 1). Pauly (2000, 53) described him as a "gifted scientific entrepreneur," always trying to expand his group and its responsibilities.

39. Fairchild headed the USDA's Division of Seed and Plant Introduction when it was first established (Taylor 1941, 18).

40. Fairchild 1903, 2–4.

of the signatories of the proposal, was another USDA employee, a botanist and entomologist, author of almost four hundred books and papers,
mostly dealing with economically important plants.[41] The kinetic theory
was yet another attempt to reinterpret Darwinism by reducing the role of
natural selection; Cook assumed that evolutionary change was caused by
a diffuse force akin to gravity that was inherent to living things, not the
result of external environmental factors. He published a series of articles
that argued that evolution was primarily driven by "spontaneous devel-
opmental tendencies," which created an evolutionary "movement" within
organisms (analogous to the way the random movement of molecules in a
gas results in pressure). Natural selection might sometimes influence the
direction of that pressure, but it did not cause it and could not therefore
be considered a creative force.[42] He illustrated his idea by analogy with a
river, whose precise course was shaped by local accidents, such as soil and
rock type, but whose ultimate cause was gravity, which made water run
downhill. The same, he believed was true of evolution: "The vital river,
when unconfined, is in motion; change is a law of organic succession;
evolution is a property of protoplasm."[43] The force that supposedly drove
life to evolve interacted with specific environmental factors to give rise to
particular kinds of animals and plants, but these accidents were not the
ultimate cause of evolution.

Cook's theory was a form of orthogenesis ("straight-line" evolution),
based on the claim that evolution is not random but has a predetermined
direction. Orthogenesis emerged in Germany in the late nineteenth century and became quite popular in the USA, especially among paleontol-
ogists who used it to explain the appearance of seemingly progressive
patterns in the fossil record (for example, the earliest fossil horses were
small, with generalized anatomy, but over millions of years they became
larger and more specialized).[44] Such patterns could also be explained by
rival theories, notably neo-Lamarckianism (also popular among US pale-
ontologists, with Edward Drinker Cope being its most prominent expo-
nent).[45] While the two ideas partially overlapped, the neo-Lamarckians
assumed that the changes were adaptive: organisms strengthened their
muscles and behaviors by using them, while those they neglected became

41. "Cook, Orator Fuller" 1953.

42. Cook first announced his theory in 1895; see Cook 1901. See also Cook 1904b;
1907; 1908.

43. Cook 1901, 973.

44. Bowler 1983 (1992), 34, 86–87, 118–26; MacFadden, Oviedo, Seymour, and
Ellis 2012.

45. Bowler 1983 (1992), 122–24.

weaker; and the resultant changes were inherited—muscles got larger and behaviors became inherited instincts. Over time, the inherited effects of use and disuse would produce better-adapted organisms (such as bigger, faster horses with a single specialized hoof). By contrast, orthogenesis tended to reject the claim of improved adaptation; evolution might improve organisms, but it could also be a relentless force driving a species to extinction.[46] The most celebrated example was *Megaloceros giganteus*, or—as Stephen Jay Gould famously called it—the poor "misnamed, mistreated, and misunderstood Irish elk."[47] These extinct fossil deer (which are neither elk nor exclusively Irish) had extraordinary, massive antlers, which were used by orthogeneticists to argue that the species' extinction had been the result of runaway evolution. Large antlers might originally have conferred some benefit on their possessors, but once the trend toward largeness set in, there was no stopping it; over the generations, the antlers got bigger and bigger until the poor creatures could barely lift their heads. Growing these unwieldy structures required enormous biological energy, so they became an ever-increasing liability, which had apparently consigned *Megaloceros* to extinction.

Cook's theory (like both orthogenesis and neo-Lamarckianism) shared the assumption that evolution was not random, but he argued that organisms "are not compelled in *one* direction, but must move in *some* direction"—evolution was the movement itself.[48] He rejected existing theories because they all explained evolutionary change by external factors ("the direct action of the environment, to selective isolation, to abrupt transformation or mutation, or to some combination of these"). As a result, they were all mistaken; his kinetic theory "interprets vital motion as continuous, gradual and self-caused, or inherent in the species." For Cook, evolution was similar to the random movements of particles in fluids known as Brownian motion.[49] He called the mysterious evolutionary force "symbasis" and acknowledged that he had no idea what caused it ("but the same might have been said of gravitation and many other

46. Bowler 1983 (1992), 17, 142–43.

47. Gould 1974. See also "The Misnamed, Mistreated, and Misunderstood Irish Elk," in Gould 1977 (1987), 79–90.

48. Cook 1903, 15, original emphasis. Peter Bowler (1983 [1992], 62–63) has noted that the common form of Lamarckianism, use-inheritance, cannot operate in plants. This point may explain why Cook, like most botanists, seems to have rejected it as an evolutionary mechanism. He wrote very little on the topic but dismissed natural selection because he believed it relied on the inheritance of acquired characters, which had been "abandoned" (Cook 1904a, 450). See also Cook 1908, 18–20.

49. Cook 1904a, 454.

properties of matter for which names have proved useful").[50] Recognizing the mysterious force's importance would prompt research into its nature, but before that could happen biologists had to give up their misplaced faith in external causes.

Cook used journals like *Science* and the fairly specialized *Popular Science Monthly* (*PSM*)—as opposed to popular, mass-circulation newspapers and magazines—to explain his ideas and criticize alternatives. He repeated the widely held view that selection only "eliminates deficient and aberrant individuals"—no one had seen natural selection creating new species in nature because it did not create them.[51] However, he was equally critical of "De Vriesian species," or mutations, which he argued could not be considered "true evolutionary species, either actual or potential" because of their low fertility (they were "obviously declining toward extinction").[52] De Vries might have explained how species sometimes enter degenerate phases, but the mutation theory was no better than natural selection when it came to accounting for the arrival of the fittest. Nevertheless, Cook was attempting to answer the same questions as de Vries, and in doing so he made similarly optimistic forecasts about the impact of understanding evolution. Once biologists accepted that the pressure to evolve was "inherent in all species," he claimed they would (in some, largely unspecified manner) find solutions to "many practical problems."[53] Researchers were wasting their time trying to induce mutations because evolutionary changes "take place spontaneously without our interference." Instead of trying to cause evolutionary change, breeders needed to focus on their power "to control and direct, to accelerate, retard, or reverse it."[54]

Just like de Vries, Cook claimed that his theory would increase human control over evolution, thus solving such practical problems as increasing crop yields. The most detailed account of the kinetic theory appeared in *Methods and Causes of Evolution*, a USDA pamphlet that carried the authoritative imprimatur of the US government. It was prefaced by a letter from Beverley Galloway, head of the Bureau of Plant Industry (see p. 59), to James Wilson, the Secretary of Agriculture, which stressed the practical aspects of Cook's work ("evolution is now being made of practical use in the solution of problems of breeding and acclimatization"). Galloway

50. Cook 1903, 16.
51. Cook 1908, 9.
52. Cook 1903, 19.
53. Campos 2015, 104–7, 118–19.
54. Cook 1908, 8.

supported these claims by citing the various experts who had read and endorsed the work, a group that included most members of the Evolution Committee and Alexander Graham Bell. The celebrated inventor of the telephone might seem an unlikely expert to be asked to comment on a work about evolution, but Bell had recently taken up experimental breeding and created a new breed of sheep (and he was also Fairchild's father-in-law).[55] Bell agreed "that natural selection does not, and can not, produce new species or varieties"; it could not explain the arrival of the fittest because it was merely a destructive force. Natural selection's "sole function is to prevent evolution," so he agreed with Cook that evolution's true cause was some "dynamic force compelling progress along other lines."[56]

Bell's endorsement added credibility to the claim that progress depended on understanding evolution, yet Cook never gave specific examples of how practical results would be achieved.[57] Breeders should abandon attempts to cause variations or mutations, but the work of selective breeding would seemingly go on unchanged. The kinetic theory mainly claimed to explain why existing practices such as Burbank's worked—and Burbank was quoted in the *PSM* as saying that "the facts of plant life demand a kinetic theory of evolution."[58] *Cosmopolitan* quoted Burbank as saying that plant-breeding was part of "A Kinetic Creation—a Universe of organized lightning" and that a breeder "does not create, but he guides nature in creating."[59] So, it is hardly surprising that in a letter to Fairchild, Burbank wrote that "of all men who tread this green earth, Professor O.F. Cook in my opinion better understands and more deeply and truly sees into the action of evolutionary forces than any other."[60] Unlike his rather awkward alliance with de Vries, Burbank's and Cook's initially seemed like a match made in heaven.

Cook and his allies hoped CIW funding would allow them to find solid evidence for the kinetic theory, and turning Burbank's grounds into a research station would have saved money. In addition, the intensity of the

55. The story appeared in various local newspapers, e.g., Carpenter 1906, but also in *Science* (Bell 1912) and the *Journal of Heredity* (February 1, 1924). For Bell and Fairchild, see Keogh 2020, 186.

56. Bell, quoted in Cook 1908, 3–4.

57. Cook 1908, 7.

58. Burbank, quoted in Jordan 1905, 225; reprinted in Jordan 1909, 77. In an unpublished manuscript, "A Kinetic Creation" (1899), Burbank railed against the notion that the universe consisted entirely of "dead material which possessed no power of its own" (Pandora 2001, 504–5).

59. Serviss 1905c, 65.

60. L. Burbank to D. G. Fairchild, December 10, 1904, in Burbank 1904–1905.

Darwinian debates meant that the marketplace for evolutionary theories was very crowded; having Burbank's name associated with one's scientific product would have been a form of celebrity endorsement. So his allies in Washington worked hard to get Cook appointed as Burbank's interpreter. Fairchild visited Burbank in August 1904, after which he wrote to Daniel Coit Gilman (then president of CIW) to press the case. He acknowledged that while Burbank did not express his ideas "in the language of a university graduate," they were nevertheless "clear-cut and definite and impressive." He emphasized the scientific nature of Burbank's investigations, adding that it was "a mistake to associate the name of Wizard with a man of his far-seeing and analytic character." He also told Gilman that Burbank had read Cook's theories and commented that "Cook's ideas on evolution were nearer to the facts than those of any other man living in the world"—even better than "those of De Vries," who Burbank had recently met.[61] And after Cook visited California a couple of months later, Burbank wrote enthusiastically to Fairchild to tell him that he and Cook had "talked evolution, variation, mutation etc. about fifteen hours a day." He had already been convinced that Cook "had arrived at about the same conclusions on these subjects" despite coming from "two different sides of the subject" (practical and theoretical). Burbank apparently saw himself and Cook as equals, with complementary expertise. He would be delighted if CIW were to employ Cook "to classify and write up this little understood subject," which again suggested Cook as a collaborator, almost an amanuensis. Burbank closed by re-emphasizing his own expertise: "It will be perfectly useless to send a man who is not well posted on current evolutionary thought as I have no inclination or time to educate a person in evolution."[62] Nor, it seems clear, did he have any interest in being merely a source of facts.

Despite the lobbying and Burbank's apparent willingness, the 1903 plan failed, partly because Burbank seems to have changed his mind about it (see below). When CIW finally awarded Burbank a grant two years later, it did not entail the institution taking over Burbank's gardens, merely sending a suitably qualified person to make accurate records of Burbank's work. However, when Woodward wrote to Burbank to announce the award

61. D. G. Fairchild to D. C. Gilman, August 25, 1904, in Burbank1904–1905.

62. L. Burbank to D. G. Fairchild, October 18, 1904, in Burbank 1904–1905. This letter was forwarded to Gilman by Fairchild, in late October, with a note explaining that he would be happy to discuss further "what I believe Mr. Burbank could do for the Institution." See D. G. Fairchild to D. C. Gilman, October 26, 1904, in Burbank 1904–1905.

and its terms, he initially informed him that Cook—"whom you already know"—would be their recorder and would visit regularly "to write up, with your cooperation, a complete scientific account of the ways, means, methods, and results which you have so successfully developed in recent decades."[63] Woodward explained the arrangement to his trustees in similar terms—"Dr. Cook was to act as an interpreter of Mr. Burbank's work," to translate Burbank into a more scientific idiom.[64] However, Cook apparently still hoped to use Burbank's data as evidence for his own theory of kinetic evolution (and had doubtless originally envisaged himself as director of the new experimental station, with Burbank as his effective subordinate).[65] Despite the apparent similarities in their thinking, and Burbank's initial enthusiasm, their differing conceptions of the relationship meant it broke down almost before it had begun. The problems were partly practical (Cook hoped to combine his full-time position at the USDA in Washington, DC, with visits to Burbank in California, which proved logistically impossible), while other difficulties are discussed below.

Kellogg briefly replaced Cook in October 1905, spending several days with Burbank (and collecting the material he would later publish as "Scientific Aspects of Luther Burbank's work").[66] However, CIW quickly found a more permanent replacement in the form of George Shull, a young geneticist from CIW's Station for Experimental Evolution on Long Island (see chapter 2), who spent several years trying—but ultimately failing—to write a scientifically acceptable account of Burbank's work (figure 4.2).[67] As Bentley Glass has shown, Shull spent much longer on the project than Cook but ultimately failed, partly because he was young and unknown, while Burbank was world-famous. The young geneticist also had more formal scientific education than the breeder—and tried to share what he knew (Burbank complained that Shull talked too much). However, the biggest difficulty was that Shull tried to present Burbank's work in a form that would be acceptable to his professional scientific colleagues (which would have differed sharply from Burbank's own understanding). Burbank was to be appropriated by orthodox science; his years of work, his observations and experiments, would be cleansed of his own idiosyncratic interpretations and be reduced to straightforward facts, ready for others

63. R. S. Woodward to Luther Burbank, December 29, 1904, in Woodward, Shull, and Burbank 1908, 2–3.

64. Woodward, Shull, and Burbank 1908, 5–6.

65. D. G. Fairchild 1903, 4–5.

66. Kellogg 1906; Largent 2000, 85.

67. Glass 1980.

FIGURE 4.2. Luther Burbank, Hugo de Vries, and George Shull—a rare moment of harmony in Burbank's garden.

to theorize over.[68] Meanwhile, Burbank believed he was well-qualified to interpret his work himself.

While Cook and others were hoping to use Burbank to revise traditional Darwinism, Jordan and Kellogg were hoping he might be used to defend it. Jordan tried to distance the Californian from claims of wizardry by stressing that Burbank's methods were simply the practical application of "the theories of Darwin and his followers" and insisted that it was "Darwin who first gave us the knowledge on which all [Burbank's] work rests."[69] When Jordan and Kellogg produced a textbook, *Evolution and Animal Life* (1907), based on their popular Stanford lecture series, reviewers noted that "Burbank's work has a prominent place" in it (partly thanks to the many photos of his plant creations that Burbank had supplied), but he

68. Glass 1980, 136–37. Mark Largent (2000, 66) notes that Burbank used the phrase "capturing Burbank for science" in his personal correspondence to describe what he saw as the purpose of the Carnegie Institution of Washington's grant.

69. Jordan 1905, 201; 1909, ix.

was used to explain good old-fashioned selection.[70] The Stanford professors acknowledged that Burbank had his own views on heredity, but these views were all-but-dismissed in a sentence.[71] Instead they commented on his "masses of valuable data" that were "being let go unrecorded."[72] Like many other scientists, the Stanford men were almost salivating at the thought of getting their hands on Burbank's data.

However, Burbank rejected the subordinate role to which CIW and other interpreters hoped to relegate him.[73] Despite his public image as an almost naive innocent, he was an adept political operator who knew how to use the power of the press. In response to the Evolution Society's proposal to take over his garden, he planted a story in his local paper, the *Santa Rosa Republican*, that claimed he had refused CIW's money on the terms that Cook and the others proposed. The story appeared on December 23, 1903 (just six weeks after the initial grant proposal had been drafted). The story claimed that Burbank had "Declined the Grant," because allowing the scientists in would have disrupted his work.[74] In fact the archives show that no formal offer of a grant had yet been made; Burbank used the newspaper to preempt CIW by making it clear that he would only accept their money on his terms—and two years later he got exactly what he wanted.

As noted, the attempts by CIW and others to interpret Burbank were part of wider dialogues over the future of Darwinism. And the question of whether or not Burbank was a man of science was part of debates over the nature of the scientific community. If, as Wharton's Professor Linyard

70. Jordan and Kellogg 1907. The book was successful enough to warrant a second edition in 1920 (Jordan and Kellogg 1920; "Science [book reviews]" 1907, 426). See also Review of Jordan and Kellogg, *Evolution and Animal Life* 1907; Grinnell 1908.

71. "Of Burbank's own particular scientific beliefs touching the 'grand problems' of heredity we have space to record but two." Firstly, that he was a strong believer in the inheritance of acquired characters, "thus differing strongly from the Weismann school of evolutionists." And that he is also convinced of the "constant mutability of species"—so was not a supporter of de Vries (Jordan and Kellogg 1907, 101–2).

72. Jordan and Kellogg 1907, 101–2.

73. Patricia Craig (2005, 42–43) noted that Burbank was not interested in exemplifying other people's ideas.

74. "Declined the Grant" 1903. The newspaper's story included several phrases repeated verbatim from the grant application and quoted a letter from Burbank to Fairchild ("one of the committee asking for the grant") in which he had explained his reasons for rejecting it. The archives at CIW contain a clipping of this story, sent in by Burbank's supporter, William Morrow (see p. 126), who called it a statement by Burbank that suggested he was now beyond CIW's "reach." See [Judge] Wm. W. Morrow to Charles D. Walcott, January 7, 1904, Burbank 1904–1905.

argued, science was no longer an "inaccessible goddess" but one who "offered her charms in the market-place," who was qualified to comment on those charms? Were the opinions of magazinists like Harwood as valid as those of Jordan and Kellogg? And what weight should be placed on Burbank's own views? Orator Cook was probably not the only one who would have agreed with the fictitious Linyard that things had been better when the scientist's audience had consisted only of "fellow-students."[75] However, these various men of science seem not to have understood the power of Burbank's public image as a scientific mystic; had he been successfully translated into orthodox scientific terms, he would surely have lost a good deal of his popular appeal.

The extraordinary volume of publicity that surrounded Burbank made him a synecdoche for the hope of remaking nature to suit human ends. So much so, that *Webster's* dictionary added a new verb *to burbank*, meaning to "modify and improve (plants or animals)" and "figuratively, to improve (anything, as a process or institution) by selecting good features and rejecting bad."[76] Burbank had come to embody American biology's achievements and, as a result, the complex and often contradictory elements that made up his public persona (the altruistic huckster, the elitist democrat, the unnatural parent, and the scientific mystic) became part of the definition of "science" for many readers. Meanwhile the mainstream scientific community (and his would-be publishers) often saw Burbank as a resource to be exploited; his unpublished results were like virgin land waiting to be profitably tilled. And at a time when media attention was becoming an ever-more-important part of science, Burbank's allure was greatly enhanced by the fact that he possessed fame and media skills that few scientists could match.

Cook was one of several scientists who tried to make use of Burbank but found the wily Californian uncooperative. Paradoxically, the failure of Cook's kinetic theory partly stemmed from the fact that its author was unambiguously part of the orthodox scientific community. He valued the scientific community's judgments more than the public's and chose to write only in reputable scientific journals, not for general-interest periodicals. There were other issues too: Cook's theory did not suggest exciting

75. Wharton 1904, 6–7.
76. There was also an entry for "burbankian . . . Produced by burbanking; resembling the act or product of burbanking," *Webster's New International Dictionary*, 2nd ed. (Springfield, MA: Merriam-Webster, 1934 [1942]): 357.

new experiments (unlike de Vries's).[77] And the kinetic theory endorsed Darwin's original conceptions about the tempo and mode of evolution. At a time when America was in a hurry to address practical problems, Cook seemed to offer very little by contrast with de Vries and, having largely failed to convince his "fellow students" of his theory's value, Cook and his ideas disappeared, leaving almost no trace.[78] However, had he succeeded in gaining Burbank's support, the story could have been very different; at the very least, the kinetic theory would have gained the attention that was becoming increasingly vital to successfully doing science in public.[79] We can imagine the alternative future that Cook's theory might have enjoyed by looking more closely at the complex role Burbank played in the actual reception of de Vries's rival theory.

Who Is a Scientist?

During de Vries's first visit to the USA (June–October 1904) he attended the opening of CIW's Station for Experimental Evolution, visited the St. Louis International Congress of Arts and Sciences, and received an honorary doctorate from Columbia University.[80] However, most of his time was spent in California, where he lectured at the state's university (which had organized his trip) and first met Burbank, making two visits to his home at Santa Rosa.[81] De Vries claimed that Burbank was the main reason he had decided to visit California, because he was "*the* man who creates all the novelties in horticulture," possessed of both "great genius" and "an almost incredible capacity for work."[82] These visits were widely reported in the state's newspapers, which were delighted to see Burbank receiving the approbation of such a distinguished visitor—and their reports in turn

77. Largent (2000, 82–83) has shown that Kellogg was, at least briefly, an exception, in that he was interested in Cook's ideas and appears to have done experiments to test them.

78. In an unpublished confidential report written a few years after the Burbank grant had been awarded, Woodward claimed that Cook had been replaced after "consultation with a number of the leading biologists," from which it had become clear that "Dr. Cook lacked their confidence" (Woodward, Shull, and Burbank 1908, 6–7; Largent 2000, 84).

79. I have been unable to locate a single published contemporary comment on Cook's theory. The only other historian who seems to have examined his ideas is Mark Largent; see Largent 2000, 78–81, n. 53.

80. As Anahita Rouyan (2017, 99) has noted, the 1904 St. Louis congress marked the start of major US press interest in de Vries and mutation.

81. Bavel 2000, 3–6.

82. De Vries 1905d, 329, original emphasis.

made the Dutch professor increasingly well-known.[83] The *San Francisco Chronicle* devoted a whole page to the "Famous Scientist" from Holland, but added that when it came to experimental plant-breeding "the name of Luther Burbank of Santa Rosa reigns supreme" (a fact that "no one more quickly acknowledged it than De Vries").[84] Their rival paper, the *San Francisco Call*, called de Vries a "great" and "distinguished" botanist, who was nevertheless about to "meet a Californian as learned as himself in the secrets of plant life—Luther Burbank."[85] De Vries showed no aversion to publicity and described his encounters with Burbank in an article for the popular Dutch magazine *De Gids* (the Guide), which was translated and published by the American *Popular Science Monthly* (and he wrote three further popular articles about Burbank over the next two years).[86] As noted, de Vries was not the first scientist who tried to promote his ideas using Burbank's fame, but the publicity that brought mutation theory to a wider audience also reinforced the public's perception that Burbank was a scientist. As a result, de Vries found it increasingly difficult to make effective use of Burbank's reputation.

De Vries's problems were exacerbated by the sheer novelty of his own theory. Theories that claimed to resolve Darwinism's apparent problems were plentiful—Burbank and Cook both offered alternatives, each of which described evolution as "kinetic." How were audiences to know which theories (or theorists) to trust? Meanwhile, the emerging field of experimental evolution did not even possess an agreed name, much less a consensus about how and where its results should be published. Burbank, de Vries, and the field of experimental evolution were all trying to establish themselves in the new mass-media market, often in the same magazines. New York's *Independent* told its readers that although de Vries was looking for "the scientific laws of growth," while Burbank's "aims are practical and commercial," the two men were "working in the same line"—of "creating new species of plants."[87] The periodical press generated publicity for science, but at the cost of blurring the differences between self-taught practical workers and university-trained theoreticians.

83. Endersby 2013; Rouyan 2017, 94.
84. Wilson 1904, 7.
85. "Distinguished Botanist Plans for Experiments" 1904.
86. De Vries 1905d, 329; 1906a; 1906b; 1907a. The first article also formed the third chapter of de Vries's Dutch-language account of his American travels, *Naar Californië*. For de Vries and popular writing, see Theunissen 1994b, 299; De Rooy 1998, 424; Bavel 2000.
87. [Hamilton Holt] introduction to De Vries 1906b, 1134.

The similarities between de Vries's and Burbank's goals made it easier to associate the Californian's fame with the mutation theory. De Vries
told readers of the New York *Independent* that Burbank worked to feed
more people, more cheaply; if he succeeded, "the earth would be transformed"; humanity would rise above destructive competition to collaborate on higher goals. "Such are Burbank's ideals; such is the aim of his
work."[88] As we have seen, de Vries had offered similar utopian hopes for
his own theory (p. 58). Despite sharing the biotopian goal of remaking
nature, de Vries tried to distinguish his science from Burbank's heuristic
wisdom. He claimed that his first visit to Burbank had been prompted by
the hope of discovering "what secret method" Burbank had used to create his famous stoneless plum, but he found that the breeder had simply
taken an old-fashioned stoneless French variety and hybridized it with a
larger, juicier one. As de Vries admitted, this had been "to a certain degree,
a disappointment," because Burbank had not contributed to explaining
the "nature and origin of new characters" (that is, the arrival-of-the-fittest
problem).[89] Since Burbank was famous for apparently creating new forms,
de Vries doubtless hoped that some of his novelties would prove to be
mutations, thus providing additional support for his theory (which relied
heavily on evidence from a single species, *Oenothera lamarckiana*, which
many of its critics claimed was a hybrid, and thus not a real species at
all).[90] It would have greatly strengthened de Vries's arguments if Burbank's
gardens had supplied a few more examples of species undergoing mutation periods, but—as far as he could judge from Burbank's frustratingly
idiosyncratic record-keeping and terminology—they had not. Burbank
simply used long-established hybridization techniques, albeit on an unusually large scale.

Despite the disappointment of realizing how conventional Burbank's
methods were, de Vries acknowledged the plant breeder as a fellow scientific worker when he first lectured at the University of California, referring regularly to his results and commenting that his "principles" were
"in full harmony with the teachings of science."[91]Nevertheless, he tried
to establish his own superiority by emphasizing the value of Burbank's
data (not his theories), and reminding his audience of what "everybody
knows"—that Darwin himself used the work of plant and animal breeders

88. De Vries 1906b, 1134. He was quoting Burbank 1904, 38–39.
89. De Vries 1905d, 335.
90. Endersby 2013.
91. De Vries 1905c, 768.

to supply much of the evidence for natural selection.[92] The implication
was that de Vries was Darwin, the great thinker, while Burbank merely
supplied the grist for his theoretical mill, an interpretation he promoted
when he told the *Salt Lake Herald* that despite similarities between his
work and Burbank's "his is practical work, while mine is theoretical. I
find the way and let others secure the results."[93] The US press (perhaps
reflecting an ethos of valuing practical know-how over highfalutin' foreign
theory) generally concurred, with the *Chicago Daily Tribune* emphasizing
that "the principles" involved in improving plants were "being studied
by De Vries," while the practical work was being done by "experimenters
like our own Burbank."[94] And the *New York Times* made a similar contrast
when it commented that Burbank was trying to make the biggest plums
and potatoes, "but de Vries is trying to discover the laws which govern
such vagaries of nature."[95]

De Vries's putative alliance with Burbank was problematic from the
outset, not least because de Vries's apparent approbation helped bolster
Burbank's scientific credentials, which encouraged him to offer his own
ideas about evolution. What would have been even worse from de Vries's
perspective, was that numerous popular articles on Burbank claimed he
had *disproved* the mutation theory. Instead of finding confirmation in California, the Dutch theorist found himself regularly trying to correct what
he regarded as mistaken or misleading accounts. He noted that "some
magazine articles and popular books" on Burbank were characterized by
"effusiveness and unconscious exaggerating." It was clear ("to the scientific
reader") that their writers were not scientists because they made misleading claims, such as that Burbank "overthrew the Mendelian laws, that he
opposes the theory of mutation," or that he had "proved the inheritance
of acquired characters."[96] These assertions had been "made by a reporter"
(Harwood, who was unnamed but quoted) "who claimed to give an

92. De Vries 1906b, 1135.

93. "Dutch Botanist Here to Study" 1906.

94. "Backward America," *Chicago Daily Tribune*, December 6, 1912: 8. The article
was reprinted in other papers, e.g., "Backward America" 1913. Similar points were
made in "Distinguished Botanist Plans for Experiments" 1904; and J. McKeen Cattell's introduction to De Vries 1905c (329). And Donald Jones, of the Connecticut
Agricultural Experimental Station, dismissed Burbank as a "doer, not a thinker"; see
Pandora 2001, 486.

95. "Noted Holland Expert Tells How to Double Our Crops" 1912.

96. De Vries 1907a.

authoritative account of the work of Burbank."[97] (And, true to form, Serviss parroted each of Harwood's claims, noting for example that the "mutation theory of Professor De Vries cannot stand in the light of Mr. Burbank's experiments."[98])

De Vries attempted to dismiss such claims because "everybody knows" that Burbank was a practical man, far too busy to attempt "purely scientific investigations."[99] He recounted his own conversations with Burbank, from which he concluded that his experimental results agreed ("in the main") with both mutation theory and Mendelism (which de Vries regarded as a subsidiary theory to his own). De Vries acknowledged that Burbank might not agree with him, but that was because his temperament was "that of an artist" whose "genius," according to de Vries, was being able to "grasp the questions from their commercial side," which allowed him to "increase the enjoyment of life for his fellowmen" (675, 681). Despite his achievements, Burbank's lack of formal training effectively debarred him from discussing the scientific implications of his own work. Yet this argument was undermined almost as it was made, because de Vries made it in *Century Magazine*, the same magazine where both Harwood's articles and Burbank's own "Training of the Human Plant" appeared. The difficulty of distinguishing his own expertise from Burbank's was exacerbated by writing for general audiences (tacitly inviting their judgment) and citing popular articles (such as Wickson's in *Sunset*), thus giving them credibility. Nor could de Vries fully adopt the persona of a popularizer, conveying established scientific truths to a less expert audience, because mutation theory was an undisciplined science about which the scientific community had yet to reach a consensus.[100] De Vries and other proponents of experimental evolution were addressing both lay and expert audiences, often in the same publications. De Vries's decision to publish in *Century* meant that for its readers (as for those of hundreds of similar publications), sober science and extravagant prophecy formed a seamless web. De Vries and Burbank became part of a single dialogue, competing prophets of a rather similar-sounding range of imaginary futures.

97. De Vries wrote: "Concerning Burbank's view of the Mendelian laws, I read the statement that 'he had disproved them over and over again years before he knew they existed.'" De Vries's quote came from Harwood 1905b (340–44) but had originally appeared in Harwood 1905d (823).

98. In a similar vein, Serviss claimed that Burbank had also proven Mendel's laws "inadequate" and had "proved the falsity of the doctrine that acquired characteristics are not transmitted" (Serviss 1905b, 169).

99. De Vries 1907a, 677–78.

100. I discuss this point more fully in Endersby 2013.

Among the difficulties faced by mutation's fans was clarifying the difference between genuine mutants and novelties like Burbank's. In 1910, US Assistant Secretary of Agriculture Willet M. Hays explained to members of the American Breeders Association why scientific breeding should be taught in schools and colleges. He claimed Mendel, de Vries, *and* Burbank had transformed breeding into "organized science" because "the potato seed from which came the Burbank potato; *and other mutants,* will stand along with the theoretic truth brought forward by De Vries."[101] Moreover, Burbank rejected the distinction that de Vries was attempting to establish, claiming that *mutation* was simply a new term for *sport* (but had become "peculiarly popular in recent years").[102] He not only conflated sports and mutations but also claimed that "everything is a sport," further undermining de Vries's theory.[103] Harwood backed Burbank, arguing that although de Vries's theory had been endorsed by "leading scientists," it was disproved by Burbank's "vast experiments." He quoted Burbank as saying that a mutation was merely the result of particular conditions, which could "be produced at will."[104]

Other writers on both sides of the Atlantic repeated the claim that the mutation theory had been disproved by Burbank, but it was not just journalists who accepted these claims.[105] In their booklet *The Scientific Aspects of Luther Burbank's Work,* Stanford's professors Jordan and Kellogg also conflated mutations and saltations, claiming that both were common and that, according to Burbank, "*these mutations can be produced at will* by any of the various means which disturb the habits of the plant."[106] They quoted his claim (as did Harwood and others) that "*mutation is not a period, but a state*" and directly contradicted de Vries by asserting that there was no "period in the life-history of the species when it is more subject to mutation than at other times" (15, original emphasis). Burbank was even quoted as having criticized de Vries's theory for a "lack of sufficiently wide experimentation," which would eventually show that de Vries's mutations were merely normal variations that the plant's environment had

101. Hays 1910, 224–25, 227, emphasis added.

102. Burbank et al. 1914, vol. 1, 80.

103. Burbank, quoted in De Vries 1907a, 678.

104. Harwood 1905d, 835. In an unpublished report to CIW's trustees, Woodward recorded that Burbank had distanced himself from "some bald statements" made by Harwood that had "aroused Prof. De Vries, as of course they must have done" (Woodward, Shull, and Burbank 1908, 15).

105. Serviss 1905b, 169, original emphasis; review of Harwood, *New Creations in Plant Life* 1906.

106. Jordan 1909, 11–12.

"carried beyond the critical point" of stability, allowing them to develop "in a new direction" (17). Finally, Burbank also rejected the claim that de Vries had been the first to witness "the actual birth of a species," arguing that "I have produced several good species by hybridization, apparently as good as nature herself has produced" (19–20). De Vries responded to such claims by noting that the Californian used the word *mutation* (as with other scientific terms) far too loosely. De Vries insisted that "jumps or leaps . . . called 'Sports' embrace a number of phenomena of which the mutations are only one instance." What distinguished mutations was that they alone produced "really new, progressive, or retrogressive [unit] characters."[107] Once Burbank's rather vague terminology was translated into more conventional language, his supposed refutation of the mutation theory simply disappeared: after several visits and hours of conversation, "I failed to discover any real difference between Burbank's opinion and my own on the main points."[108]

Despite repeated attempts, de Vries found it impossible to confine Burbank to a purely practical role, partly because (as Cook and Shull had already discovered), Burbank saw himself as a scientist with his own theoretical insights to offer (a view that was reinforced by many who wrote about him). However, there were two more significant issues at stake. Firstly, Burbank's mystical and scientific sides were crucial to the fame that made him such a potentially valuable ally for science, but his unorthodoxy created unease for some scientists. Burbank opened the wider question of whether science could address questions of purpose or meaning, which preoccupied many Americans. The second, related point, was that deciding *who* might properly address scientific questions also required deciding *where* the discussion should be published. Burbank was a subject for almost every genre of publication, which exacerbated the complex issues of scientific trust and reliability that Wharton had highlighted so effectively. How were audiences to know whom to believe when attempting to evaluate Burbank (or his critics) and, even more importantly, when evaluating any new, supposedly scientific claim (whether made by mutationists, Mendelians, Lamarckians, or proponents of kinetic or spiritual evolution)? These two problems met in the problem of interpretation: some sought to translate Burbank into clear scientific terms, while others tried to translate de Vries into more familiar language (for example, by substituting *sport* for *mutation*)—and both attempts were resisted. The

107. De Vries 1907a, 678.
108. De Vries 1907a, 676.

questions of where and how to talk about science were central to deciding who was a scientist.

Wharton's discussion of Linyard and his fictitious book highlighted some of the issues that real-life science faced in the reverberating age. The mutation theory had made its US debut as media science "in the market-place"—precisely the kind of science that Linyard (and many of his real-life counterparts) despised.[109] Yet even as Linyard was complaining that scientific discussions were no longer confined to serious scientific publications, he took advantage of popular interest to place his would-be satire, *The Vital Thing*, with a general publisher, not a specialized scientific one. Once published, the book took on a life of its own, being read and interpreted in ways its author had never imagined. And Linyard's publisher noted that—far from being a drawback—the book's ambiguity would ensure a large sale and a long shelf life because "you fit in everywhere—science, theology, natural history" (adding, "Good as fiction? It's better—it'll keep going longer"). And although de Vries never wrote anything remotely similar, his book *Plant-Breeding* combined material addressed to very different audiences, which created mild uncertainty about its intended audience (but may also have expanded its appeal).

Plant-Breeding (one of the titles the *Atlantic Monthly* included in its roundup of the latest biology; see p. 19) was de Vries's attempt to clarify his relationship to Burbank. It summarized the experiments of both Burbank and Hjalmar Nilsson (a noted Swedish plant breeder) and concluded that their work proved that evolutionary theory was being transformed by the theory that species originate "by means of sudden mutations" (figure 4.3).[110] Unfortunately, the evidence was negative; neither man had discovered new mutations—each simply relied on hybridization to create new combinations of existing characters (as in Burbank's slightly disappointing stoneless plum). De Vries took this evidence as proving that hybridization cannot make new species. De Vries concluded that "the results of Burbank and others wholly agree with the theory of mutation"; the absence of genuinely new characters was "a fact of high scientific significance"—it

109. Wharton 1904, 6–7.

110. De Vries 1907b, v. Nilsson worked at the Svalöf Seed Association, an experimental research station founded in the 1880s that relied on a mixture of private and state funding. See Staffan Müller-Wille, "Plantbreeding at Svalöv: Instruments, Registers, Fieldwork," *The Virtual Laboratory: Essays and Resources on the Experimentalization of Life* (Max Planck Institute website), 1908, http://vlp.mpiwg-berlin.mpg.de/essays/data/art69.

A New Book by Prof. De Vries

PLANT BREEDING

Comments on the Experiments of

NILSSON AND BURBANK

BY

Hugo De Vries, Professor of Botany in the University of Amsterdam

A scientific book in simple language. Intensely interesting as well as instructive.
Of special value to every botanist, horticulturist and farmer,
Pp. XV + 360.   Illustrated with 114 beautiful half tone plates from nature.
Printed on fine paper, in large type.   Cloth, gilt top.   Price, $1.50 net.
Mailed, $1.70.
Supplied by your dealer; or direct, on receipt of your order with the mailing price.

THE OPEN COURT PUBLISHING CO.

1322 Wabash Avenue, Chicago, U. S. A.

*Readers may secure copies of this book of booksellers in Eastern and Southern states
through the Baker & Taylor Co., 33-37 East Seventeenth Street, New York.*

FIGURE 4.3. De Vries's book *Plant-Breeding* was published by the New Thought publishing house, Open Court, which promoted its connection to Burbank.

supported his contention that only mutation could explain the arrival of the fittest.[111]

However, if de Vries hoped that *Plant-Breeding* would settle the matter finally, he was to be disappointed. One reason was that the book itself was a form of bricolage. It was a compilation of previously published papers addressed to different audiences, which included university lectures, scholarly papers, and a straight reprint of a popular article about Burbank that de Vries had written for the New Thought magazine *Open Court*.[112] And while that chapter was sober and cautious, it retained the original article's acknowledgement of the value of Wickson's articles in *Sunset*, thus effectively endorsing Wickson's descriptions of Burbank as a "gifted seer" or "prophet," who moved "above the pathway of the contemporaneous scientists" (644).[113]

By the time *Plant-Breeding* appeared, de Vries was becoming well-known in the USA, which—combined with Burbank's name—helped the book get widely reviewed, yet reviewers were unclear as to audience and message. One American reviewer ignored de Vries's stated goal of promoting the mutation theory and saw the book as intended to correct "America's Wrong Notion of Luther Burbank's Achievement" (which had been distorted by journalistic excesses); far from being a "wizard," "Luther Burbank has not enlarged human knowledge of anything essential in the science of plant-breeding."[114] By contrast, a British reviewer ignored Burbank almost completely, but "confidently commended" the book to a fairly expert audience consisting of both "the practical plant-breeder" and "students of science."[115] The American *Nation* magazine saw the book as intended primarily for a popular audience, praising its "graphic and un-technical language" and recommended it as "an interesting and safe guide to amateurs" ("safe" perhaps implying a contrast with more popular, but *un*safe accounts such as Harcourt's).[116] The specialist *Botanical Gazette* also argued it should be read by general audiences, both because it gave a "true estimate of Burbank" and, even more importantly, it provided "a compact and popular presentation" of de Vries's ideas on mutation.[117] The same message emerged in a very different publication, the Chicago-based literary magazine the *Dial*, which also interpreted the book as primarily

111. De Vries 1907b, v, 16.
112. De Vries 1906a.
113. Wickson 1902; De Vries 1907b, 164.
114. "America's Wrong Notion of Luther Burbank's Achievement" 1907.
115. Review of de Vries, *Plant-Breeding* 1907a, 243.
116. Review of de Vries, *Plant-Breeding* 1907b.
117. Coulter 1907, 147–49.

popular ("clearly and pleasantly written" and "may be read with satisfaction and profit by all"). It was also "one of the most interesting volumes of the year for speculative science" (a description that could easily have applied to *The Vital Thing*).[118] The *Atlantic* (see p. 19) also imagined *Plant-Breeding*'s audience as a nonexpert one, eager to learn from "the world's first authority in his field" (but potential readers were assured that it was "much the briefest and least technical" of his major works).[119] The sheer range of publications that reviewed the book was testimony to de Vries's growing reputation and the widespread interest in the topic, but also to the diversity of its imagined audience (in addition to those mentioned, it was reviewed in the *Journal of Education*, the *Literary Digest*, the *Living Age*, the *Journal of Philosophy*, the *Scientific American*, and others).

The analogy with Linyard should not be pushed too far, but de Vries's modest celebrity was slightly boosted by the fact that reviewers had some difficulty assigning his book to a well-defined genre, and the fate of *The Vital Thing* illustrated what might happen. As noted, its publisher had initially assumed the book was a serious religious one, and some of its imaginary reviewers made the same mistake. The book was so successful that his publisher suggested Linyard write a sequel and recommended he turn his hand to the "now-for-the-best" genre, the optimistic, progressive evolutionary epic "which is so popular just now" (23). As publications like the *Open Court* and *Success* showed, there was a definite market for claims about the spiritual or higher meaning of evolution. And while de Vries never wrote anything of this kind, his connection to Burbank may have assisted interpreters who tried to connect the mutation theory with broadly spiritual interpretations of evolution (see pp. 48–49). The Californian's scientific-mystic persona relied on the sense that he had surpassed orthodox science. Harwood described how de Vries, during his visit to Burbank, had commented on "the unreliability of Nature," to which the breeder responded, "You are all wrong; Nature never lies." According to Harwood, "the great botanist sat some time in silence, and then gravely nodded his head."[120] In a similar vein, Serviss claimed that Burbank's experiments had "flowed all around certain conceptions of formal science, leaving them like islands in the stream, and thus revealing their inadequacy and the partial character of such truth as they do contain."[121] Perhaps Burbank knew more than book-taught scientists, or at least different things? Serviss recounted

118. "Plant Breeding and the Origin of Species" 1907.
119. Brewster 1908, 121.
120. Harwood 1905b, 130–31.
121. Serviss 1905c, 64.

asking Burbank whether he ever felt that he was "exerting a psychic force upon these plants," so that they were "in some way, not yet expressible in scientific terms, . . . following the suggestions of your imagination?" To which Burbank replied simply, "Why not?"[122]

Burbank's work was frequently interpreted as transcending conventional science, a claim that, ironically, had been boosted by the Carnegie grant, which raised Burbank's scientific status and generated a flood of ever-more-imaginative press attention. As a result, one of the difficulties the Carnegie Institution faced was justifying their extraordinary funding for Burbank without endorsing the media's more extraordinary claims about him. He brought them welcome publicity (Carnegie himself told CIW's trustees that "nothing you have done has attracted such general attention as your aid to Mr. Burbank," which was "a gain to the Institution").[123] Nevertheless, the publicity was a mixed blessing, and the institution's president, Robert Woodward, told the trustees that they had been "placed on the defensive" because of the "popular expectations of sensational and spectacular results" arising from the press reports. Despite all the hyperbole, Woodward asserted that there was "nothing mysterious or occult in the work of Mr. Burbank" and his methods were "neither unique nor unknown" (meanwhile, Shull's efforts to translate Burbank were revealing that his work was "of less value to biological science than supposed," partly because of his haphazard record-keeping).[124] The various members of the scientific community who were trying to utilize Burbank's fame faced similar problems: only certain aspects of his work could be translated into established scientific terminology, but once shorn of his idiosyncratic language and almost mystical claims, his work would appear more conventional—and Burbank would lose a good deal of his publicity value.[125]

Scientific attempts to interpret Burbank were most likely to founder when they used the mass media to make their case; the scientists were competing with many other writers, each offering their own interpretations. Some simply accepted de Vries's writing as more authoritative, but other readers might have noticed that Burbank's claim that mutations could be induced at will was strikingly similar to those being made by respected

122. Serviss 1905b, 168.

123. Carnegie was quoted in Woodward 1908 (13–14). See also CIW Trustees minutes, December 12, 1905: 474, quoted in Craig 2005 (44, 258).

124. Woodward 1908, 13–14; Reingold 1979, 321.

125. Some saw Burbank's apparent eccentricities as proof of his genius (Reingold 1979, 320–21).

orthodox scientists like Charles Davenport and Daniel MacDougal—and de Vries himself (chapter 1). Meanwhile, shared goals, such as inducing mutations and creating new species in order to feed the world, ensured that Burbank's work became part of the wider excitement generated by experimental evolution, and the media attention encouraged Burbank to offer his own interpretation of his work's future implications. He argued that if what he knew about plants were to be applied to people, the eventual result would be "the finest human product ever known." Such claims meshed seamlessly with visions of blooming deserts, where thornless cacti supported ever-larger populations, which were in turn strongly reminiscent of the predictions made when CIW's Station for Experimental Evolution opened, that the journey from "four-toed guinea pigs" to producing "superman" was well underway (see p. 70). And, of course, the Carnegie Institution was funding both the Cold Spring Harbor Laboratory *and* Luther Burbank; their grant was the most prominent piece of evidence for the claim that—despite the epithet "Wizard"—the key to Burbank's magical future was a new kind of science. It might, of course, prove to be a highly unconventional form of science that shattered existing laws and transformed the orthodox scientific community's understandings of nature, but that prospect only added to its appeal.

What Burbank offered was not merely better plums, but control over evolution itself. As Serviss wrote, thanks to Burbank, humans "hold a master hand in the game of evolution"; *Sunset* claimed that Burbank's spineless cactus had taken him "ten years to evolve"; the *World To-Day* summarized Burbank as a "scientist, inventor, discoverer, the Edison of plant life, a disciple of Darwin in the evolution of plants and a master workman in the learning of his craft"; and Wickson argued that the spineless cactus had been created by Burbank "in his process of plant evolution."[126] And of course, Burbank himself always used the term, explaining for example that the main goal of his work was to "to provide short cuts for the process of evolution."[127] He even claimed that he had been unable to complete the US census form in 1910, because none of the occupations listed adequately described him. So, he devised his own: "My business is that of New Plant Originator or Evolutor of New Forms of Plant Life for Economic

---

126. Serviss 1905c, 63; "The Sage of Santa Rosa" 1905, 542; Wright 1905, 282; Wickson 1908, 159.

127. Luther Burbank, quoted in *Luther Burbank's Spineless Cactus* 1912 (1); Burbank 1921, vol. 8, 225.

Purposes."[128] The noun *evolutor*—like the verb *to burbank*—is a reminder of how new these ideas still were. There were as yet no words to explain what Burbank or experimental evolution were going to do. It proved impossible to translate Burbank into fully orthodox scientific terms without discarding the paradoxical qualities that made him so appealing. That untranslated residue contributed to his lasting popular reputation, but it also left its mark on biotopianism, which helps explain why it—like the mutation theory itself—persisted as a popular imaginative project long after the scientific community lost interest.

Burbank's claim to be an "evolutor" is also a reminder that the meaning of *evolution* was (like *Darwinism*) being actively debated at this time. Burbank and his interpreters inadvertently collaborated in creating a new language, redolent of science fiction, a babel of voices speculating about the possibilities of controlling and accelerating evolution, of making new species to order, reclaiming deserts, and doubling the earth's population. Much of this language was shared with other proponents of experimental evolution (the adjective *experimental* itself being part of an attempt at redefining evolution). The new language promised a biological revolution even more transformative than the industrial one of the nineteenth century, particularly because a burbanked biotopia would allow individuals to profit as they reshaped nature. Burbank's vision was as radical as those of the later British biotopians who are discussed in the next chapter; he argued that applying what he knew about plants to people should produce "a more or less modified form to the human being."[129] Those future humans might even evolve "a sixth sense" (he was a firm believer in clairvoyance).[130] Small wonder that he concluded, "Man has by no means reached the ultimate. The fittest has not yet arrived."[131]

For historians of science, there are obvious similarities between Cook's theory and that of de Vries: both were interventions in the Darwinian debates, born of skepticism about the power of natural selection; both were

128. "Burbank's Report to the Census Enumerator," 1910, George Harrison Shull papers, American Philosophical Society, Philadelphia, PA. Quoted in Largent 2000, 65–66.

129. Burbank 1906, 127.

130. Burbank 1906, 138. The botanist David Fairchild (1938, 264) recorded in his autobiography: "I was surprised and nonplussed to find that Burbank believed in clairvoyance." Elsewhere, Burbank (1923, 68) described intuition (which he was famous for) as one of the signs of a more advanced kind of person. Katherine Pandora (2001, 505–6) argues that "The Training of the Human Plant" was Burbank's attempt to bypass the "arbiters of scientific decorum."

131. Burbank 1906, 138.

published by formally qualified scientists in recognized scientific journals; each garnered some support from among recognized scientists—at
least for a time; both were ultimately rejected by the scientific community
(and, as a result, neither gets much attention from historians). However,
historians do not need to follow Cook's example and privilege the opinions of scientists over those of their publics. Let us imagine for a moment
that Cook *had* become Burbank's main scientific interpreter; with much
greater public interest, his could have been the theory that left a permanent
mark on popular conceptions of evolution. We will never know, but the
counterfactual possibility helps clarify the roles that various publics have
played in the long-term, historical impact of scientific theories—including
those that the scientific community has written off. Unlike Cook's kinetic
theory, mutation continued to shape ideas about future evolution long after the scientists had lost interest, partly because it escaped from scientific
journals into the paraliterature that Cook scorned—from newspapers, science textbooks, and feminist utopias to pulp science fiction. De Vries had
tried to purify his work of popular (mis)interpretations, but its impurities
ensured mutation theory would continue to reverberate for many decades.

❋ 5 ❋

Perverse Futures

In sharp contrast to the American horticulturalist Liberty Hyde Bailey, who saw evolution as essentially harmonious ("Adaptation and adjustment mean peace, not war"), the British biologist Thomas Henry Huxley described it as violent. Natural selection—the "cosmic process" as he called it—was not merely indifferent but actively hostile to human goals.[1] Any social progress would depend upon "a checking of the cosmic process at every step and the substitution for it of another, which may be called the ethical process." And ethical guidance, he argued, could only be found in human traditions; naturalistic ethical codes like Bailey's, which put their faith in nature's goodness, were mistaken. Despite the fact that Enlightenment rationality had celebrated both nature and human reason (embodied in scientific progress), Lorraine Daston has argued that, as the power and authority of science grew, conventional sources of moral guidance such as custom and tradition were increasingly seen as the opposite of nature. One result was that some nineteenth-century thinkers, including Huxley, came to regard nature as amoral or nonmoral; it thus became as powerless as God when it came to offering moral guidance.[2]

Many Victorians, like Huxley, either lost or questioned their faith, whether in God or nature, and asked instead whether science could tell humans how they ought to live.[3] As we shall see, Huxley's initial confidence in nature's laws had largely failed him by the time he wrote his celebrated essay *Evolution and Ethics* (1894), in which he ultimately failed to find an alternative, universal ethical code that would fulfil the roles apparently abandoned by God and nature. However, the problems that

1. Paradis 1989, 8.
2. Daston 2014, 585–86.
3. Turner 1993; Brooke 1991; Brooke and Cantor 1998; White 2003, 166–69; Dixon 2008b.

Huxley left unsolved were taken up by others, including his former student
H. G. Wells, whose fictions vividly explored the ethical and political implications of fin-de-siècle science. Between them, Wells and Huxley inspired
a trio of twentieth-century scientists—J. B. S. Haldane, John Desmond
Bernal, and Hermann J. Muller—to develop biotopianism into its most
discomfiting and challenging form as they each rejected nature in favor of
the pleasures of what they openly called a perverted future.[4]

Among the things that united these writers was their use of gardens
as a metaphor with which to explore the relationship between nature and
culture. As noted earlier, Francis Bacon's *New Atlantis* (1627) was the first
scientific utopia, built on the assumption that nature was in some way
deficient. His celebrated assertion that scientific knowledge was power
was predicated on acquiring the ability "to generate and superinduce on
a given body *a new nature* or new natures."[5] Bacon's imaginary research
institution, Solomon's House, inspired his seventeenth-century successors to found London's Royal Society and, nearly 250 years later, Bacon's
work was claimed as a model for Carnegie's Station for Experimental Evolution at Cold Spring Harbor (see p. 70). Like Bacon, Huxley and the
biotopians were fascinated by gardens, by the paradoxical complexity of
an artificial space that hovered between re-creating and improving upon
nature, where a new artificial nature would be created (partly by eliminating everything that failed to meet human standards). The biotopians
accepted that humans were inescapably part of nature, but they saw it as
neither inherently good nor bad. As twentieth-century biology began to
imagine remaking human nature as easily as it created new plants, these
scientists imagined the future as an unnatural, scientific garden populated
with newly invented plants—and with people whose ethical codes would
be as malleable and artificial as their plants.

The Ethical Garden?

Huxley began *Evolution and Ethics* with a reminder of the almost infinite vastness of time. From his window he could see the native plants
of Britain's south coast, but beneath them was the chalk, over a hundred
meters thick yet built from the minute fossil remnants of countless dead
creatures.[6] As Huxley had explained in his earlier lecture, "On a Piece of
Chalk," the white cliffs that symbolized England also encapsulated the

4. Kumar 1987, 175–77.
5. Bacon 1620 (2000), 102, emphasis added.
6. T. H. Huxley, Paradis, and Williams 1989. All page references are to this edition.

central, most frightening fact of Victorian geology—time was a bottom-less abyss.[7] As the Victorians dug like moles—mining and quarrying to construct their canals and railways—the accumulating mounds of fossils confirmed that the Earth was unimaginably ancient but also (and more alarmingly) that most of the creatures God had presumably designed were already extinct. Even worse, the recognition gradually grew that "all the history of civilized men is but an episode," a trivial moment overshadowed by the thousands of silent centuries during which the earth had been devoid of humans—and human purpose (59–60).

Yet the view from Huxley's window also provided a hopeful contrast to the pitiless cosmic processes that had shaped the landscape:

> Three or four years have elapsed since the state of nature, to which I have referred, was brought to an end, so far as a small patch of the soil is concerned, by the intervention of man. The patch was cut off from the rest by a wall; within the area thus protected, the native vegetation was, as far as possible, extirpated; while a colony of strange plants was imported and set down in its place. In short, it was made into a garden. (67)

Huxley's garden embodied the Enlightenment hope that nature could be improved by human work. It contained the fruits of empire—exotic plants from all over the world that could not survive "except under conditions such as obtain in the garden." As a result, the plants were "as much works of the art of man" as greenhouses were. Gardens were as much civilization's monuments as cathedrals, and both—like every human achievement—were relentlessly battered by wind and rain as nature endeavored to "reclaim that which her child, man, has borrowed from her" (the image of humanity as a rebellious, naughty child was one of several that Huxley bequeathed to the biotopians). Like many other writers, Huxley saw gardening as morally improving, but not just because it provided fresh air and healthy exercise; for Huxley, gardening was applied ethics—active resistance to natural selection. As they planted, weeded, and watered, humans were opposing the pitiless, amoral "cosmic process" with their human values; Huxley referred to this as opposition as either the "ethical process" or the "horticultural process" (69, 93–94).

Unlike many of his contemporaries, who feared that degeneration would result from restricting natural selection, Huxley argued that

7. First published in *Macmillan's Magazine* in 1868; collected in Huxley 1870 (174–201). Stephen Jay Gould (1990, 44) coined the phrase "geology's most frightening fact" to describe the impact of this revelation. See also Gould 1988, 1–19.

civilization depended on curtailing it.[8] Nevertheless, he acknowledged the Malthusian assumption that civilization's comfortable conditions would cause catastrophic overpopulation and admitted that a new, artificial Eden might need regular weeding, so that only "the strong and the healthy . . . would be permitted to perpetuate their kind" (79). This weeding was eugenics, a term that Francis Galton had coined just a few years earlier in 1883. Huxley noted that many of his contemporaries were discussing the possibility that humans should take control of their own breeding—selecting and destroying as necessary, just as pigeon breeders improved their birds—but he confidently asserted that a "rigorously scientific method of applying the principles of evolution to human society hardly comes within the region of practical politics" (81–92), not least because he doubted that anyone had the skills to identify the most valuable future members of society among a crowd of children.

Huxley not only doubted the practicality of eugenics but also condemned it as immoral. A eugenic state (a "pigeon-fanciers' polity"), would require the "active or passive extirpation of the weak, the unfortunate, and the superfluous," which would destroy the "natural affection and sympathy" which provided the moral basis of society (a view his friend Charles Darwin largely shared). Adopting nature's ways as an ethical guide (by emulating natural selection) would create a world with "no conscience, nor any restraint on the conduct of men" (94–95). Huxley clearly saw eugenics as committing the naturalistic fallacy—those who attempted to derive moral guidance from factual evidence were making a disastrous mistake (138–39). One of his targets was those, like his friend Herbert Spencer, who urged that all state intervention in society was an unwarranted interference in the natural order.[9] By contrast, Huxley averred that society, including science, had to be guided by an independent moral code which would determine when and how people should interfere with nature's workings. However, Victorian anthropology (a science in which he was deeply interested) was demonstrating that humans lived by many different religions and philosophies, raising the question of *which* moral code an advanced, scientific society should adopt. Huxley utilized his gardening metaphor to show why even such apparently robust standards as

8. Malthus's opposition between population and production was, of course, between nature and culture. And both Huxley and Malthus rejected a social idealism that claimed to find benevolence in nature. Malthus was the first to naturalize the concept of original sin (Paradis 1989, 10–11). For a more comprehensive discussion, see Hale 2014.

9. Jones 2004, 12–13. See also Richards 1987; White 2003, 168.

the ethical Golden Rule ("Do as you would be done by"), were unworkable. He invited his readers to put themselves in the place of a person who had robbed them; they would find they had a strong desire not to be punished. As he put it, "What would become of the garden if the gardener treated all the weeds and slugs and birds and trespassers as he would like to be treated, if he were in their place?" (89–91). The Golden Rule could not hold in a garden; human sympathy could not be extended to animals and plants who resisted cultivation's civilizing process.

It was, of course, no accident that the garden—the epitome of civilization—was described by Huxley as a "colony" from which the natives had been "extirpated."[10] Huxley (along with Darwin and most of their contemporaries) was convinced that the people they called savages were anything but noble; indeed, they were in urgent need of civilizing.[11] And when indigenous people or plants resisted European notions of cultivation (or were judged incapable of improving) they were liable to "extirpation," a word whose original meaning was horticultural (to clear land of stumps; from *stirps*, the stem or stock of a tree). Thousands of years of cultivating their small island had made the British particularly prone to seeing gardens as the opposite of wilderness—few Victorians idealized uncultivated nature.[12] Huxley argued that ethics began when people "revolted against the moral indifference of nature." Resistance to nonmoral nature was a thread running through most human ethical codes.[13] That common thread gave them a degree of universality and coherence, which could potentially allow humanity to unite with a common purpose—to protect the ethical garden (117).

In Huxley's well-run colony, the uncultivated natives—whether plants, animals, or people—would all be conquered and then replaced by "an earthly paradise, a true garden of Eden," in which everything served "the well-being of the gardeners." The ideal colonial administrator would abolish the struggle for existence by replacing nature with "a state of art; where every plant and every lower animal should be adapted to human wants." He explicitly compared the process to empire-building: the gardener was

10. My thinking on this point is much indebted to the work of Richard Drayton, who explored the links between agricultural/horticultural improvement, enclosure, and colonization in depth (2000).

11. For Darwin, see his comments on the people of Tierra del Fuego, e.g.: "I could not have believed how wide was the difference between savage and civilized man: it is greater than between a wild and domesticated animal, inasmuch as in man there is a greater power of improvement" (1845, 205).

12. Hoyles 1991; Paradis 1989, 23.

13. Gould 1983b.

a colonizer who must struggle to prevent the "native savage" from destroying "the immigrant civilized man" (70–71, 75). Despite condemning eugenicists for being willing to contemplate "the active or passive extirpation" of those they regarded as "superfluous," Huxley expressed no concerns about the same processes being applied to non-European peoples; his garden colony would, he argued, be an "ideal polity" (77–78). He not only ignored the ways colonization undermined the ethical foundations of human society, he virtually defined civilization as the creation of a racist, imperialist state based on biology.

Huxley's seemingly hypocritical embrace of imperialism while rejecting the application of eugenic policies to his fellow Britons was the result of his inquiry into the problem that preoccupied many of his contemporaries: how could humans avoid degenerating under the "artificial conditions of life" which they were creating (81, 91–92)? For Huxley, the simple answer was that they could not; at least, not permanently. The underlying problem was human nature, which Huxley explained in terms of ancestral heredity, the burden of the past. He believed that everyone sought to maximize pleasure and minimize pain with no thought for the needs of society as a whole: "That is their inheritance," he declared, "from the long series of ancestors, human and semi-human and brutal," whose evolutionary success was "largely indebted to those qualities which he [Man] shares with the ape and the tiger."[14] Victorian anthropologists saw non-European people as less evolved than themselves, thus providing a glimpse into humanity's early history.[15] These assumptions led to indigenous people being judged by Eurocentric standards, with horrifying consequences for those deemed primitive. Huxley assumed that those he called savages exhibited the violent and selfish "innate tendency to self-assertion" which our ancestors had bequeathed to modern humans. This tendency was, he argued, "the reality at the bottom of the doctrine of original sin" (85, 109–10).

However, if the least attractive aspects of human nature were to be explained by evolution, so too were our better angels; we had evolved the potential for cooperation, sympathy, and above all for developing a conscience, which Huxley tellingly called our "artificial personality." These traits gave humans the ability to live and work together, to overcome nature—as long as people used their higher traits (like reason) to

14. Huxley was referring to Tennyson's *In Memoriam* ("Move upward, working out the beast, / And let the ape and tiger die"). Apes signified the primitive origins of humanity, and in the nineteenth century, tigers were a symbol of wanton ferocity, believed to kill for pleasure and not necessity. See Qureshi 2017.

15. Stocking 1987, 25ff, 142, 237.

master their lower, more bestial impulses (88–89). Huxley mocked those who believed that our moral instincts were evidence of "an eternal and immutable principle" implanted in each of us by our Creator. No doubt, he suggested, that is how a philosophical bee would account for its species' achievements, but a human biologist could analyze the steps between solitary and hive bees and show how the relevant instincts had evolved gradually. Huxley assumed that human social and moral instincts (the foundation of the "ethical process") had arisen in exactly the same way. Despite the apparent altruism epitomized by instinctive sympathy or artificial conscience, both were simply adaptive traits which had allowed our ancestors to successfully band together and overcome brute nature (83–84). The opposition between nature and culture crystallized in Huxley's description of the history of civilization as "building up an artificial world within the cosmos" (140–41). Once again, the garden epitomized civilization. The only practical approach to what he called the "gardening of men" would be to create the artificial conditions that would allow the best to thrive; only "the garden of an orderly polity" could produce "the finest fruits humanity is capable of bearing" (101–13). Just as the enclosed plot outside his window had been designed to favor attractive flowers and tasty vegetables while keeping out weeds and wild animals, so human society must be engineered to encourage the qualities humans chose to prize in themselves (140). Untamed nature beyond the wall could never be allowed to dictate what belonged within it, a claim that supposedly justified the brutality of imperialism against those deemed not to have made sufficient progress.

While Huxley accepted the natural origins of human ethical codes, he was too astute to try and derive a human ethical system from observations of the nonhuman natural world. It might seem that the social instincts, having been vital to human evolutionary success, must therefore be morally "good," suggesting that the authority of Nature could be invoked to encourage behaviors such as cooperation. But for Huxley this conclusion was a fallacy; to equate "fittest" with "highest" was to mistake past for future utility. The physical world had changed constantly in the past and would do so in future, so that the very traits which enabled an organism to thrive (such as thick fur during an ice age) would become a fatal liability when the earth became tropical again. In the same way, traits like aggression might have been vital to the success of our earliest ancestors (and were assumed to remain vital in so-called primitive cultures) but were disastrous in a more civilized society. And nobody could predict what future humans might need. Human moral sentiments were, of course, the result of evolution (like every feature of our minds and bodies), but "the immoral sentiments have no less been evolved." So although evolution explained the origins of both the "good and evil tendencies of man," science itself was "incompetent" to provide any reason why "what we call good is preferable to what we call evil" (137–38). Hence Huxley's conclusion that "the ethical progress of society depends, not on imitating the cosmic process, still less in running away from it, but in combating it." We could neither eliminate our animalistic qualities, nor turn to them for moral guidance. Instead, humans had to create their own "artificial personality"—moral codes adapted to the artificial conditions of life they were creating.

Although Huxley made no attempt to develop a comprehensive ethical system, he nevertheless listed some of the ethical principles he had in mind: self-restraint (instead of "ruthless self-assertion"); respecting and assisting others; helping them become fit so that "as many as possible . . . survive"; and above all, recognizing society as a collective achievement (each person should be "mindful of his debt to those who have laboriously constructed it"), so that people would be reminded of their "duty to the community" and reject "fanatical individualism." These were all principles, as he readily acknowledged, that were compatible with many of the world's ethical systems and religions (including Christianity); since humans were a single species, it was unsurprising to find so many commonalities in their evolved moral instincts. He summarized his view as pursuing "a course of conduct which, in all respects, is opposed to that which leads to success in the cosmic struggle for existence" (139–41). In effect, Huxley told his readers to ask themselves, "What would nature do?"—and then do the opposite.

For Huxley (rather like Burbank), ethical progress was the "gardening of men." Just as plants were "trained" to grow in desirable shapes and forms, humans must "modify the conditions of existence" to try and "change the nature of man himself" by utilizing our intelligence and moral principles to reinforce desirable behaviors, until they became instinctive (143). Huxley's view was partly built on Darwin's Lamarckian conception of instincts as learned behaviors, reinforced by repetition until they became completely automatic.[16] A key example was the social insects, the philosophical bees whose instinct to serve their community had subdued each individual's selfishness to the point where most members of the hive had even given up reproducing. However, Huxley had only limited faith

16. This conception was one of the Lamarckian aspects of Darwin's thinking, used to explain how successful behaviors became instinctive (Paradis 1989, 30). See also Richards 1987.

in Darwin's view.[17] Progress was possible—better behaviors might be devised, taught, learned, and reinforced. The ceaseless struggle to resist the cosmic process would prevent degeneration, but he was ultimately skeptical of the Enlightenment hopes that reason and education could transform people. Human nature was "the outcome of millions of years of severe training" in the harsh school of natural selection. It was an in-grained biological legacy that could not be erased by a "few centuries" of retraining. The cosmic process might be held at bay by instilling moral behaviors in humans, but seemingly cultivated man still harbored the ape and the tiger—and the beast beneath the skin would always reassert itself. The burden of evolution was innate selfishness and aggression ("the reality at the bottom of the doctrine of original sin") and these traits would be "a tenacious and powerful enemy as long as the world lasts." And, like most Victorians, Huxley believed that the world would not last forever, not least because contemporary physics seemed to prove that the heat death of the universe was inevitable.[18] As a result, even if evolution progressed upward "for millions of years," eventually "the summit will be reached and the downward route will be commenced" (85). The universe, the Earth, and humanity would all be over long before our slow gardening could cleanse human nature of its darker elements.

Huxley's former student H. G. Wells dramatized the arguments of *Evolution and Ethics* in his first successful novel, *The Time Machine* (1895).[19] (Wells sent Huxley a copy, but the older man died in the same year, apparently without ever having read it.)[20] In the novel, Wells described the Earth in the year AD 802,701 as one in which "the whole earth had become

17. Huxley added a note after "nature of man himself": "The use of the word 'Nature' here may be criticised. Yet the manifestation of the natural tendencies of men is so profoundly modified by training that it is hardly too strong. Consider the suppression of the sexual instinct between near relations." This note clearly acknowledged that the goal of education and moral training is the inculcation of taboos so strong they seem instinctive. Such taboos would mitigate or conceal humanity's more bestial instincts but could not ultimately eradicate them (1894 [1989], 14 n. 23).

18. For the cultural impact of thermodynamics, see G. Beer 1999a. The bleak vision of the end of the universe was, of course, shared by Huxley's onetime pupil H. G. Wells, who invoked the heat death of the universe in the finale to *The Time Machine* (1895 [2001]).

19. Wells (1896a, 594) defined sin as the "conflict of the two factors," which emerged as humans tried to perfect their artificial personalities—"as I have tried to convey in my *Island of Dr. Moreau*." Wells's novel is usually interpreted as a retelling of *Frankenstein*, but it also makes sense as an imaginative reworking of Huxley's *Evolution and Ethics*.

20. Hale 2010, 29. For *The Time Machine* and Huxley, see Hughes 1977; McLean 2009; Parrinder 1981; 1995; Smith 1986, 3–25; Stover 1990.

a garden." As a result, "the air was free from gnats, the earth from weeds or fungi; everywhere were fruits and sweet and delightful flowers. . . . Diseases had been stamped out." The Time Traveller saw this state of affairs as the inevitable outcome of the scientific agriculture of his day, which resulted in "a new and better peach, now a seedless grape, now a sweeter and larger flower, now a more convenient breed of cattle." The Traveller assumed that as science progressed, greater triumphs would follow: "things will move faster and faster towards the subjugation of Nature. In the end, wisely and carefully we shall readjust the balance of animal and vegetable life to suit our human needs."[21] Huxley's dream of an artificial Eden had been achieved, but degeneration had not been avoided. The Time Traveller had found "humanity upon the wane" and was witnessing "the sunset of mankind" (90–91); thanks to scientific breakthroughs, people had given themselves comfort and security, but this "too perfect triumph of man" had left humanity with no problems to solve or dangers to face. It was thus dissolving into a homogeneous mass, each of whom possessed the "same soft hairless visage, and the same girlish rotundity of limb," flabby and childlike, "indolent" and "easily fatigued" (92–94, 86, 88–89). And this physical decline was matched by equally sharp moral decline: the Eloi (the vaguely effeminate aboveground species, whom the Traveller speculated must have descended from the Victorian upper classes) were indifferent to each other's suffering and made no effort to save a woman from drowning, while the Morlocks (the subterranean species, presumably descended from the working classes) had degenerated into cannibalism, farming the childlike Eloi and eating them.[22] And in the novel's final, bleakest section, the Traveller witnessed the heat death of the universe, a stark reminder of the ultimate futility of humanity's struggle against the cosmic process.

The Time Machine offered no more hope than *Evolution and Ethics* had, but Huxley had concluded that—despite the futility of the struggle—humans must struggle continuously against the cosmic process, "cherishing the good" while facing evil "with stout hearts set on diminishing it" (143–44). His call for what sounded suspiciously like a good Christian life, would doubtless have had many a Victorian preacher nodding in agreement. Humanity could never defeat the cosmic process, even if "the whole human race be absorbed in one vast polity, within which 'absolute political justice' reigns"—people would still struggle in vain with human nature and the problems it caused, such as of overpopulation. The only hope, he suggested, would be if humanity's "dose of original sin could be

21. Wells 1895 (2001), 90–91. All page references are to this edition.
22. McLean 2009, 23–25.

rooted out by some method at present unrevealed."[23] However, in the final decade of the nineteenth century, such a prospect seemed unimaginable.

Finding Utopia

The pessimistic outlook that Wells and Huxley shared in the nineteenth century's final decade was far from unusual. Another of Huxley's students, E. Ray Lankester, argued (in *Degeneration: A Chapter in Darwinism*, 1880) that the apparent moral and cultural decline of the fin-de-siècle had a biological basis; if an animal's conditions of life changed, so as to "render its food and safety very easily attained," it would "lead as a rule to Degeneration." Lankester offered parasites as a straightforward (and emotive) example: they adapted to their easier conditions of life by shedding "legs, jaws, eyes, and ears" until an "active, highly-gifted crab" had degenerated into "a mere sac, absorbing nourishment and laying eggs" (precisely the change pictured in *The Time Machine*'s final section). Other animals showed a similar regression; Lankester argued that the Ascidians (sea squirts), whose simple structure suggested they were more like vegetables than animals, were in fact the much-degenerated descendants of higher organisms. He made the moral of his story clear: such creatures were like "an active healthy man" who "degenerates when he becomes suddenly possessed of a fortune," and gives up honest toil for a life of sensuous self-indulgence. His initial, British audience were the rulers of the largest empire the Earth had ever seen, so he reminded them that "Rome degenerated when possessed of the riches of the ancient world" (the luxuriant decadence of the Roman Empire was proverbial, frequently offered as a terrible warning to the nineteenth century's leading imperialists). It was, he concluded, only "an unreasoning optimism" that allowed "the white races of Europe" to assume that they were "destined to progress still further." "Possibly," he speculated, "we are all drifting, tending to the condition of intellectual Barnacles or Ascidians." He urged his audience to remember that they were subject to "the general laws of evolution, and are as likely to degenerate as to progress."[24]

Lankester's ideas were an important influence on his friend Wells, as is obvious from *The Time Machine* (and Lankester would later be a key adviser to Wells while he was writing his best-selling *The Outline of History*, 1920).[25] A decade after *Degeneration* was first published, Wells popularized

23. Huxley 1894 (1989), 102.
24. Lankester 1880, 33, 58–60.
25. Barnett 2006; McLean 2009, 25; Hale 2010, 27–28.

Lankester's ideas in the *Gentleman's Magazine*, whose well-heeled and educated readers were berated for their "excessive self-admiration" and "invincibly optimistic spirit," both of which resulted from their scientific ignorance. They misinterpreted evolution as meaning that the "great scroll of nature has been steadily unfolding to reveal a constantly richer harmony of forms and successively higher grades," a comforting version of stadial history that Wells called "Excelsior biology" but that was unsupported by scientific evidence.[26] With Lankester's fear about the drift to becoming "intellectual Barnacles" clearly in mind, Wells concluded that evolution "has been fitful and uncertain; rapid progress has often been followed by rapid extinction or degeneration." Humans would doubtless continue to change in the future, "but whether that will be, according to present ideals, upward or downward, no one can forecast."[27]

The Time Machine reflected these biological ideas, and its extraordinary success suggests that it caught the mood of many readers. It reflected the fear of declining and falling empires, particularly the specifically biological anxiety that the sharp pruning knife of natural selection had been blunted by civilization. As the feckless and unfit bred and multiplied in their dark, almost subterranean hovels, their supposed betters luxuriated in an ease that would turn them into beautiful but futile Ascidians. As the Austro-Hungarian social critic Max Nordau expressed it (in his own book, also titled *Degeneration*, which first appeared in English in the same year as *The Time Machine*), the century's final decade was characterized by "fearful presage and hang-dog renunciation. The prevalent feeling is that of imminent perdition and extinction." And he forecast that the result would be "the unchaining of the beast in man" and "the trampling under foot of all barriers which enclose brutal greed of lucre and lust of pleasure."[28]

Such dark predictions made utopia impossible. Wells's later book, *Anticipations* (1902), argued that "all dreams of earthly golden ages must be either futile or insincere, or both, until the problems of human increase were manfully faced," as had been proven by Malthus's *Essay on Population*—perhaps the most "shattering book" ever published.[29] Wells concluded his book with a summary of the grim eugenic views he held at the time. He asked what would become of the "swarms of black, and brown, and dirty-white, and yellow people," who would surely prove

26. Wells 1891, 246–47.

27. Wells 1891, 253.

28. Max Nordau, *Degeneration*, trans. W. F. Barry (New York: D. Appleton, 1895), trans. Entartung, 1892. Quoted in Wells 1895 (2001), 210.

29. Wells 1902a, 312; Hale 2010, 27–28.

unable to meet future's "needs of efficiency." "Well," he concluded, "the world is a world, not a charitable institution, and I take it they will have to go. . . . it is their portion to die out and disappear."[30] Yet—despite Malthusian fears, the gloomy fin-de-siècle consensus, and the devastation of World War I—by the 1920s, Wells had not only embraced utopia but, according to the *New York Times*, had persuaded many people that they might also "arrive there, land and take possession." In a review of Wells's new novel *Men Like Gods* (1923), the *Times* argued that in the past, to call something utopian was to dismiss it, but in the present "scientific age. . . . Utopia is arousing almost as much interest as contempt," because *Men Like Gods* was only the most recent of a series of books in which Wells had "made Utopia constructively visible and almost respectable."[31] However, utopia was rehabilitated not because Wells had changed—but because biology had. Experimental evolution might soon do precisely what Huxley had claimed was impossible—remake nature, including human nature, and even eradicate original sin.

Like many of his religious contemporaries, Huxley had seen original sin as an ineradicable part of human nature. Although his explanation was evolutionary rather than conventionally religious, his secular sermons shared with their more orthodox counterparts the conviction that there was no "prospect of attaining untroubled happiness." At the time he wrote, new socialist parties and their trade union allies were promising the workers utopia, which struck him as one of the most misleading illusions to have ever been "dangled before the eyes of poor humanity" who could never hope for a "state which can, even remotely, deserve the title of perfection" (102).[32] Huxley's rejection of utopianism is easier to understand if we examine the nineteenth century's most successful utopia, Edward Bellamy's *Looking Backward: 2000–1887* (see p. 23), published just a few years before *Evolution and Ethics*. Bellamy's book told the story of Julian West, a contemporary Bostonian who slept for over a century and woke up to find himself in a socialist America (described in supposedly persuasive detail). Despite the novel's deficiencies as a work of literature, it was enormously successful—selling more than a thousand copies a day at some points in 1889, and becoming only the second American novel to sell over one million copies (Alfred Russel Wallace was one of many who credited the book with making him a socialist, while in the USA, Bellamy

30. Wells 1902a, 342.
31. Forman 1923, 1. Wells's best-known utopia was Wells 1905.
32. White 2003, 141–47.

clubs developed into a full-blown political movement dedicated to making Bellamy's vision a reality).[33]

As Mark Pittenger has shown, Bellamy's politics were based on a radical environmentalism (partly influenced by Herbert Spencer's Lamarckianism) that assumed that if society stopped rewarding competitive and selfish behavior, these qualities would die out and be replaced by nobler ones. Spencer is now commonly regarded as the prophet of a brutal, laissez-faire social Darwinism, but as Robert Richards has shown, he had been a utopian socialist in his youth, and he never abandoned the hope that evolution would make a better society inevitable.[34] Many early twentieth-century socialists shared Spencer's view that human nature was redeemable via social engineering, but—as Pittenger notes—the result would not be a new, more highly evolved, human nature, but a restoration of what Bellamy called the "natural nobility of the stock."[35] This phrase comes from a section of the novel where West listens (via telephone, as part of an audience of 150,000 people) to a sermon by a Mr. Barton, one of the future's most popular preachers. Bellamy frequently used advanced technologies to explain the rapid tempo of future change; West discovers that it had taken little more than a century for utopia to evolve, but in Bellamy's version of stadial evolution there had been no revolutionary upheaval—no change in evolution's mode—merely a collective, rational decision to replace an inefficient system with a seemingly better one.[36] The novel's mass-media preacher, Barton, used West's emergence from the distant past as an opportunity to explain the changes: the apparently miraculous improvements have a "simple and obvious explanation"—they are the impact of "a changed environment upon human nature." Instead of founding society on the "pseudo self-interest of selfishness," and appealing "solely to the anti-social and brutal side of human nature," the new society is based on a "rational unselfishness," which appeals "to the social and generous instincts of men." As a result, "it was for the first time possible to see what *unperverted* human nature really was like." In Barton's view, it was hardly surprising that nineteenth-century people had sunk to the level they had. They had lived "through numberless generations" under conditions that "might have perverted angels," yet even these conditions "had not been able to alter the natural nobility of the stock." Once the environment changed, human nature "like a bent tree, . . . had sprung

33. Roemer 1976, 2–3; J. R. Durant 1979; Lipow 1982; Hamlin 2014b, 157–58.
34. Richards 1987, 246.
35. Pittenger 1993, 9–10, 17–20, 67.
36. Williams 1978, 206.

back to its normal uprightness." Humanity under capitalism was like "a rosebush planted in a swamp," growing in conditions so foul that the resulting plant appeared to be "a noxious shrub, fit only to be uprooted and burned."[37] The claim was clear: nature, including human nature, was inherently good—evil merely the result of perversion, of turning against nature.

Bellamy—like many socialists—expressed optimism about the innate goodness of human beings, which explained inequality, cruelty, and injustice as the "perverting" effects of capitalism. The English socialist William Morris loathed *Looking Backward* (not least because it denied the need for revolution), and he wrote *News from Nowhere* (1890) as a response to it. Nevertheless, Morris held a similarly optimistic view of human nature; his novel opened with a vision of the Thames restored to fresh, salmon-rich beauty once Victorian capitalism and its pollutants had disappeared. The healthy, unaffected boatman who conveys the time-traveling William Guest on the first leg of his journey embodies the same argument as Bellamy's: rescue human nature from the dismal swamp of capitalism and it would spring back into its "natural" (wood carving, wallpaper designing) form.[38] By contrast, Huxley—like many of his contemporaries—rejected this optimistic view and saw their thriving capitalist society not as a perversion, but as an expression of humanity's competitive nature, honed by natural selection to ensure survival.

Judith Shklar has argued that utopias invariably entail rejecting the notion of original sin, because it relegated "natural human virtue and reason" to the status of "feeble and fatally impaired faculties."[39] Her insight helps us understand Huxley's pessimism. Malthus's gloomy forecasts were a response to the Enlightenment's hopes of human perfectibility and were a key source for Darwin's theory of natural selection. Malthus assumed original sin was ineradicable—left unchecked, the human instinct to reproduce must inevitably result in misery.[40] Huxley saw science as a key tool for creating a much-improved human society, in which "the cosmic struggle for existence, as between man and man, would be rigorously suppressed" (76). Yet there could be no heaven on Earth because nothing

37. Bellamy 1888 (1917), 276, 287, emphasis added.

38. Morris 1890 (1998); Morton 1952 (1978), 202–36; Manuel and Manuel 1979, 759–64, 768–72.

39. Shklar 1965, 370.

40. Spengler 1971. Krishan Kumar (1987, 12–13) has noted that Christianity has a long, contradictory tradition concerning the perfectibility of man, which some took to say that men could become like gods (a Neoplatonic idea rooted in Hellenism), while others interpreted it as simply meaning that we should strive to live as good a life as is possible for us, given our fallen natures.

could stop the weeds from overrunning the garden; as he put it, "ethical man admits that the cosmos is too strong for him" (135). The struggle might be morally essential but was ultimately futile.

For Huxley, the ultimate problem was the "nature of man himself," the tendency to compete and breed, which would lead to overpopulation. Wells built on Huxley's claim by imagining a world that had indeed become a garden, from which the struggle for existence had been excluded, but, of course, the result was not remotely utopian: aboveground, humanity had declined into "a mere beautiful futility," while nothing was beneath the subterranean Morlocks, who—despite being hardworking and industrious—had regressed to the savagery of their earliest ancestors.[41] Yet Wells's later novel *Men Like Gods* depicted an almost identical garden world created by similar means (particularly plant- and animal-breeding), whose people were neither indolent nor cannibals but "like Gods." The key difference was that *The Time Machine* was a nineteenth-century novel, which understood heredity as the burden of the past and depicted evolution as a hostile, cosmic force, far too slow to complete its work of improvement before the world ended. *Men Like Gods* was imbued with the twentieth-century optimism of the experimental evolutionists, equipped with the tools to speed up evolution and bring it under human control.[42]

Men Like Gods centered on Mr. Barnstaple, a disillusioned journalist who finds himself (along with an assortment of other "Earthlings") in a parallel earth (explicitly named Utopia) into which they have accidentally been pulled by some Utopian physicists. As the Earthlings explore, they discover the usual Wellsian utopia: an enlightened anarchism achieved through a vague, gradual scientific education, which has resulted in a state without coercion or scarcity of any kind.[43] However, most earlier utopias (*Looking Backward* is a typical example) had been based on progress in the physical sciences, combined with reforms to political, economic, and

41. Huxley 1894 (1989), 143, 85; Wells 1895 (2001), 119.

42. Piers Hale (2010) argues that Wells became increasingly anti-utopian and pessimistic post-1895, but his argument barely addresses the biological optimism of *Men Like Gods* and *The Science of Life*. The same absence is apparent in Nate (2000), despite his coming to the opposite conclusion to Hale.

43. Kumar describes *Men Like Gods* as a kind of sequel to Wells's earlier book *A Modern Utopia* (1905). The society described in the earlier work has continued to evolve and the Samurai state has withered away, a vision—as Kumar comments—that is as much Marxist as Darwinian (Kumar 1987, 190).

similar codes—all designed to manage or minimize human shortcom-
ings.[44] In *Men Like Gods*, biology played the central role, epitomized in its
cultivated landscape. "Plants and flowers, always simpler and more plastic
in the hands of the breeder and hybridiser than animals, had been enormously changed in Utopia," whose scientists had forced the plants "to
make new and unprecedented secretions," such as essential oils, "of the
most desirable quality"—desirable, of course, from a human perspective.
And the same desire had eliminated all infectious organisms and most
pests, including mosquitoes and midges.[45] As with many other biotopian
visions, the Utopians' work had begun by exploiting the "plasticity" of
plants but had been successfully extended to other animals (the visitors
meet a "naturally" tame leopard)—and even to themselves, an achievement Wells (and others) hoped would eventually be emulated in reality.

The *New York Times* was not alone in praising Wells's novel, nor in
arguing that the future it promised was tantalizingly close (figure 5.1). The
British scientific journal *Nature* described it as depicting a world in which
"life has been subjected to a similar control" to that already exercised over
machinery. Wells had simply extended processes that "the biologist sees
so obviously on its way," and so the book should "excite no surprise." The
reviewer was the biologist Julian Huxley (a zoology lecturer at Oxford and
Thomas's grandson), who argued that since "our knowledge of genetics"
was increasing rapidly, "Mr. Wells's wonderful flowers and trees are almost
there already: we will not worry about them." (He even recommended
some additional nonfiction reading to accompany *Men Like Gods*, including Punnett's *Mendelism*.) He astutely identified as "the most radical and
inevitably the most provocative of our author's imaginings" the fact that
his Utopians were not (as in most utopias) people just like us, just with
better laws or gadgets. What made *Men Like Gods* extraordinary was "not
the alteration of things in relation to a constant human nature, but the
alteration of that human nature itself."[46] Huxley had hit the biotopian
nail crisply on its head. His review led directly to his collaboration with
Wells on their epic textbook, *The Science of Life* (1929–1930), an extended
celebration of biology's power to remake the future (discussed in the next
chapter).[47]

44. Howard Segal (2005) describes *Looking Backward* as the most successful example of a wider trend which he named "technological utopianism."
    45. Wells 1923 (2002), 85–86.
    46. Huxley 1923, 592.
    47. Anker 2002, 111–12; Clayton 2016, 880.

FIGURE 5.1. As he left Utopia, Mr. Barnstaple got instructions from Urthred on how to make the Earth more utopian. The scene is depicted here in one of the illustrations that accompanied the *Hearst's International* serialization of *Men Like Gods* (see p. 274).

Radical Indecency

As we have already seen, Wells's newfound optimism about biology's possibilities was far from unusual. But while most American readers first read about new theories of heredity in the nonspecialist press, in Britain the same enthusiasm was more likely to be communicated by scientific intellectuals. Wells was the most famous, but in the 1920s the boldness of his biological theorizing would be more than matched by several others, beginning with the British biochemist and geneticist John Burdon Sanderson (J. B. S.) Haldane. He first presented his controversial vision of the future as a talk at Cambridge University in February 1923, and it was published the following year as *Daedalus; or, Science and the Future*. The book initially sold seventeen thousand copies (making it one of the most successful titles in the newly launched series, To-Day and To-Morrow); nevertheless, it was a highbrow book for an elite audience quite unlike the readerships of US mass-market magazines.[48] By the 1920s, Wells was so famous that Haldane had to acknowledge that "a word on Mr. H. G. Wells" was required, since the "very mention of the future suggests him." However, Haldane seemed unaware of *Men Like Gods* (which had not appeared when he gave his initial talk) and focused on Wells's earlier *Anticipations* (1902). He dismissed its predictions as "singularly modest"; Wells was "a generation behind the time," having been raised when flying, radio, and telegraphy were scientific problems. "Now these are commercial problems," Haldane opined, "and I believe that the centre of scientific interest lies in biology."[49]

The claim that biology was the defining science of the twentieth century was the heart of Haldane's argument. He claimed that the "chemical or physical inventor is always a Prometheus" (the mythical thief of fire, who epitomized the achievements of the physical sciences), whose inventions were often "hailed as an insult to some god":

> But if every physical and chemical invention is a blasphemy, every biological invention is a perversion. There is hardly one which, on first being brought to the notice of an observer . . . would not appear to him as indecent and unnatural. (44)

According to Haldane, "a sentimental interest" in Prometheus had distracted people from "the far more interesting figure of Daedalus," whose monstrous hybrid of human and cow, the Minotaur, was utilized by Haldane to symbolize the allure of perversion—the ability to make new organisms with no regard for supposedly natural barriers. Haldane explicitly linked the mythical monster to the achievements of twentieth-century genetics; if only, he joked, "the housing and feeding of the Minotaur [had] been less expensive," Daedalus might "have anticipated Mendel" (but Haldane had to admit that science funding bodies would probably consider an annual sacrifice "of 50 youths and 50 virgins . . . excessive as an

48. Saunders 2019, 339.
49. Haldane 1924, 9–10. All subsequent page references are to this edition.

endowment for research" [46–47]). Nevertheless, the modern biologist was heir to Daedalus, able and willing to pervert nature to serve human needs and so, despite being "a poor little scrubby underpaid man," Haldane was convinced that "the biologist is the most romantic figure on earth at the present day" (77). The biological inventor's perversions were positively desirable; Haldane celebrated the "profound emotional and ethical effect" on humanity of such inventions as the domestication of plants, animals, and fungi. Some of these were so familiar that he suspected his readers would no longer notice just how "indecent and unnatural" they were, so he invited them to consider the dairy industry, built on stealing a cow's milk so that it could be "drunk, cooked, or even allowed to rot into cheese." As he observed, "We have only to imagine ourselves as drinking any of its other secretions, in order to realise the radical indecency of our relation to the cow" (45).

Haldane dramatized his forecasts by presenting them as a report from the future. At the center of his book was an undergraduate essay supposedly written in 150 years' time, which described the series of historical breakthroughs that had shaped the future. The compressed temporal frame was vital to the book's impact: the future was close and the history that had created it began with the achievements of real scientists (including Haldane himself). Such details made Haldane's future so plausible as to seem almost inevitable. Among the breakthroughs the student essay described were drugs that modified human moods, artificial foods synthesized directly from inorganic chemicals, and coal and oil being replaced by renewable energy, generated by huge windmills (24).

However, while physics and chemistry played their parts, it was the biological inventions that took center stage. *Daedalus* forecast that "synthetic food will substitute the flower garden and the factory for the dunghill and the slaughterhouse" (39)—the sights and smells of traditional rural life would be replaced by a cleaner, more efficient, and artificial world. That replacement of nature with a human-made landscape was also a feature of *Men Like Gods*, in which Wells (like his teacher, Huxley) used the image of a garden to dramatize the conflict between nature and culture.[50] On his first morning in Utopia, for example, Barnstaple meets two gardeners tending extraordinary and beautiful roses. Yet the flowers are distinctly unnatural: they are described as "monsters," with flowers that "dwarfed" their leaves. Their petals are like "drifting moths," "snaky-red" and resembling blood—a description that blurred the lines between

---

50. The link between Wells and this essay of Huxley's is noted in Hughes 1977 (53) but not explored in detail.

plant and animal, imbuing the roses with a slightly sinister air. Barnstaple learns that some Utopians consider it too much trouble to cultivate these double-flowered roses in the mountains because they are growing unnaturally, in conditions to which they are not suited. (An obvious contrast to the views of those, like Liberty Hyde Bailey, who argued that gardening required learning "how to adapt one's work to nature"). For Barnstaple the rose growers prove that Utopia's people were willing to "work and struggle for loveliness," and he sees their slightly unnatural work as a microcosm of the utopian project: "He knew enough of Utopia now to know that the whole land would be like a garden, with every natural tendency to beauty seized upon and developed and every innate ugliness corrected and overcome."[51] Wells's utopians no longer regarded nature as a guide but merely as raw material for the human imagination to work on.

Denying any normative role for nature was a central feature of each of the texts considered in this chapter. For example, *Daedalus*'s claim that biology would eventually be able to reshape both the human body and psyche directly inspired Haldane's younger contemporary, the physicist and crystallographer John Desmond (J. D.) Bernal, to write his own scientific forecast, *The World, the Flesh and the Devil: An Inquiry into the Future of the Three Enemies of the Rational Soul* (hereafter, *The World* . . .), which appeared in 1929, also as part of the To-Day and To-Morrow series.[52] Like Haldane, Bernal gleefully discarded any normative role for nature, and imagined using science and technology to reshape both human beings and their environments until the organic and synthetic were fully hybridized. The result would be a "new life . . . more plastic, more directly controllable and at the same time more variable and more permanent than that produced by the triumphant opportunism of nature" (46). In Bernal's view, progress largely consisted of "the replacement of an indifferent chance environment by a deliberately created one" (62). And similar attitudes were apparent in the American geneticist Herman Muller's *Out of the Night* (1936), which despite being published a decade later brought together various essays and speeches that Muller had first produced in the 1920s and that reflected many of the same enthusiasms as Wells, Haldane, and Bernal.[53] The *New York Times* called it a "stimulating little collection

---

51. Wells 1923 (2002), 155.

52. Bernal 1929 (1970); all subsequent page references are to this edition. Bernal was a member of the Cambridge University Heretics club, where Haldane first presented what would become *Daedalus*, but it is not known whether Bernal was present (Brown 2005, 72–73).

53. For example: Herman Muller, "Revelations of Biology and Their Significance: An Address Read to the Peithologian Society of Columbia University," 1910, Muller

of lectures and essays" that was "both a popular exposition of genetics and a prophecy," while Britain's Manchester *Guardian* also described it as "a daring, lively, provocative essay."[54] Muller predicted that genetics would eventually allow "quite new forms of plant life [to] be invented and evolved, exotic growths whose characters we have not yet imagined."[55]

Haldane offered the first, clearest and most influential summary of the full-blown biotopian attitude to nature: it was to be corrected, improved upon, and remade according to human standards. His undergraduate narrator described various biological inventions, of which the most remarkable was a laboratory-built alga that fixed nitrogen so efficiently that global food gluts resulted. The artificiality of the invention was strongly emphasized (the student essayist explained that the alga "could not, of course, have been produced in the course of nature, as its immediate ancestors would only grow in artificial media and could not have survived outside a laboratory" [59–60]). The imaginary history of this discovery described how a strain of the engineered alga had escaped into the ocean the Atlantic, which "set to a jelly" for two months ("with disastrous results to the weather of Europe"), but the accident eventually led to fish stocks booming to become the world's main source of protein, finally abolishing hunger. Haldane acknowledged that many of his readers would dismiss his speculations as "improbable or indecent," because they were "perversions," violations of the natural order. And yet he was convinced that in a generation's time, the ideas he put forward would "appear as modest, conservative, and unimaginative as do many of those of Mr. Wells to-day" (10–11). In time, physical inventions as unnatural as flight became commonplace, and the same happened with biological inventions. In passing, the essayist noted that it was after the synthetic alga's escape that "the sea assumed the intense purple colour *which seems so natural to us*, but which so distressed the more aesthetically minded of our great grand-parents who witnessed the change" (59–62, emphasis added). Such tongue-in-cheek asides were the heart of Haldane's challenge to the very concept of "natural." Anything, he asserted, can become natural in time, from eating cheese to purple oceans, because "natural" is a cultural category—and if anything can *become* natural, nothing is inherently natural (or unnatural). In similar terms, Barnstaple described the Utopians' remaking of nature

---

archives, Lilly Library, University of Indiana, Bloomington, IN, Quoted in Carlson 1981 (34–35); Muller 1936, 5. Also, Muller, "The Promise of Biology," speech, University of Texas, Austin, 1924 (Garrett 1924).

54. Kaempffert 1936; Stocks 1936.

55. Muller 1936, 94. All subsequent page references are to this edition.

(including their own natures) as "the most natural and necessary phase in human history."[56] If nature was no longer "the norm," only human standards—utilitarian or aesthetic—could be used to judge the desirability of various perversions.

Men Like Gods offered biological solutions to the world's problems. Barnstaple described Utopia approvingly as "a world where ill-bred weeds, it seemed, had ceased to thrust and fight amidst the flowers," because of Utopian eugenics.[57] The analogies with the real world were obvious; the Earthlings learn that in Utopia's earlier history, medical progress had led to rapid population growth that threatened a Malthusian catastrophe. The Utopians had decided to restrict the population to the number able to enjoy "a fully developed life" by instituting eugenic measures, which had initially included "weeding" their population. However, by contrast with the forms of eugenics that most of Wells's contemporaries were offering (and in which he had firmly believed just a few years earlier), the utopia he described no longer seemed to have any unfit to be eliminated. Earlier policies had presumably worked, but the book also dramatized the expectations aroused by speculative heredity, including new ways to increase the numbers of the fittest. Early in Wells's book, the visitors were treated to a brief history of Utopia's earlier "Age of Confusion," partly caused by the "great masses of population" who had simply "blundered into existence, swayed by damaged and decaying traditions and amenable to the crudest suggestions." Under these conditions, it was better—the Utopians concluded—for such people never to born at all.[58] And so they created a world in which "every child is born well," of perfectly healthy mothers who consciously decide when to have children, using artificial birth control.

When this history was recounted, one the Earthling visitors, Father Amerton, a Catholic priest, is appalled by the revelation of birth control and eugenics. ("The human stud farm," he cries, "Refusing to create souls! The wickedness of it!")[59] Although Wells satirizes the priest's attitudes,

56. Wells 1923 (2002), 87.

57. Wells 1923 (2002), 32.

58. Wells 1923 (2002), 155, 63–64, 87.

59. Wells 1923 (2002), 81–82, 63–64. One contemporary reviewer suggested that Amerton had been modeled on "the late Father Vaughan" (review of Wells, *Men Like Gods* 1923). Bernard Vaughan (1847–1922) was an English Roman Catholic clergyman, a Jesuit, and author of *Sins of Society* (1906), a book based on popular sermons he gave in London. He was often regarded as publicity seeking. Geoffrey Holt, "Vaughan, Bernard John (1847–1922)," *Oxford Dictionary of National Biography* (Oxford: Oxford University Press, 2004); online ed., January 2011, http://www.oxforddnb.com/view/article/36631.

the description of Utopia as a human stud farm is an accurate one; humanity is farming itself by choosing the best to breed from. However, the next generation's *nurture* gets at least as much attention as their nature; education and childcare are crucial to maximizing the quality, rather than the quantity, of utopian children. And improving the population takes many forms; the Earthlings learn that there are no punishments for antisocial behaviors in Utopia, because nobody admires such behavior. The indolent or selfish individual will "find no lovers, nor will it ever bear children," thus eliminating undesirable genetic traits.[60] The utopians relied on Darwin's theory of sexual selection, which ascribed a vital role in evolution to mate choice (it was also central to several feminist utopias; see chapter 7).[61] The result is a startlingly beautiful population (particularly evident because of the Utopian custom of nudity). The Utopians, as naked and unashamed as classical deities, dramatize the many ways in which *Men Like Gods* envisaged a radical reorganization of sexual morality. To utopia's biologists, minds and morals had become as malleable as plants.

Each of the books discussed here reflected the fear of degeneration, but—as in Wells's utopia—each went beyond conventional, mainstream eugenics in their efforts to overcome it. In *Daedalus*'s historical-essay portion, the undergraduate narrator argued that civilization would have collapsed very rapidly "owing to the greater fertility of the less desirable members of the population" (66–67). However, this fate had not been avoided using conventional eugenics, which Haldane dismissed with characteristic wit by describing the widely proposed eugenic official as a compound of "the policeman, the priest and the procurer" (40–41).[62] *Daedalus* identified artificial contraception (a standard—albeit controversial—aspect of mainstream eugenic policy), as an important biological invention that would become ever more effective and widely used in future. However, Haldane believed more radically indecent, perverse approaches would be needed; his future assumed babies grown outside human bodies in artificial wombs (ectogenesis), thus completing the disconnection of sex and reproduction that had begun with contraception. In the future, a handful of the supposedly best men and women would be selected to donate their gametes and thus become parents of all future generations

---

60. Wells 1923 (2002), 72–73.

61. Milam 2010.

62. Maurizio Esposito (2011, 42) notes that Haldane rapidly lost his limited faith in eugenics but remained committed to a bioengineered future for humanity; see *The Inequality of Man and Other Essays* and *The Causes of Evolution* (both 1932).

(63–66).[63] Haldane's knowledge of genetics persuaded him that the forms of eugenic selection proposed in his day were, among other things, far too slow to work effectively, whereas ectogenesis would allow the mass production of whatever kinds of people the future wanted. The *New York Times* gave the book a lengthy review that acknowledged that "unnaturalness" was "inseparable from the very nature of biological inventions." As a result, such inventions were bound to be controversial, "since they are innovations bearing on the most intimate phases of life." While the idea of creating better people seemed obviously desirable, the reviewer believed that the prospect of giving some officials the power to decide which types were desirable enough to be multiplied, "thus elevating them into as many little gods," was "anything but cheerful."[64]

Haldane's book is, of course, best-remembered today because it was a key source for Aldous Huxley's *Brave New World* (1932). Huxley's deeply ambiguous book is usually—and rather simplistically—considered as a straightforward dystopia, yet it retained traces of Haldane's original utopian mood. *Daedalus's* report from the future explained that ectogenesis allowed for the scientific selection of the best men and women as parents, with "very startling" results, including "decreased convictions for theft" and (perhaps with tongue in cheek) "increased output of first-class music." Haldane also forecast that in the post-natural world, aging and disease would be controlled to the point where death becomes "a physiological event like sleep," causing both the fear of death and the resulting "desire for an afterlife" to fade away. Meanwhile, drugs, hypnosis, and hormones would modify our behavior and multiply our sources of pleasure.[65] Ectogenesis was considerably more humane than conventional eugenics, and it was not the only utopian idea *Brave New World* took from *Daedalus*.

Muller shared several of Haldane's views and also argued that it was vital to break the link between love and reproduction before true eugenic policies would be possible. Thanks to modern biology it was, he argued, finally possible to "unyoke the two" so that each could fulfil "its already distinct function" (138–40). Once this unyoking happened, Muller proposed that artificial insemination be used to spread the best genes. He also cast doubt on the prescriptions of existing eugenics (noting, for example,

63. The term *ectogenesis* was first used to describe structures that organisms created outside their bodies (1909), but Haldane's usage, referring to human reproduction using artificial wombs, rapidly became the dominant one. See *Oxford English Dictionary Online*, s.v. "ectogenesis, n." June 2019, https://www.oed.com/dictionary/ectogenesis_n?tab=factsheet#5974942. See also Squier 1994.

64. Bagger 1924, 24, 27.

65. Haldane 1924, 66, 71–73.

the difficulty of deciding who should be considered defective), but continued, "Fortunately, nature does not limit us to a particular fixed standard of sanity; her own standard has been ever changing, and she allows us always to make for the horizon" (102). The expanding horizon of expectation created by new biological ideas allowed nature's standards to be abandoned in favor of human ones. Muller proposed a form of positive eugenics based on human seed-banking (the sperm of great men would be stored after their death and made available to future generations, since "mankind has a right to the best genes available"). The *Washington Post* found Muller's proposals repugnant and noted that his "genetic principles" were no different to those of the slave owner "who loaned or hired his healthy big bucks to neighboring plantations."[66] However, Muller was convinced that "many women, in an enlightened community devoid of superstitious taboos and of sex slavery, would be eager and proud to bear and rear a child of Lenin or of Darwin!" (152–53). The technology to achieve this vision did not yet exist, but Muller was confident it was close (although he found it much harder to imagine a technology that would allow great women to spread their genes more widely).

As Elof Carlson has argued, despite important similarities in their views, Haldane and Muller had strikingly different views of the roles that women might play in their projected futures.[67] Haldane's attack on conventional morals and socially endorsed norms did not extend to considering (much less, challenging) supposedly natural gender roles; by contrast, Muller embraced an overtly feminist agenda, arguing that a woman should refuse to be "either a queen-bee or a galley-slave," but instead join the "growing strike against child-bearing . . . until the dire age-old grievances have been removed!" (129). He argued that equality for women was a precondition for "real eugenics." There would need to be free, universal birth control, effective education, and legal, regulated abortion. Childbirth must be completely voluntary and a good deal more comfortable (he noted that male doctors tended to regard women's pains as "obligatory, or even sadistically look upon them as desirable"). He also emphasized nurture over nature, promoting better health care for new mothers and their babies, supplemented by proper social support (such as socialized food production and laundering) to relieve women of these unnecessary burdens. Finally, he advocated more flexible working arrangements to allow

66. Flury 1936.
67. Carlson 1995, 92–93.

women to return to work after birth (129–31).[68] Muller summarized his own proposals by observing, "What an admirable eugenic system is all this, how accurately calculated to inspire women of the highest type of intelligence to enter upon the career of motherhood!" (132). Despite its grating, self-congratulatory tone, he attempted to offer a progressive political agenda. Bernal also saw human morals and behaviors as raw material for the imagination to work on. However, as Andrew Brown has shown, just as Haldane rejected eugenics for being too slow, Bernal rejected ectogenesis for the same reason and also argued that it would tie humanity too tightly to its biological origins. Despite Bernal's commitment to Freudianism (during this period of his life, at least), *The World* . . . did not regard human instincts as immutable but assumed they could be altered as easily as bodies.[69] "The sexual instinct in particular," Bernal noted, "which still finds considerable direct gratification, would be unrecognizably changed" as humans modified their artificial bodies (61). Bernal imagined a more radical reconstruction of human bodies and minds, including their sexuality and morality, than even *Daedalus* had offered.

In each of these books, new approaches to sex and reproduction were presented as an essential component of a better world (not least because they would permit the supposedly healthier and more rational approach to reproduction that even speculative biotopian eugenics required). Wells also argued for the need to discard old taboos; when Father Amerton attempts to enlist Barnstaple in a crusade against Utopian morals, Barnstaple responds that he believed Amerton embodies all that is "wrong and ugly and impossible in Catholic teaching."[70] The Utopians seem to agree, since their chief spokesman, Urthred, says of Amerton that he is "afraid even to look upon men and women as they are," and that as a result, "this man's mind is very unclean. His sexual imagination is inflamed and diseased. . . . Tomorrow he must be examined and dealt with." (This comment, Barnstaple has to admit, created "a queer little twinge of fear" in him.)[71] In general, however, Barnstaple's reaction to Utopian eugenics was entirely accepting, and once again, horticultural metaphors were used to persuade the reader of its desirability. Utopian eugenics was described by Barnstaple as "revision and editing," the "weeding and cultivation of the kingdoms of nature by mankind. . . . 'After all,' he said to himself, 'it was

68. Many of these ideas were repeated in the so-called Geneticist's Manifesto, which Muller largely wrote (Crew et al. 1939).

69. Brown 2005, 72–75.

70. Wells 1923 (2002), 110.

71. Wells 1923 (2002), 97, 82–83.

a good invention to say that man was created a gardener.' And now man was weeding and cultivating his own strain" (87). No doubt weeds also felt a "queer little twinge of fear" at the gardener's approach, but Barnstaple was persuaded that was an acceptable price for advancing the utopian gardener's goal of "cultivation," nurturing the best that grew.

Barnstaple's single, brief reference to 'weeding' is the only hint of negative eugenics in his book. The more familiar, darker side of eugenics—measures to prevent the less-fit from breeding—was almost entirely absent from all these texts. When eugenics was first mooted by Galton, he famously used the "convenient jingle of words"—*nature and nurture*—to summarize the influences that shape human character and ability.[72] (And, as Lorraine Daston notes, Galton's phrase epitomized the newfound opposition between nature and culture that characterized nineteenth-century modernity.)[73] A majority of eugenicists accepted his conclusion that nature (inheritance) dominated nurture (cultivation), hence the need for harsh negative eugenic measures to eliminate undesirable traits. However, Galton's argument was rooted in the nineteenth-century view of ancestral heredity. By contrast, Wells's Utopians were committed to twentieth-century speculative heredity, which led them to emphasize nurture. The same was true of the other biotopians: Haldane rejected negative eugenics; Bernal noted that eugenicists might improve people, but "they do not touch the alteration of the species" (32); and Muller based his future on giving women the power to choose the best genes for their children. The almost-complete absence of conventional, negative eugenics in these books tells us as much about the nature of these biological utopias as the presence of positive eugenics.

The way *eugenics* was used by the writers discussed in this chapter illustrated that, as with *evolution*, a contest was underway over exactly the what word meant, how it should be used, and who was entitled to use it.[74] Although "eugenics" was an aspect of each of the utopias considered in this chapter, there was no common concept behind the word because early twentieth-century biologists were questioning the malleability of "nature," which for Galton had been the epitome of stability. As old words acquired new meanings, new futures became possible.

---

72. See Kevles 1995 for a summary of Galton's views and the subsequent history of eugenics.

73. Daston 2014, 586.

74. Lene Koch (2006) has used Koselleck's work to trace similar shifts in the meaning of *eugenics* in twentieth-century Scandinavian encyclopedias. See also Esposito 2011.

Artificial Ethics

Unsurprisingly, some reviewers objected to the various sexual revolutions proposed by the biotopians. While many greeted *Men Like Gods* as something of a return to form, the London *Bookman*'s reviewer took exception to Utopian eugenics, commenting that—however great the supposed benefits—"I do not like the prospect of letting the doctor usurp the office of the irresponsible Cupid."[75] Another reviewer described *Daedalus* as an "unpleasant but provocative little book" and was particularly appalled by Haldane's vision of a "dehumanized humanity . . . wholly mechanical, with love and religion outgrown."[76] Haldane predicted that once reproduction was "completely separated from sexual love mankind will be free in an altogether new sense" (68), and the reviewer grasped—and opposed—his point that unlinking sex and reproduction would threaten the various systems of morals that regulated sexual behavior and much of the social structure. Haldane himself dismissed the very idea of permanent ethical values; not only would scientific inventions force people to abandon traditional ethical systems, but science itself would shape the future's morality. The claim that new, artificial plants were only the first step on the way to a new, artificial morality was among the most radical aspects of biotopianism.

Haldane accepted that separating sex and reproduction would transform family and personal relationships, but he showed little interest in revolutionizing supposedly natural gender roles (it would be left to a series of women writers to take up the radical possibilities he had raised.[77]) Instead, he argued that, among other things, science provided the tools for "man's gradual conquest, first of space and time, then of matter as such, then of his own body and those of other living beings, and finally the subjugation of the dark and evil elements in his own soul" (the passage that inspired Bernal's title, *The World, the Flesh and the Devil*).[78] Banishing demons such as guilt and superstition would enable people to reflect rationally on moral questions, but Haldane did not believe that our choice of ethical systems would therefore become arbitrary.

Haldane argued that scientific progress, in physical and biological sciences, would tend to "bring mankind more and more together, to render life more and more complex, artificial, and rich in possibilities—to

75. A.R. 1923, 32.

76. Ellis 1925.

77. E.g., Brittain 1929; Firestone 1970 (2015); Piercy 1976 (2000). See also Squier 1994.

78. Haldane 1924, 71–77, 82.

increase indefinitely man's powers for good and evil" (20). And although scientific progress offered no guarantee of moral progress, new scientific knowledge would impose new moral obligations. He averred that as human power over nature grew, it would "render actions bad which were formerly good," as improving medical knowledge demonstrated: in the past, when illness had been largely inexplicable, people had to stoically accept the suffering it caused (perhaps by interpreting it as divine punishment), but modern medicine "transformed resignation and inaction in the face of epidemic disease from a religious virtue to a justly punishable offence." It was, Haldane argued, morally indefensible to refuse to treat a communicable disease, so a scientific discovery *imposed* a new moral obligation. Haldane's view went beyond the claim that things change or that moral codes need to be updated; he argued that moral virtues were "essentially quantitative"—with increasing power over nature comes a growing obligation to use that power (89–90). He made the same point a couple of years later when he argued that prior to the advent of the telegraph and steam engine, news of a famine in China imposed no moral duty on Englishmen, since there was nothing they could do about it. But with great (electrical and steam) power came great responsibility, and in the twentieth century such news constituted an ethical problem.

Haldane also argued that when facing moral choices, scientific, statistical data were needed to decide both the best course of action *and* the degree of responsibility.[79] With the same logic, Haldane shocked many of his contemporaries (particularly on the political left) when he wrote *Callinicus: A Defence of Chemical Warfare* (1925).[80] He noted that poison gas was, in most people's opinion, "an innovation as cruel as it is unsoldierly," but dismissed such views as "sentimental." He compared war to epidemic disease, arguing that pious hopes could not abolish either, only scientific research. Many gases disabled rather than killed soldiers and most of those affected recovered completely (as Haldane had seen at first hand, having served in the trenches during World War I). By contrast, the wounds caused by conventional shells and bullets were much worse and more likely to permanently maim or disable survivors. Given the evidence, he argued, poison gas was morally preferable to conventional weapons.[81]

79. Haldane 1928, 12–13, 31–32, 39–40.

80. In a note, Haldane explained that "Callinicus" means "he who conquers in a noble or beautiful manner"; it was the name of the Syrian general credited with inventing Greek fire in the eighth century CE, whom Haldane credits with saving much of "Christendom" from domination by Islam (1925, viii, 6).

81. Haldane 1925, 2–3, 19–22.

Haldane went on to note explicitly that what was true of chemical warfare was true of all science: a combination of respect for tradition and fear of the unknown tended to make us fearful of novelties, but in most cases once "we have got over our first not very rational objection to them, [they] turn but to be, on the whole, good." And Haldane argued that embracing such novelties would become much easier if people would accept that the world itself is "neutral and indifferent to human ideals"; we should abandon both naturalistic ethics and conventional religion and try instead to tackle moral questions using scientific evidence.[82]

In *Daedalus*, Haldane had argued that because medicine was already prolonging life and creating new moral obligations, it had already changed Western Europe as much as the industrial revolution had done (and even more profound changes were to come).[83] As the link to the industrial revolution illustrated, Haldane was already beginning to draw on Marxism, which he (like many of his contemporaries) interpreted as a technologically determinist philosophy of history; new modes of production (the economic base) created new social, political, and cultural superstructures.[84] Haldane acknowledged that the relevant sciences (including psychology) were as yet too underdeveloped to predict the future in any detail, but was confident in predicting "that no beliefs, no values, no institutions are safe."[85] Haldane's argument implied that morals were (at least to some extent) *determined* by science: new sciences produced new technologies, which created new modes of production, which produced new cultural, political, religious, and ethical forms.[86] Marx famously argued that "the hand-mill gives you society with the feudal lord; the steam-mill society with the industrial capitalist"; to which, Haldane might have added, the ectogenetic "people mill" would give us another social form, whose ethical values would be as different to today's as nineteenth-century

82. Haldane 1925, 81–82.

83. Haldane 1924, 89, 53–54.

84. This interpretation is arguably a caricature of Marxism that some of his contemporaries would have rejected; however, it was a dominant view in parties affiliated to the Stalinist Third International at this time (McLellan 1998, 148–51; Sheehan 1985 [1993], 316–21).

85. Haldane 1924, 89, 86–87. Haldane reiterated the point about the relatively underdeveloped state of the sciences in several later essays, including Haldane 1928 and "Science and Politics," in Haldane 1927 (1930) (182–89).

86. Sheehan (1985 [1993], 323) notes that Haldane was a rather uncritical and enthusiastic Marxist whose thinking was often rather crude, e.g., "in assuming mind mirrors matter only in a straightforward and uncomplicated way, thus failing to take the possibility of distortion into account."

FIGURE 5.2. J. B. S. Haldane became increasingly well-known for his uncompromising politics during the 1930s, particularly after his well-publicized visit to Madrid in January 1937, when he advised the Republican government on air-raid precautions.

capitalism's were from the moral economy of the early-modern crowd.[87] He argued that "biological progress will prove to be as incompatible with certain of our social evils as industrial progress has proved to be with war or certain systems of private ownership" (figure 5.2). Confronted by breakthroughs such as ectogenesis, the family (like the state), would simply "wither away" (86–87).[88] As Diane Paul has commented, these writers shared a Marxist confidence that in "transforming nature, we also transform our capacities and sensibilities"; from artificial limbs and senses to artificial morals—everything was plastic.[89]

---

87. Karl Marx, *The Poverty of Philosophy* (1847), quoted in E. P. Thompson 1993. Some recent commentators on Haldane suggest that he argued that "ethical progress" was needed to keep pace with scientific progress (Dronamraju 2016, 68; also Dronamraju and Haldane 1995, 17; Dyson 1995, 56). However, this view overlooks the determinism implicit in Haldane's Marxism, which meant ethical systems were dependent upon more fundamental changes, and thus could not simply be "updated" as new inventions were made.

88. This idea was taken up by US feminist Shulamith Firestone (1970 [2015]).

89. Paul 2004, 134.

Given its fully developed biotopian perspective, it is not surprising that *Daedalus* demurred from Thomas Henry Huxley's claim that "traditional morals were sacrosanct and impregnable" to the challenge of science (90). Like his fellow biotopians, Haldane believed that the morals of the Minotaur-like, self-created humans of the future could be neither predicted nor determined by the traditions of the past. A similar point could be made about the consciously feminist agenda that Muller argued should guide the application of biological knowledge. However, as Carlson noted, Haldane assumed that scientific developments would direct future ethical systems, whereas Muller was concerned to place evolution under political control.[90] Nevertheless, both agreed that neither nature nor tradition could offer guidance—ethics were a purely human concern. T. H. Huxley's image of conscience as an "artificial personality" had become a key component of the biotopian ideal of the artificial person, whose morals would be as synthetic as the ectogenetic womb from which they had been decanted.

This Old Hag, Our Mother

When Wells's Utopians explained their society using analogies from gardening, some of the Earthlings objected to it as strongly as Amerton objected to Utopian morals. Among them Rupert Catskill, secretary of state for war (a barely fictionalized portrait of the young Winston Churchill), who ironically congratulated the Utopians for having "tamed the forces of nature and subjugated them altogether to one sole end, to the material comfort of the race."[91] However, he was convinced that the result was too safe, too easy, and simply too boring to be bearable. Without terror and danger, he asserted, the Utopians' joy must also be muted; life on Earth is "titanic" while Utopia "is merely tidy." The Utopian policy of exterminating all irritating insects was a small example of the wider loss of "the intensity of life." Catskill claims that "all the energy and beauty of life are begotten by struggle and competition and conflict" and had clearly absorbed the lessons of the degeneration theorists, since he argued that without struggle, Utopia must degenerate, and that it already possessed only "Autumnal glory! Sunset splendour!" (recalling the "sunset of mankind" in *The Time Machine*). He was also concerned about the fall of empires, and he argued that the experiment that had brought the Earthlings into this parallel universe would doubtless be duplicated elsewhere, so

90. Carlson 1995, 94.
91. For the identification of Catskill and Churchill, see Toye 2008.

that the peaceful Utopians would eventually find themselves invaded by more warlike, fitter worlds.[92] (Wells's readers would doubtless have been reminded of the marauding Martians from *The War of the Worlds* [1898].)

The Utopians dismiss Catskill's claims. Positive eugenics prevented degeneration ("the indolent and inferior do not procreate here") and competition survived since everyone worked to do their best for the world (like the rose growers). As a result, they were sure their science would always prove sufficient to close the inter-dimensional doors on barbaric worlds. Science has made the Utopians feel secure, confident of their ability to manage their planet according to their own standards. By contrast, Urthred complains, the Earthlings' minds "are full of fears and prohibitions" that hold them back from shaping their future. Humans half-realize that "they may possibly control their universe," but the prospect is too frightening "for them to face," and so they prefer to leave the universe's management in the hands of Amerton's traditional God, or of Catskill's updated one—competition. Upon hearing this, Barnstaple comments "Evolution was our blessed word," to which Urthred responds that "Evolution" is only another name for "a Power beyond your own which excuses you from your duty."[93] The Utopians instead urge humanity to put aside childish things and face full, adult responsibilities.

Perhaps the most interesting of the Earthlings' objections focused directly on the unnatural results of Utopian gardening. It was presented by Catskill's secretary, Freddy Mush (a "gentleman of serious aesthetic pretentions"), who objected to the fact that there were no swallows in Utopia, because the "Balance of Nature" had been disturbed.[94] The reader learns that swallows (and all other insectivorous birds) had become very rare because of "an enormous deliberate reduction of insect life," which had largely destroyed the birds' food supply. The Utopians explained how they had engaged in "a systematic extermination of tiresome and mischievous species . . . from disease-germ to rhinoceros and hyena." Some poisonous plants survived because they produced chemicals "too costly or tedious" to synthesize. Only a few "noxious insects" had been retained, such as

92. Wells 1923 (2002), 90–95; Cantor and Hufnagel 2006.

93. Wells 1923 (2002), 96. Strictly speaking, Barnstaple doesn't "speak," since the telepathic Urthred overhears his thought.

94. Wells based the character on Eddie Marsh, secretary to Winston Churchill (Toye 2008, 150). Marsh (1872–1953) was an art collector and patron of painters and poets (Hassall and Pottle 2004). Several reviewers took Wells to task for his portrayal of "Mush," with the *New Statesman* commenting that he was "inartistically, unfairly, and spitefully represented as a thousand times less intelligent than his prototype must certainly be" (Mortimer 1928, 695).

caterpillars whose butterflies were particularly beautiful. However, in every case, the implicit assumption was that a death sentence ("final and complete") was automatic unless the organism's existence could be justified by human values. Such culling was the "natural and necessary" work of weeding and cultivating nature, which created a world so reminiscent of the crepuscular Earth depicted in *The Time Machine*, free from gnats or illness—but also of swallows.

Despite Utopia's achievements, Mush still objected to human standards being allowed to disturb the balance of nature—"I hold by the swallows," he explained—their loss was too high a price to pay, even for a utopia. In response, Urthred told the Earthlings to take a long, hard look at "Mother Nature":

> She is purposeless and blind. She is not awful, she is horrible. She takes no heed to our standards, nor to any standards of excellence. She made us by accident; all her children are bastards—undesired; she will cherish or expose them, pet or starve or torment them without rhyme or reason. She does not heed, she does not care. She will lift us up to power and intelligence, or debase us to the mean feebleness of the rabbit or the slimy white filthiness of a thousand of her parasitic inventions. There must be good in her because she made all that is good in us but also there is endless evil. Do not you Earthlings see the dirt of her, the cruelty, the insane indignity of much of her work? (98)

This argument is a powerful one for rejecting Thomas Huxley's relentless cosmic force as a guide to human life (as Mush commented, after hearing it, "Phew! Worse than 'Nature red in tooth and claw,'" [99]). The Utopian responses to the Earthlings' various objections exemplify biotopia's most important claim: there are no sources of moral guidance outside humanity. Any self-conscious, intelligent species has to stop putting its trust in imaginary, external Powers (God, Competition, Evolution, or Nature) and face its full responsibility. Either humanity takes on the role of gardener or we let the cosmic process run unchecked.

Urthred explained the consequences of the choice that humanity faced. Before the Utopians "took this old Hag, our Mother, in hand," more than half the species on the planet were "ugly or obnoxious, insane, miserable, wretched, . . . helplessly ill-adjusted to Nature's continually fluctuating conditions"—as was still the case on Earth, where humans and other animals suffered because of misplaced faith in the goodness of nature. To eradicate these problems, the Utopians had been forced to suppress Mother Nature's "nastier fancies" to wash and comb her and teach her. As a result they are no longer "the beaten and starved children of Nature, but her free and adolescent sons. We have taken over the Old Lady's Estate. Every day we learn a little better how to master this planet" (99).[95]

As we have seen, the same arguments about Mother Nature neglecting her children were made in connection with Burbank (p. 110), and the language of male mastery over "Mother Nature," derived in part from Bacon, has, of course, been a long-standing feature of Western science. Science has often been described in terms of gendered violence—active and arrogant men interrogate or even torture a passive, feminized nature in order to extract knowledge from "her" (usurping God's power in the process).[96] Given the time when these texts were published, the presence of such gendered imagery is as unsurprising as the repeated use of colonialism as a metaphor for conquering nature (Barnstaple describes utopia as "intensely militant, conquering and to conquer" [113]). Nevertheless, these familiar images both naturalize gender relations and link texts from apparently different genres into a single way of describing the nature/culture relationship. For men like Wells, male mastery of nature—including the systematic destruction of numerous species—*was* utopia. Their schemes capture the Baconian paradox at the heart of the biotopian approach: nature is both exalted and manipulated. As Bacon put it, *natura parendo vincitur* (nature is mastered by obeying), an almost oxymoronic claim exemplified by Barnstaple's description of "editing" nature as "most natural." Instead of adapting themselves to nature's constraints and limited resources (as Bailey had argued a good gardener must), the inhabitants of these new utopias would force nature to adapt to them. The language of gardening, of "cultivating" a plot of waste ground to make it beautiful and fertile, collapses the distinction between nature and culture, and the loss of that distinction explains a key difference between Huxley's view and fully biotopian ones.[97] For Huxley, resistance to the cosmic process produced an inverted naturalistic fallacy (nature was still a source of moral guidance, but only a negative one); for his biotopian successors, nature

95. The phrase "free and adolescent sons" was undoubtedly a tribute to Wells's friend Lankester, whose 1905 Romanes lecture had described humanity as "Nature's insurgent son" (Lankester 1907, 7, 26–27, 31–32).

96. For a fuller discussion, see Merchant 1982; Harding 1993; Wertheim 1997; Pesic 2008.

97. My thinking on this point is indebted to Philip J. Pauly's (2000, 8–9; 2007) discussion of "culture," which—he notes—meant "tillage" before it meant art and literature. Both meanings retain a sense of improvement through the activity of cultivating; plants and people are similarly improved by careful cultivation. See also Ortner 1972; Haraway 1991.

could no more be a negative norm than a positive one (as Urthred said, Mother Nature made "all that is good in us" but also "endless evil"). In the hands and minds of the experimental evolutionists, "nature" and its meanings became increasingly fluid and flexible, raw material for humans to sculpt, neither a set of ineluctable forces nor one of inherited burdens.

Daedalus can be considered the first fully developed biotopia partly because Haldane not only predicted that biology would finally erase the supposedly inviolable barrier between nature and culture but did so in a book whose literary style and commercial success challenged the boundary between elite and accessible science. Haldane's efforts to write for laypeople (albeit, highly educated ones) implicitly invited them to take part in the work of imagining the futures that biology offered. His book also contributed to the creation of a strange, hybrid genre that would soon become known as "science fiction," which borrowed equally from both the "scientific romances" of writers such as H. G. Wells and from the language of dry, scientific papers to establish its plausibility.[98] In addition to bringing the richly impure language of biotopia to fresh audiences, *Daedalus* utilized a familiar narrative device (often used in both utopian and science fiction) and presented its predictions in the form of an undergraduate essay written "150 years hence" that described a much perfected future world.[99] The format imagined the future as already familiar—compressing time and space in ways that made its predictions all too plausible in the eyes of some reviewers. And, as noted, Haldane added to its plausibility by constructing a history that linked present-day scientists to their imaginary successors (and by making this section of his book as brief, underreferenced, and laden with generalizations as a real undergraduate essay).

Biotopianism emerged from the participatory culture that grew up around new theories such as mutation, which, according to some of its fans, might provide the tools needed to eradicate original sin. However, biotopianism's wider impact and optimism arose from its proponents' confident rejection of both traditional religious and naturalistic codes of ethics (including so-called evolutionary ethics). In their most radical (and, arguably, autocratic) moments, the biotopians also rejected ethical relativism, offering instead the prospect of a synthetic, human-made code of ethics based on scientific principles and data. Haldane's key argument—which

98. See "The Scientific Romance and the Evolutionary Paradigm" (Luckhurst 2005, 30–49); see also Parrinder 2011, 246–49; Saunders 2019, 9.

99. The same device was used by Jack London in *The Iron Heel* (1908). For utopian narrative strategies, see Suvin 1979 (2016); Moylan 1986; Kumar 1987, 230–32; Ruppert 1986; Ferns 1999.

was endorsed to varying degrees by Wells, Bernal, and Muller—was that both human biology and ethics would prove as malleable to the scientist as the color of the oceans. "We can already alter animal species to an enormous extent," Haldane noted, and "it seems only a question of time before we shall be able to apply the same principles to our own" (69). As Mark Adams has argued, the themes of *Daedalus* ran though all Haldane's work; he consistently imagined and predicted a future in which humanity had gained complete control of its own biology and would use science to modify *Homo sapiens* into numerous, distinctive new species.[100] Haldane developed this idea further in *The Last Judgment* (1927), which depicted several artificial subspecies of humans, each reengineered to enable them to colonize a different planet (an idea that inspired a number of science fiction writers, particularly Olaf Stapledon).[101] Humans would evolve themselves until they became as unnatural as minotaurs, and this vision appalled many, particularly the popular Christian writer C. S. Lewis, who attacked Haldane's vision in his science fiction trilogy (see p. 344).[102]

The optimism that marked the dawning of the twentieth century was particularly strong in the USA, one factor that would eventually lead many to identify the twentieth as the *American* century. Yet none of the biotopian authors discussed here were Americans (the one apparent exception, Herman Muller, was always conscious of his immigrant background, and he felt so out of step with the USA that he was in self-imposed exile in the USSR at the time *Out of the Night* appeared). The books considered in this chapter were philosophical and rather abstract. They might be considered (and were sometimes reviewed) as offering kinky fun for jaded European intellectuals, which could be safely enjoyed because of their ambivalence—perhaps the writers were not really serious. By contrast, homegrown American biotopianism had a severely practical aspect; you could buy yourself a utopian garden from a mail order catalogue, published at the experimental gardens in Santa Rosa California, of Luther Burbank. And while writers like Haldane and Bernal were offering their philosophical musings in Britain, the biotopian mode was demonstrating its genre-bending qualities by reverberating through the USA not in futuristic utopias but in school and college textbooks.

100. Adams 2000, 460–69.
101. This essay was included in Haldane 1940.
102. Haldane 1946; Adams 2000, 483; Parrinder 2011, 255.

6

Textbook Futures

In the sweltering summer of 1925, hundreds of US newspapers and magazines covered the forthcoming trial of Tennessee schoolteacher John T. Scopes, who was to be prosecuted for violating the state's law against teaching evolution. As Adam Shapiro has shown, what became known as the Scopes "Monkey Trial" concerned such issues as America's continuing arguments about science, faith, and expert authority, but it also illuminates another way in which experimental evolution achieved its lasting influence.[1] When the *New York Times* previewed the trial, it forecast that "Erudite headlines will confront the nation," as "Creationism" and the book of Genesis were to be pitted against evolutionary champions—with de Vries being listed alongside Darwin and Weismann. The evidence for evolution would include both celebrated protohuman fossils ("Pithecanthropus, Australopithecus, Heidelberg, Piltdown, Cro-Magnon"), but also "the mutations of the evening primrose."[2]

A quarter-century after its first publication, the mutation theory and its founder were still considered famous enough that a leading US newspaper could assume they needed no introduction. By contrast, any geneticists who read the *Times* might have been surprised to see de Vries and his theory mentioned (especially as Mendel's name was conspicuously absent). By this period, geneticists had mostly abandoned de Vries's original theory because it had become apparent that *Oenothera* (which had provided its most compelling evidence) possessed a number of unusual genetic features that were not shared by most plants.[3]

Given the scientific community's loss of interest, the abiding fame of de Vries and his theory seems perplexing, but the Scopes Trial reveals the

1. Shapiro 2008; 2013.
2. "Natural Law in the Dock" 1925.
3. Endersby 2013. See also Endersby 2007, 128–69.

importance of school and college textbooks in ensuring that mutation retained its place in the public's understanding of evolution. The trial was intended to test the legality of Tennessee's recently passed Butler Act, which prohibited the teaching of evolution in schools. As Scopes himself said:

> I don't see how any man can teach biology without bringing in the theory of evolution. . . . For instance, there is the book that I teach here in the Rhea County High School—"Civic Biology," by George W. Hunter. I don't see how I can teach it without bringing in a lot of evolutionary talk.[4]

Civic Biology, from which 90 percent of Tennessee's students learned their biology, did not merely include a lot of "evolutionary talk," it was one of many textbooks that continued to describe the mutation theory as current science.[5] Hunter's text claimed that "*recently* a new method of variation" had been announced by Hugo de Vries, who discovered that new species "arise suddenly by 'mutations.'" These forms were called "mutants" and they "at once bred true," making them of "immense value to breeders."[6] When the book had first been published in 1914, the mutation theory might just have been considered "recent," but not by 1925 when the trial occurred.[7] (And even when it was revised as *New Civic Biology* in 1926, the mutation theory still featured prominently.)[8] This chapter will demonstrate that Hunter's textbook was not unusual; mutation theory was still being featured in textbooks and taught as up-to-date biology in schools and colleges in the mid-1930s, two decades after most expert geneticists had abandoned it. Historians of science have not always given textbooks the attention they warrant.[9] In this case, they not only reveal why de Vries's ideas retained their public authority long after scientists had moved on, but also illustrate another way in which the promises of experimental evolution reached audiences beyond the classroom. In theory, textbooks gave readers a more permanent and authoritative version of the claims made by periodicals, but in practice they created imaginative links

4. Milton 1925, 659.

5. Shapiro 2013, 66.

6. Hunter 1914, 253, emphasis added.

7. Shapiro 2008, 412–14.

8. Hunter 1926, 385; Shapiro 2013, 1–20.

9. Some exceptions include Rudolph 2002; Shapiro 2012; Kohlstedt 2010. Ron Ladouceur's blog, *Textbook History* (https://textbookhistory.com/), made me realize how valuable textbooks could be for researching the long-term impact of de Vries. See his post "Where'd Hugo Go?" (December 14, 2013, https://textbookhistory.com/whered-hugo-go/). Also Ladouceur 2008a; 2008b.

FIGURE 6.1. W. J. V. Osterhout (in straw hat), Jacques Loeb (on de Vries's right), Hugo de Vries (facing camera), and others inspect *Oenothera* plants growing at the University of California, 1905.

between the supposedly sober and objective world of laboratory science and the more speculative worlds of political utopianism and the fledgling genre of science fiction.

Textbooks are usually assumed to be the exact opposite of idiosyncratic essays and opinion pieces; they are expected to exemplify a particularly strong form of objectivity.[10] They may not always live up to that ideal, but are nevertheless regarded as particularly trustworthy; it was precisely because students were assumed to believe what they were taught that opponents wanted evolution banned in schools. As William Jennings Bryan (leading light of the Scopes prosecution team) argued: "The objection to Darwinism is that it is harmful, as well as groundless. It entirely changes one's view of life and undermines faith in the Bible."[11] He spoke for many who feared that young people would believe their textbooks instead of the Bible and forsake their parents' religion.

10. Lorraine Daston and Peter Galison, *Objectivity* (New York: Zone, 2007): 17. Quoted in Shapiro 2013 (42).

11. Bryan 1922, 1.

Textbooks' supposed trustworthiness made them a crucial site where elite science was interpreted for the benefit of inexperienced readers. The rapid expansion of compulsory schooling in the twentieth century's early decades led to intense competition between textbook publishers, particularly in the large biology market. Biology teaching was more widespread in American schools and colleges than in Britain, and the larger US market increased each publisher's need to differentiate their books from those of their competitors. One result was a diversity that makes it difficult to define *textbook*, but they all had to be accessible—in price and content. As a result, they were often where the expert's world intersected with those of the enthusiast or the autodidact. Given the diversity of those potential readerships, and the fact that this was a relatively new market, the textbook's form was still fluid, and many different titles could be considered as textbooks.[12]

This chapter analyzes over 150 texts (both American and British, and many in more than one edition; see the appendix): the majority were explicitly addressed to students (from high/secondary school to advanced university level), but many were also explicitly addressed to "beginners," the "educated layperson," or the "general reader."[13] Some titles described themselves as being intended for teaching (even including sample lesson plans and classroom experiments), while others were aimed at a broader audience (usually written in a more informal style, they avoided technical terms, included numerous illustrations, and omitted any obvious pedagogical apparatus). All kinds of readers who wanted to know more about recent biology turned to textbooks; Edith Wharton read such titles as Vernon Kellogg's *Darwinism To-Day* and Robert Heath Lock's *Variation, Heredity and Evolution* (see p. 20), and other autodidacts doubtless did the same. In order to approximate the range of sources that a nonexpert reader might have turned to if they wanted to understand the promise—or threat—of the new biology, this chapter considers some books that were not formally identified as textbooks but nevertheless shared the goal of taking accredited scientific expertise and interpreting it in an accessible but authoritative way.

Textbooks became another arena in which to debate the question of who was entitled to interpret biology's implications. The nineteenth-century

12. Shapiro 2013, 20.
13. A full analysis of US biology textbooks would require a book in its own right (probably more than one), but I hope this preliminary analysis of their role in promoting biotopianism will encourage others to look at them more closely, particularly now that digitization has made them so much easier to find and use.

conflicts over Darwinism can be understood as a contest over cultural authority, over whether society's values were to be shaped by traditional (often religious) authorities or by the newer (often secular) expertise embodied in the man of science, the engineer, and the public-health official.[14] The Scopes trial offers a microcosm of the way similar debates were replayed for twentieth-century America. Textbooks allowed proponents of new biologies to reopen issues of cultural authority, particularly because the public sphere had been dramatically enlarged—not least by the growth in public education. Many of the younger men (and they were mostly men) who wrote the twentieth century's first textbooks were more than happy to tackle the "big questions" of order, progress, meaning, and whether intelligence and free will played a role in life's drama.[15]

In the USA, large-scale high school biology teaching began in 1902 in New York but spread rapidly. In 1910, only 1.1 percent of pupils in public secondary day schools took biology, but that figure had risen to 14.6 percent by 1934, by which time it was more popular than any other science (indeed, most of the other sciences declined over the same period). And the total school population expanded dramatically in that period; in 1900 the USA had just over half a million secondary school pupils—by 1934 there were 5,669,000 (so biology textbooks had about eight hundred thousand readers, compared with around five thousand in 1902).[16] By the mid-1920s, what was being called "civic biology" was being taught all over the country. As Philip Pauly argued, it was "both ostentatiously objective and intensely value laden" and encapsulated key goals for many progressive US educators, which was precisely what prompted the Scopes Trial. Civic biology's key goal was to teach the children of immigrants to be clean, healthy, productive Americans (a project that parallels Burbank's campaign to discipline unproductive plants). The teachers' approach was rooted in their understanding of biological evolution; experiment and adaptation were the tools that would produce rational progress toward a better future. Lamarck and Spencer were at least as important as Darwin for the high school biology movement, which hoped to replace old-world diets and dogmas (often associated with the religious and cultural traditions of immigrant communities) with a modern, secular humanism based in objective, scientific evidence (figure 6.2). Hence the term *civic* biology, which combined health and hygiene lessons (including human physiology and reproduction) with recent advances in zoology and botany. A

14. Heyck 1982; Turner 1993; Collini, Whatmore, and Young 2000; Hale 2014.
15. Clark 2001; Moore 2001; Pauly 2000, 195.
16. Barabba 1975, vol. 1, 377, 368.

Compare the unfavorable artificial environment of a crowded city with the more favorable environment of the country.

FIGURE 6.2. George Hunter's *Civic Biology* (1914), the textbook at the heart of the Scopes trial, invoked a long-standing contrast when it invited its readers to compare the "unfavorable artificial environment of a crowded city" with the supposedly more natural one embodied in the classic American farmstead.

progressive curriculum tailored to urban students led textbook authors to replace traditional nature study with a focus on biological processes, including evolution, invariably imagined as progressive, dynamic change.[17]

Like other civic biology texts, Hunter's reflected both the ambition and optimism of experimental evolution. It reflected the broader biotopian hope of reshaping nature (including human nature) by utilizing tools that ranged from plant-breeding to improved sex education and eugenics.[18] Alongside the discussion of de Vries and mutation, mentioned above, *Civic Biology* used Burbank's work as an example of what could be done, and it also emphasized Mendel's discoveries as being "of the greatest importance in plant and animal breeding" (253–54, 259). Hunter claimed it had been "necessary for men like Charles Darwin or Gregor Mendel to prove their theories before men like Luther Burbank . . . could benefit mankind by growing new varieties of plants" (261–65, 398). These sections were followed by one titled "Improvement of Man," which stressed personal hygiene (take exercise and avoid alcohol), and then sections on eugenics and "euthenics," or "the betterment of the environment" (a nurture-based variation on classical eugenics that illustrates the continuing debates over the meaning of eugenics; see pp. 27–28).

Just as Darwinism had done in nineteenth-century Britain, twentieth-century biological ideas touched on matters of personal morality, thus challenging established religious faiths. However, there was more to these debates than is captured by the simplistic contrast between science and religion. Just like their nineteenth-century precursors, the early twentieth-century biologists had been emboldened by recent scientific advances and were claiming the right to remake the culture and morality of their fellow citizens. However, most Victorian evolutionists had been concerned with separating church and science (Huxley's *Evolution and Ethics* typified the strategy of rendering unto the church those questions which were moral, while rendering unto science those things which concerned facts). As we saw in the last chapter, this compromise was based on the widely held conviction that human nature could not be changed; whether one blamed our apish ancestry or original sin, humans were too flawed to take responsibility for their own morality. By contrast, some twentieth-century biologists increasingly believed they had the power—and therefore the obligation—to reshape society and its moral codes.[19]

17. Pauly 2000, 171; Shapiro 2008.
18. Shapiro 2008, 67–74, 409–10.
19. Pauly 2000, 172.

An often-utopian faith in biology's power took many forms. Civic biology's proponents were just one example; Henry Fairfield Osborn, a traditional museum-based paleontologist and taxonomist, had ideas about biology that were quite distinct from those of either the experimental evolutionists or the promoters of civic biology.[20] Nevertheless, when Osborn was asked for his response to the Scopes Trial, he told the newspapers that America "should quit haggling" over "established evolutionary laws." He insisted that it was time for people to study and master "the evolutionary processes of today and tomorrow" and use them "to guide the human race to a new Utopia." Achieving this goal would require a new "liberal and enlightened religion" combined with a "true understanding of the forces of evolution, which are ennobling when rightly understood."[21] Despite his differences with them, Osborn shared the optimism of many of the laboratory biologists, emboldened by their understanding of speculative heredity and the better future it predicted.

Among their defining features, textbooks were expected to be up to date, which encouraged authors to discuss the latest ideas. However, as Hunter's *Civic Biology* demonstrated, it was often easier to get new theories into a textbook than it was to remove them. The economic realities of the textbook market meant that its products were often a force for conservatism, repeating stale ideas long after active researchers had abandoned them. As Stephen Jay Gould argued, textbook authors are particularly prone to copying one another's work. One symptom was arcane analogies being retained in later texts; linguistic remnants that functioned like the vestigial organs of living species and allow the lineage's history to be traced.[22] Such copying also created the textbook genre, which—like every genre—both shaped and was shaped by readers' expectations; early twentieth-century readers (whether teachers, students, or state officials in charge of textbook purchasing) clearly expected their textbooks to include forecasts of better living through biology. Those who produced textbooks were sensitive to readers' expectations, and one result was that Burbank was often used to illustrate the exciting practical implications of up-to-date science. He could also provide a familiar face for unfamiliar science, add scientific credibility to a general-interest book, or lend a bit of popular

20. For more on Osborn and his role in the evolution debates, see Clark 2008, 16–40.

21. Reported under the headline, "Battlecry of Peace Times Should Be 'Toward the New Utopia,' Says Scientist" (Cohn 1925). Constance Clark (personal communication, via email, July 10, 2020) found the same article in the *Poughkeepsie Enterprise*, so it was clearly syndicated.

22. Gould 1991, 155–67.

appeal to a more technical one. However, references to Burbank often reflected the way textbook writers drew on items from periodicals and hence gave permanent form to what might otherwise have been forgotten. Many of the examples of Burbank coverage in earlier chapters appeared in 1905, when the Carnegie Institution's grant brought intense media attention to Burbank and his work. Harwood's *New Creations* began as little more than an attempt to cash in on the media frenzy. However, a decade later Hunter included Harwood's book in the bibliography of *Civic Biology*, extending its shelf life for many more years while adding substantially to its authority. And *New Creations* was cited in at least a dozen of the books discussed in this chapter; as late as 1923, *Biology of Home and Community* was still recommending Harwood's title as one of its "reference books for the high school library."[23]

Being less ephemeral than periodicals, textbooks kept ideas in circulation for longer, particularly because competition ensured that new and revised texts were often closely modeled on their successful precursors. At the same time, new titles aimed to improve upon their predecessors by being more up to date, which could lead to comparatively short-lived scientific fads becoming fossilized, permanently embedded in the textual bedrock. New books could also aim to be more comprehensive than their rivals, by omitting nothing while adding new material. As a result, even potentially outdated ideas such as mutation theory might be retained to ensure that the textbook had not omitted anything its competitors had included. The rather contradictory goals of being both up to date and inclusive ensured the mutation theory survived much longer in textbooks than in the faster-moving world of scientific journals and conference papers. (The tendency of historians of biology to rely on sources such as specialist scientific journals and conference proceedings has probably contributed to the misleading claim that interest in the mutation theory had ended by 1915.)[24]

Readers might reasonably have expected scientific textbooks to offer a considered judgment on issues that periodicals had long since forgotten about, but that was seldom the case (in reality, they were just as likely to ensure that what had begun as filler in an ephemeral 10-cent magazine acquired lasting historical impact). Behind the textbook's impersonal covers, debates over rival theories and their adherents continued, sometimes with

23. Trafton 1923, 607.
24. I.e., when the Columbia-based *Drosophila* researchers demonstrated that Mendelian factors were in fact physical entities located on an organism's chromosomes. See T. H. Morgan, Sturtevant, Muller, and Bridges 1915.

undiminished vigor (and theories that were no longer even mentioned in current scientific journals continued to be presented and discussed in textbooks). The patterns of copying, cross-citation, and influence between different textbooks provide another example of the dialogue within which the claims of the new biology were made and assessed; the plausibility and authority of particular biotopian claims, for example, would have been affected by the frequency with which they appeared in textbooks. And, of course, almost all textbooks contained bibliographies and suggested further reading; the way in which authors cited and recommended each other's work further expanded the readerships for these books, as did reviews in periodicals. While sales figures are all-but-impossible to find for these books, the number of reviews and the variety of periodicals in which they appeared provide a general sense of a book's impact. In addition, the number of writers and publishers who produced such books suggests a buoyant market (as does the fact that many of the titles considered here went through numerous editions). And while nobody would have followed up every citation or recommended reading, such references would have shaped their sense of which ideas were considered current and who was to be regarded as authoritative.

Recent Opinions

The ways a textbook could shape the interpretation of biology become clearer if we analyze Liberty Hyde Bailey's *Plant-Breeding*, a typical and successful example. As we have seen, Bailey was a proponent of both traditional and progressive values who hoped to see agriculture scientifically modernized. His "biocentric" philosophy (that nonhuman nature can offer moral guidance), was the antithesis of some of the biotopians' key claims. Nevertheless, his textbook promoted the claims of experimental evolutionists, so it is likely that many of his readers have struggled to distinguish his version of biological optimism from theirs.

Plant-Breeding first appeared in 1895 and rapidly became a classic, running through numerous editions over many decades (it was in print until at least 1936). Its success was partly due to the fact that Bailey was so well-known, being both a Cornell professor and a popular writer of everything from poetry and philosophy to pioneering works on environmentalism. He also addressed many different audiences, including schoolchildren (he had helped found the nature study movement with *The Nature-Study Idea* [1903]; see p. 75).[25] All his works sold well; some went through twenty

25. Kohlstedt 2010, 77–110.

editions and were still selling decades after their initial appearance.[26] *Plant-Breeding* was based on Bailey's ever-changing college-level lectures, and successive editions provide a snapshot of changing ideas about the subject's possibilities. The preface to the first edition observed that when gardeners noticed an apparently new type of plant, they usually dismissed it as an accident, but he urged them instead to recognize that the new plant might enable them to contribute to "the philosophy of organic evolution."[27] Plant-breeding was far more significant than most people realized.

By the time *Plant-Breeding* reached its third edition (1903), Bailey noted that there had been "great changes in our attitude toward most of the fundamental questions" since he had first published. In less than a decade, the view "held by Darwin and the older writers," based on time-honored natural-history practices of observation and collecting, had been eclipsed by a new one "arising from definite experimental studies." Work was still underway, but there was increasing confidence that the art of breeding followed "definite laws." So, Bailey added a chapter called "Recent Opinions" that summarized "the investigations of de Vries, Mendel and others."[28] His chapter had originally been a lecture, but by the time it appeared in *Plant-Breeding* it had been revised and corrected with the help of de Vries himself, who also contributed a short section "On Hybridization." Bailey's text thus became one of the first authoritative accounts of de Vries's views to appear in book form, and was read by successive generations of American students.

Bailey's scientific credentials ensured that each new edition of his book was reviewed by specialist publications, including gardening and botanical ones (for example, the Ecological Society's journal, *Plant World*, recommended it as one that "must be in the hands of every intelligent devotee of this fascinating subject").[29] His lucid style also made the book suitable for a wider range of readers, as reviewers repeatedly stressed. The *Botanical Gazette* called it a "compact simple statement" of recent experimental biology, thus "a boon to the general reader," while the widely read *American Naturalist* agreed that Bailey's book met "the needs of a wide circle of readers, included the general public."[30] Thanks to Bailey's comparative fame, his book was also reviewed in the *Congregationalist and Christian World*, while the *Journal of Education* recommended it as a "fascinating"

26. Banks 1994.
27. Bailey 1902, v–vi.
28. Bailey 1906, viii–x.
29. F.H.K. 1904.
30. [Cattell] 1904; W.E.C. 1905.

introduction to plant-breeding (the most tantalizing of "all the mysteries in this mysterious world").[31]

Reviews helped ensure that *Plant-Breeding* required many editions and frequent reprints (for example, the fourth edition was reprinted five times before the fifth appeared). As result, successive generations of readers over many years would have read Bailey's claim that—thanks to de Vries—the "experimental method has finally been completely launched and set under way." A photo of de Vries's experimental garden was used to emphasize the point (welcomed by one reviewer for bringing "the reader into a more living touch with this renowned investigator").[32] In Bailey's view, earlier approaches—including the observational one used by Darwin—were being replaced "by experiments . . . under conditions of control" (179). And, of course, the ongoing debates over Darwinism's future shaped the book's reception (one reviewer praised Bailey's "sincerity and honesty" in admitting that the "views of Darwin, are passing away").[33]

Bailey presented de Vries's views in a more permanent form and to a much wider audience, yet he doubted that the mutation theory was quite as important as its originator believed. It would "have a profound and abiding influence on our evolution philosophies," he allowed, but he believed that even if some species were produced by mutation, most "originate by means of natural selection" (155). Yet despite such doubts, Bailey not only retained the discussion of mutation in successive editions, he almost doubled its size: the 1906 (4th) edition contained ten pages on de Vries, with another dozen written by de Vries himself; the 1915 (5th) edition contained an entire forty-page chapter titled "Mutations," which added a portrait of de Vries. And the whole chapter appeared unchanged in the 1936 edition—and still included no mention of the various criticisms of mutation theory that had appeared in the intervening decades. For more than three decades, one of America's most popular textbooks not only stressed the importance of de Vries's work but continued to describe it as the latest science. Bailey's was typical of numerous textbooks that ensured that the mutation theory continued to be taught as an important aspect of supposedly up-to-date biology.

Given that Bailey's textbook also brought Mendel's work to a wider audience, one might assume that it would have played a similar role in communicating genetics to students. However, Bailey presented de Vries

31. *Journal of Education* 63, no. 21 (May 24, 1906): 581; *Congregationalist and Christian World*, July 14, 1906, 54.

32. Lloyd 1904, 110.

33. *Journal of Education* 63, no. 21 (May 24, 1906): 581.

and Mendel in rather different ways. Both were described as experimental evolution's founding fathers, but de Vries was credited with banishing "gradualism and slowness" from evolution, and as having produced the "most pronounced counter-hypothesis" to Darwinian orthodoxy (145). In addition, de Vries was not only still alive (and regularly covered in the US press), but had contributed a section to Bailey's text, so he and his work dominated Bailey's chapter. By contrast, Mendel had merely "found uniformity and constancy of action in hybridization" and proposed a theory to explain it (157). Many textbooks paired de Vries and Mendel, and de Vries was often the more prominent.

By 1915 (largely thanks to the *Drosophila* researchers), Mendel was rapidly eclipsing de Vries among elite geneticists, but outside the fly rooms and academic journals de Vries often retained the limelight—partly because of his continued prominence in textbooks. Bailey's claim that de Vries's "counter-hypothesis" was "also the newest" appeared unchanged as late as the 1936 edition.[34] Meanwhile, Mendel's work was consistently described as secondary because Bailey (like many others at the time) saw Mendelism as dealing only with the hybridization of existing characters. While it would help breeders and biologists understand precisely how traits were passed on, Mendelism—like natural selection—seemed unable to explain the appearance of entirely new traits or species. Natural selection only destroyed failures (it was "not a constructive or augmentative agency. It merely weeds out" [145–46]). And as we have seen, it was precisely this problem—explaining the arrival of the fittest—that de Vries was widely believed to have solved.

Ever the traditionalist, Bailey was somewhat skeptical about mutation, but he had even stronger reservations about Mendelism. He could not see how Mendel's work would "greatly modify our plant-breeding practice" (169–70) and decried what he called the wild "prophecies" that it had prompted, such as growers eventually being able to synthesize new organisms in the way that chemists synthesize new compounds (Bailey demurred: the comparison was "fallacious and the conclusion unsound" [174–75]). Several reviewers of *Plant-Breeding* shared Bailey's assessment of Mendelism. The *Botanical Gazette* commented that "it is a satisfaction to see that Professor Bailey has not been swept off his feet by the swelling tide of Mendelism. The wild prophecies that the application of Mendel's law will reduce plant breeding to a science of mathematical precision find him waiting for proof."[35] By contrast, Bailey explained that de Vries's work

34. Bailey and Gilbert 1915, 52–53.
35. [Cattell] 1904, 471.

suggested the existence of "true progressive mutations . . . upon which the progress of the plant race depends," and that in such cases, "there can be no Mendelizing," since the new mutations bred true from their first appearance (177–78). Mutation could account for new forms, but Mendelism only explained how they were inherited. His opinion of the respective importance of mutation and Mendelism was similar to de Vries's own views, unsurprisingly, given de Vries's active involvement in the book (his contribution restated Mendel's laws in a revised form "proper to the mutation theory").[36]

When it came to illustrating biology's practical implications, Bailey naturally turned to Burbank; the Californian's achievements were described in fairly measured tones; nevertheless, Burbank's inclusion added to his authority. In the book's fourth edition (1906), Bailey wryly observed that the "editor of one of the great magazines" had recently asked him whether Luther Burbank was "the only plant-breeder in the country"— and acknowledged that anyone "who has read the current Burbankiana can well understand why the question was asked." Bailey tried to offer a more balanced view, patiently explaining that Burbank was "a plain, modest, sympathetic, single-minded man." And he attempted to curb the wilder enthusiasms of the "current Burbankiana" by describing how Burbank pursued traditional methods on an unprecedented scale but cared "little for the scientific method"; he was thus "a plant-lover rather than plant-breeder" and definitely "not a wizard" (238–46).

Despite Bailey's attempts, the Burbankian hyperbole persisted, which may explain why Bailey gave less space to Burbank in later editions. And he noted that while some of Burbank's productions, "may be the starting-points of strong and noble lines of evolution," he felt that some "that have been much heralded are of doubtful economic value."[37] Nevertheless, Bailey concluded with the statement that there "is a new kind of pleasure to be got from gardening, a new and captivating purpose in plant growing. It is a new reason for associating with plants" (323). These were final words of the book's main text, so (despite the brevity of Bailey's treatment), readers might well have concluded that Burbank's ability to remake plants for pleasure was the ultimate goal of plant-breeding.

Even though Bailey expressed skepticism about some of experimental evolution's claims, his book shared its optimistic, utopian mood. *Plant-Breeding* was written to satisfy the demands of the textbook market, so it provided an accessible, topical, comprehensive summary of an ongoing

36. De Vries, in Bailey 1906 (197).
37. Bailey and Gilbert 1915, 323.

scientific revolution that promised to have significant practical implications. Humanity's future associations with plants would be characterized by rapidity and precision—de Vries's and Mendel's complementary gifts to the plant breeder. These were the defining characteristics of experimental evolution, a tool that would transform the slow, laborious processes of plant-breeding into a kind of precision engineering, a distinctively American kind of know-how for the dawning twentieth century. And plant-breeding was, for Bailey and for many others, a microcosm of evolution; it too was to be transformed from a slow nineteenth-century waltz to a lively twentieth-century Charleston, a rapid process that promised a brighter future for a century in a hurry.

Mutation's Long Afterlife

The way Bailey's *Plant-Breeding* reported the mutation theory was typical of textbooks in this period, not least in that (like Hunter's *Civic Biology*) it was still describing mutation theory as an exciting new science long after geneticists had moved on. As late as the 1930s, some books were still explaining that "mutations" showed how "new species may in some cases have arisen at a bound instead of by gradual steps."[38] Despite their ostensible goal of educating the public, textbooks often fueled an increasing divergence between expert and nonspecialist understandings. From its initial home in popular magazines, the mutation theory soon migrated to books. Some were technical monographs, while others grew out of accessible lectures and were commonly addressed to "those who would like a brief introductory outline of this important phase of biological theory."[39] Since de Vries's experiments had all been conducted with plants, his claims were particularly prominent in botanical textbooks; Bailey's *Plant-Breeding* was one of the first to address mutation theory and Osterhout's *Experiments with Plants* (see p. 46) was another, which complemented Bailey's by offering simple, practical experiments in a form readers could emulate, inviting both students and laypeople into the experimental garden.[40] Reviewers singled out the practical qualities of Osterhout's book,

38. Latter 1930, 130–31. Other examples include Thomson and Geddes, who explained in 1931 *Oenothera*'s "mood is all for mutation; it is always producing something new" (Thomson and Geddes 1931, vol. 2, 961). And Alfred Kinsey (1938, 414) was still using a picture of de Vries to illustrate the concept of mutation in 1938.

39. Metcalf 1904, vii. Other monographs included Baldwin 1902, viii; Bateson 1902, 8–14; Lloyd and Bigelow 1904, 52, 138; Morgan 1903b, 287–99; Weismann 1904, vol. 1, 317–19.

40. Osterhout 1905, 442–53.

commenting that the apparatus needed for the experiments was so simple that "any handy, intelligent boy can readily make it" (147). (Osterhout's biographer mentioned having been taught from the book himself, which was still in use at Harvard twenty years after it first appeared; the equipment included "a homemade balance, sensitive to one-tenth of a gram, made from umbrella ribs!")[41] And Osterhout's book was successful: in addition to at least nine reprints (the last in 1923), it was translated into both Dutch and Russian and was widely reviewed on both sides of the Atlantic in both botanical and educational publications, from the *New York Times* and the New York *Independent* to London's *Athenaeum* and the *Speaker* (which described it as "that rarity among school-books—one which is a pleasure to read").[42] Like Hunter's *Civic Biology*, Osterhout's became a state-approved textbook—one that was still being recommended by the Washington Academy of Sciences as one of their hundred best "Popular but accurate books" on biology more than fifteen years after it first appeared.[43]

Osterhout tried to make his book both up to date and authoritative, but the novelty of experimental evolution forced him to rely on an eclectic range of sources that included popular journalistic writing. As noted previously (p. 46), his book recycled ephemeral material from periodicals into a long-lived reference work. He told readers who wanted further information on "the general subject of plant-breeding" to consult various magazine articles about Burbank (including one by Harwood from *Scribner's* and others from *Cosmopolitan, Country Life in America*, and the *World's Work*, as well as Wickson's series in *Sunset Magazine*; see p. 110). Similarly, anyone curious about the mutation theory was recommended to read articles by or about de Vries in magazines (including *Everybody's Magazine*, the *Popular Science Monthly*, and the *Independent*). These recommendations were, literally, the last words in the book, so readers would have been left with the impression not only that de Vries and mutation were the future but that magazines provided a trustworthy source of information about that future.

41. Blinks 1974, 218.

42. Specialist reviews included one in *American Naturalist* (Hus 1906, 147) and another in *Plant World* 8, no. 7 (July 1905). See also *New York Times*, April 29, 1905: BR285; and *Independent*, August 3, 1905: 270. British reviews included Review of Osterhout, *Experiments with Plants* 1905 and "School Science" 1905.

43. See *Los Angeles Herald*, June 25, 1905: 6, report of the school board. Also *Washington Herald*, October 19, 1921: 4; and December 12, 1921: 4. See the Library of Congress's *Chronicling America* project database at https://chroniclingamerica.loc.gov/.

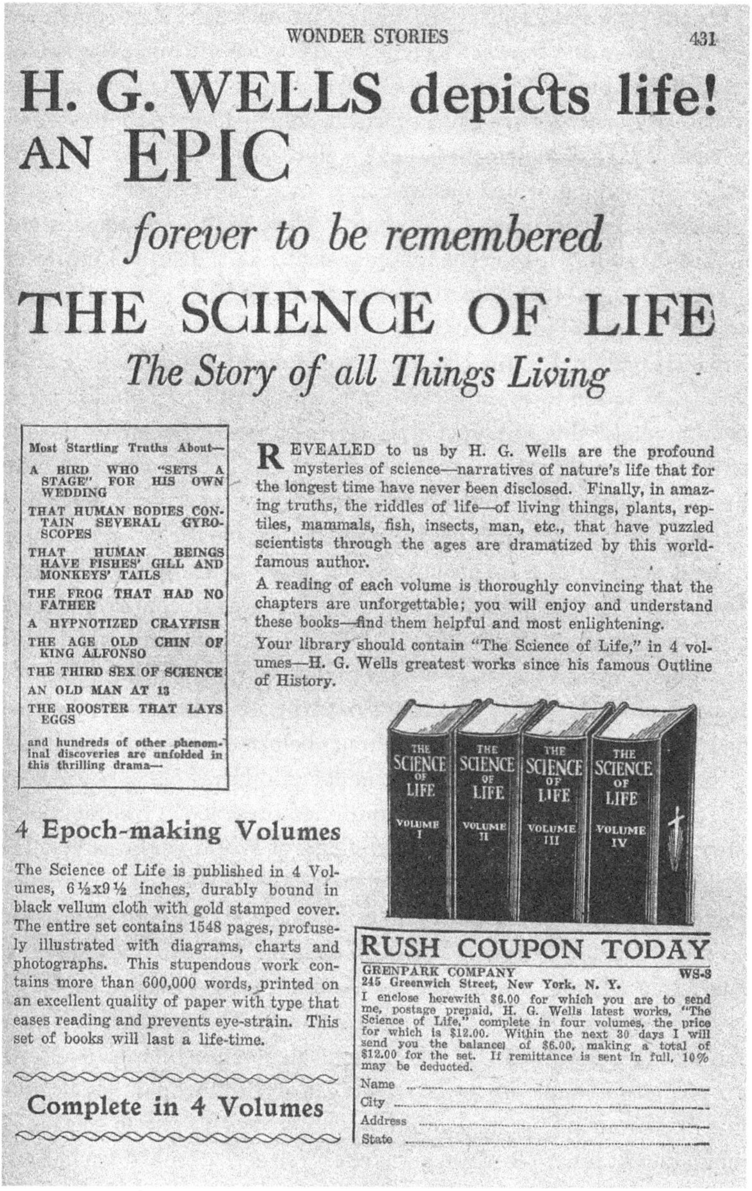

FIGURE 6.3. *The Science of Life* was initially published in inexpensive serial form and the books were advertised to the widest possible audience—this ad appeared in the US pulp science fiction magazine *Wonder Stories* (August 1931).

Osterhout's sources illustrate Donna Haraway's claim about public science: "the boundary between technical and popular discourse is very fragile and permeable."[44] When a professor at the University of California used articles by journalists to write a practical, college-level textbook (which was intended to complement Bailey's well-regarded text), a new hybrid form was emerging, which was welcomed by reviewers (the widely read *American Naturalist* praised Osterhout's "easy, almost colloquial style" and described his tips for creating new plants as "a distinct innovation as far as botanical text-books are concerned").[45] The book epitomized several aspects of biotopianism: the blending of de Vries's theory with Burbank's practicality, the eclectic mix of sources, the accessible style, and the invitation to everyone to join in. Above all, it emphasized how plant-breeding "adds, in a superlative degree, to the permanent wealth and increased material happiness of a nation." In Osterhout's view, the most important aspect of the mutation theory was that it had launched a new approach; de Vries's discoveries "demonstrate conclusively that evolution can be studied experimentally in a manner hitherto unsuspected," and the theory raised the "important question whether we can control evolution and so produce species at will" (441, 452–53). This paragraph was quoted by London's highbrow literary weekly the *Athenaeum*, which emphasized the possibility of "making new kinds of plants," thanks to both de Vries and Burbank.[46] Textbooks like Osterhout's helped experimental evolution reach new audiences and gain permanent credibility.

Despite the mutation theory's continued presence in textbooks, scientists had largely lost interest by the end of World War I, largely because what was increasingly being called "genetics" was making so much progress. The expert biologists' shift from mutation theory to Mendelism was exemplified by Thomas Hunt Morgan, who had been one of the first American academics to embrace the mutation theory. He had explained his doubts about orthodox Darwinism in an accessible article for *Harper's Magazine* (see p. 46), and his critique appeared in more permanent form as *Evolution and Adaptation* (1903). Morgan argued that natural selection (just like a human breeder) could only improve an existing species by eliminating its less well-adapted members (so selection had "an entirely negative value"). If selection were as powerful as Darwin and his modern

44. Haraway 1989, 14–15.

45. Hus 1906, 147. The review was by Dutch-born American botanist Henri Theodore Antoine de Leng Hus, who was one of Osterhout's graduate students (Blinks 1974, 219).

46. Review of Osterhout, *Experiments with Plants* 1905.

followers believed, breeders should be able to "go on indefinitely in the same direction, and produce, for instance, pigeons with legs five metres long." In reality, natural selection could not create. However, Morgan argued that Darwin had created his own problem by assuming adaptation was always slow and gradual, which led him to overlook the large, discontinuous variations that explained the origins of new traits (and thus, ultimately, of new species). Morgan concluded that, despite the usefulness of Mendel's recently discovered work, it was de Vries who had transformed evolution by showing that "Nature's supreme test is survival. She makes new forms to bring them to this test through mutation, and does not remodel old forms through a process of individual selection."[47]

Morgan was one of the first established scientists in the English-speaking world to embrace mutation theory and then explain why in detail. As a result, his book was widely reviewed: Britain's *Manchester Guardian* welcomed it as a "useful account . . . of such important matters" as the work of de Vries and Mendel's Law ("things which it is not easy to find gathered together elsewhere").[48] The book received even more attention in the USA, where the specialized journal *Science* welcomed it as "timely" and "the first non-technical work of its kind." As a result, it would "find its place on the general bookshelves" (and was specifically recommended as complementary to works by such celebrities as Spencer and Wallace).[49] And the *American Naturalist* recommended it because the "general reader will find in it a convenient summary" of both "older views" and "the new point of view," especially that of de Vries, as a result of which a "real obstacle to the older ideas about evolution has thus been removed."[50] Even critics acknowledged Morgan's work (thus giving it extra publicity).[51] When Orator Cook addressed the Biological Society of Washington to promote his kinetic theory of evolution (see p. 128), he cited Morgan's book as the best summary of what he regarded as the mistaken mutation theory.[52] And when James Walsh attacked Darwinism in the *Catholic World*, Morgan's

47. Morgan 1903b, 464, 91–97, 103, 123–24, ix–x, 261, 273–78.
48. Review of Morgan, *Evolution and Adaptation* 1904.
49. Dean 1904, 222.
50. W.E.C. 1904, 398–99. The reviewer was presumably Harvard-based geneticist William Earnest Castle. Similar points were made in "Concerning the Mode of Evolution" 1904, 678; Cockerell 1904, 197; "Reviews of New Books" 1903.
51. Merriam 1906, 244.
52. Cook 1904b, 549.

book was quoted at length to support Walsh's claim that he had "never in the last five years met an enthusiastic Darwinian."[53]

Four years after *Evolution and Adaptation*, Morgan published *Experimental Zoology* (1907), another accessible, introductory text based on lectures. He was still committed to the mutation theory (asserting that de Vries's work had "opened a new era in the study of evolution").[54] The book received fewer reviews than its predecessor, but they were even more positive. For example, New York's *Independent* called it "the best, indeed the only up-to-the-moment" account of "experimental investigation" into evolution. Although it was "not primarily" for general readers, they should choose it over any "more popular and less reliable account."[55] Thanks to his widely reviewed books, Morgan was probably the best-known scientific supporter of the mutation theory in the USA, yet by 1907 he had started to redefine the meaning of mutation. In *Experimental Zoology*, he wrote that "the term *mutation* will be used in the following chapters in a very general way," and specifically that "it is *not* intended that the word shall convey *only* the idea which De Vries attaches to it" (emphasis added). Instead, mutation was defined "as synonymous with *discontinuous and also definite variation* of all kinds," including what Darwin had called "single variations" or "sports."[56] By retaining the word de Vries had made popular, Morgan helped create endless confusion—maddening for biologists, but a rich source of imaginative interpretations for its wider publics.

The details of the work that Morgan and his team at Columbia did with the fruit fly, *Drosophila*, are too well-known to need repeating here.[57] In brief, Morgan's lab found numerous new forms of flies appearing in their breeding bottles and soon realized that the visible changes (such as a fly with white eyes, rather than the more common red ones) corresponded to specific genetic changes. As the new types multiplied, Morgan initially believed he was witnessing a de Vriesian "mutation period" in an animal (and had perhaps managed to induce one artificially). That idea was quickly abandoned, because the new forms did not appear to be new species, as de Vries's theory had led Morgan to expect. Instead, the rapidly

---

53. Walsh 1905, 505.

54. Morgan 1907, 214, 232–33.

55. "Biology a Science of Experiment" 1907, 219. See also Review of Morgan, *Experimental Zoology* 1907; "The Experimental Method in Biology" 1907, 229; Child 1907, 829; Conklin 1908, 139–40.

56. Morgan 1903b, 340.

57. For a full account of the *Drosophila* research program, see Carlson 1974; Allen 1975; 1978; Roll-Hansen 1978; Kohler 1994. For an overview, see Endersby 2007, 170–208.

multiplying flies allowed complex breeding experiments to be conducted quickly, so the Columbia geneticists were able to elaborate Mendel's laws and piece together many of the key components of what became classical genetics. Thanks to *Drosophila*, the initially rather vague notion of Mendelian "factors" was transformed into the concept of genes, located at specific points on the flies' chromosomes.

While *Drosophila* was proving to be an inexhaustible source of fresh ideas and publishable results, the search for other organisms (particularly animals) that displayed the same behavior as *Oenothera* had proven largely fruitless. As a result, interest in de Vries's original claims faded rapidly. Nevertheless, Morgan and his team retained the term "mutation" to describe their new forms, despite abandoning most of the original mutation theory. This new meaning became widespread and was soon understood by experts in the field, but it created ambiguity once it started being used more widely. For example, when Julian Huxley described Morgan's *Drosophila* work to a nonspecialist audience in 1926, he commented that roughly four hundred "variations, or mutations, have cropped up spontaneously in his stock."[58] No doubt some of his readers would have assumed these were de Vriesian mutations, implying that hundreds of new species had arisen rapidly. The distinguished biologist Ernst Mayr recalled that when he was at university in the 1920s, he and his fellow students assumed exactly that, that Morgan and his team were using "mutation" in the de Vriesian sense (and recalled that "we used to make fun of the 'new species' of Drosophila" that the Columbia geneticists imagined they were discovering).[59]

As Arlin Stoltzfus and Kelehave Cable have shown, Morgan's flexible definition of mutation was becoming more common at this time. Britain's William Bateson argued that even though he was "doubtful of the validity of the superstructure which de Vries has created" (that is, the mutation theory as a whole), he was one of several geneticists who started to use *mutation* to mean inherited variations, as opposed to those that were the result of environmental changes.[60] Stoltzfus and Cable cite numerous examples from both sides of the Atlantic of expert geneticists using mutation in this new, slightly confusing sense.[61] As noted, this shift in usage led to an increasing divide between expert understandings of mutation and those of their wider audiences, who continued to think of the word

58. Huxley 1926a (1933), 18. Huxley's piece originally appeared in the popular science magazine *Discovery*.
59. Mayr 1980 (1998), 20.
60. Bateson 1909, 287.
61. Stoltzfus and Cable 2014, 507–9.

in the way that it had originally appeared in the media—as describing new species appearing overnight, in a single leap.[62]

Morgan's later works continued to reflect his skepticism about orthodox Darwinism (in 1916 he was still noting that natural selection "has not produced anything new"), but de Vries's name became less prominent as Mendelism become the main framework for explaining evolution.[63] Morgan claimed that natural selection only acted "after a beneficial mutation has occurred"; it was therefore mutation, not selection, that accounted for the arrival of the fittest—but for Morgan and his team, *mutation* no longer meant a de Vriesian mutation but simply any heritable change to a gene (194). This point was obvious to expert reviewers; for example, William E. Castle (writing in the *Journal of the Washington Academy of Sciences*) noted that "Morgan's use of the term mutation is very different from that of DeVries, its original form. To Morgan, mutation as illustrated in Drosophila is simply change by a unit-character."[64] However, Morgan's latest book was too technical to be widely reviewed in general publications, and none of those that did examine it commented on the changed meaning of *mutation*. Expert and lay understandings of mutation were diverging rapidly, but it is unlikely that most lay readers would have noticed.

By the time Morgan published *The Physical Basis of Heredity* (1919), he had abandoned the mutation theory completely, defining mutation entirely in the light of a new understanding of Mendelian genetics. (De Vries was acknowledged as having inadvertently made crucial contributions to several relatively esoteric genetic questions.)[65] Once again, the few reviews the book received in general-interest publications made no mention of the shift in the meaning of mutation.[66] Nevertheless, as the specialists moved toward a new definition of mutation, general textbooks that appeared in the second and third decades of the twentieth century began to reflect the scientific community's growing skepticism about the

---

62. Ernst Mayr (1980 [1998], 8) noted, "As late as the 1930s and 1940s systematists referred to 'de Vriesian mutations,' many of them quite unaware of the fact that the term mutation had since been transferred by Morgan to a different class of phenomena."

63. Morgan 1916 (1919), 154.

64. Castle 1917.

65. Morgan 1919, 269, 265.

66. For example, the *Nation* described it as a "college text-book" that examines heredity as "conceived by the recent experimenters along the lines initiated by Gregor Mendel." The book presumed a high level of prior knowledge, "and wholly lacks a summary or any general statements which would make it understandable by the intelligent layman" (Review of Morgan, *The Physical Basis of Heredity* 1920, 829).

mutation theory. For example, Benjamin Gruenberg (*Elementary Biology*, 1919) described Mendelism as crucial, with de Vries being reduced to a distinctly secondary role.[67] And Babcock and Clausen (*Genetics in Relation to Agriculture*, 1918) argued that de Vries's *Oenothera* results resulted from either chromosomal abnormalities or hybridization. They proposed that the term *mutation* should only be used in its new specialist sense (to mean "those changes in specific factors, which result in the appearance of new Mendelizing characters")—which meant that de Vries's original *Oenothera* mutations were no longer to be considered mutations at all.[68] Babcock and Clausen's definition represents what might be called the "post-*Drosophila*" (or Mendelian) definition of mutation (for example, any heritable change to a gene), a definition that soon began to be widely used by geneticists. That would have created confusion for readers who were not up to date with the latest developments in biology, which would have been compounded because many of the earlier texts discussed here were still in print, being used in schools and colleges, or available in libraries.

During the 1920s and early 1930s, most scientists adopted the Mendelian definition of mutation, but there were still devotees of de Vries's earlier ideas even among the geneticists, some of whom continued their advocacy in textbooks, particularly those that offered a general overview of evolution and biology.[69] For example, the successful four-volume *Outline of Science* (1922; it went through ten printings in its first year alone), which was aimed at "the intelligent student-citizen, otherwise called 'the man in the street,'" included details on evolution that might have been repeated from a textbook published fifteen years earlier, with the evening primrose and *Drosophila* being described as organisms "which are at present in a sporting or mutating mood." The phrase "mutating mood" evoked de Vries's original concept of mutation periods, and the sentence implied that the new forms observed in *Oenothera* and *Drosophila* were essentially the same.[70] The book also described "the creations of Mr. Luther Burbank," including the spineless cactus and the Shasta Daisy, as "striking instances of what is always going on" in evolution "before our eyes."[71] Nor

67. Gruenberg 1919, 462. See also Hodge and Dawson 1918, 338.

68. Babcock and Clausen 1918, 282–86.

69. Hagedoorn and Hagedoorn-Vorstheuvel la Brand 1921; Gager 1920; Coleman 1921, 148.

70. A similar claim was implied by Gruenberg (1929, 184–85) when he described "the breeding experiments of de Vries and of Morgan with the mutations which they studied."

71. Thomson 1922, vol. 1, 187–89. Individual authors were not given for most chapters, but E. Ray Lankester and Julian Huxley were among those credited.

was the retention of older examples, theories, and definitions restricted to popular books. In 1921, Horatio Hackett Newman (zoology professor at the University of Chicago) produced a successful collection of *Readings in Evolution, Genetics and Eugenics,* which was reprinted in its first year and revised in 1925, and which had been through at least six printings by 1931. Its third, updated edition (1932) retained all the original material on the mutation theory (including a lengthy extract from de Vries, complete with multiple *Oenothera* illustrations). Three decades after they first appeared, de Vries's original descriptions of his *Oenothera* work had been elevated to "classic" status, and his claims—despite being debated and criticized—were still prominent in this successful textbook.[72] And the original sense of *mutation* was also invoked occasionally to explain one of the great evolutionary mysteries: the apparently rapid emergence of large-brained modern humans. The progressive educationalist Henry Linville (one of the founders of the civic-biology movement), suggested that the "particular variation or mutation which gave man his start was the sudden appearance of a larger brain than his ancestors had"—modern humans appeared abruptly, perhaps in a single step.[73] (This claim became one source for a long-standing science fiction trope of big-brained humans and aliens representing highly evolved intelligence; see p. 286.)

The original de Vriesian meaning of *mutation* persisted both because older books were still in circulation and because supposedly up-to-date texts continued to use *mutation* in ways that retained a flavor of the original theory. H. G. Wells and his collaborators explained mutation as an instance when "an individual with an entirely new feature, a 'sport' appears" in their massively successful *The Science of Life* (discussed in more detail below).[74] The book's tens of thousands of readers would have been left with a sense that *mutation* still implied the abrupt appearance of something new.

The ways in which words retained earlier meanings as they acquired new ones was a focus of Raymond Williams's *Keywords,* which offered historical analysis to argue that what made particular words so significant was that they acquired new meanings without shedding older ones as they were put to new—contested—uses.[75] Later scholars (including Williams himself) showed how the meanings of words were not fixed by dictionaries but by the ways specific communities of speakers or writers

72. Newman 1921, viii.
73. Linville 1923, 165. For Linville, see Pauly 2000, 177; Shapiro 2013, 67.
74. Wells, Huxley, and Wells 1929–1930, 380.
75. Williams 1976 (1983); Durant 2006, 10–11; Helmreich and Roosth 2016.

used them—and how both communities and usages change over time.[76] This insight reinforces the earlier point about Koselleck's *Begriffsgeschichte* (see p. 6)—their changing meanings remind the historian that words do not necessarily denote stable concepts. *Mutation* was not a word Williams considered, but it underwent the same process of accretion, modification, and expansion that his well-known examples, such as *culture*, exhibited. As Stefan Helmreich and Sophia Roosth argue about a similar nineteenth-century term (*life form*) it came to embody what they call a "capacious doctrine," precisely because it had a "constitutive incompleteness, ready for use in working out fresh problems."[77] The same was true of mutation; the theory's initial incompleteness was an invitation to interpretation, and the way mutation persisted in textbooks shows the lasting appeal of its ambiguity. Textbooks provided a key path through which de Vries's original sense of *mutation*, augmented by the new senses that its fans had given it, slipped almost unnoticed into the future, where—as the next two chapters show—it would dramatically change the possible futures that could be imagined.

The Science of Life

It is relatively easy to trace the waxing and waning of various ideas about heredity and evolution in biology textbooks, since they usually included sections on such topics as "mutation" or "Mendel" (see the appendix). By contrast, the biotopian mood is more elusive; it can be detected in references to experimental evolution and was sometimes apparent in discussions of biology's future impact, but tracing references to Burbank often provides the clearest evidence, since descriptions of his work often gave books an optimistic, futuristic flavor.[78] For example, Bailey, who generally propounded rather traditional values, sounded rather like a prophet of a biologically engineered future when he wrote of Burbank that "plants are plastic material in his hands. He is demonstrating what can be done. He

76. Williams 1961. John Gumperz and Stephen Levinson, *Rethinking Linguistic Relativity* (Cambridge: Cambridge University Press, 1996); and Anna Wierzbicka, *Understanding Cultures through Their Key Words* (Oxford: Oxford University Press, 1997). Both cited in Durant 2006, 14. See also Hall, Slack, and Grossberg 2016.

77. Helmreich and Roosth 2016, 21.

78. However, similarly optimistic claims about biologically created futures can be found in textbooks that made no mention of Burbank, e.g., Newman 1921 (39–42); and Metcalf 1911 (176).

is setting new ideals and novel problems" by beginning "strong and novel lines of evolution."[79]

Textbook discussions of Burbank also regularly used the accessible language of the magazines (for example, Bailey's book echoed the tone of his popular article for the *World's Work* [see p. 95] by reinforcing the image of Burbank as an intuitive, untaught genius).[80] One strategy textbook writers used to compete in the marketplace was to combine journalistic and scholarly tones (and sources), thus creating another form of the hybrid language that characterized biotopian writing.[81] And, of course, the mere fact that Burbank was mentioned regularly in science textbooks would have persuaded many of their readers that he was an important member of the scientific community.[82] When Henry Linville (see p. 214) wrote *The Biology of Man* (1923), he noted that human control over nature extended "even to the creation of new varieties of plants and animals" and asked "Who has not heard of Burbank's wonderful results in the creation of new kinds of plants out of older kinds?"[83] And Herbert Eugene Walter (associate professor of biology at Brown University) wrote a textbook that listed "the method of Burbank" alongside "the method of Mendel" as examples of modern approaches to plant-breeding.[84]

When textbook writers made Burbank into a scientist, they added to the difficulties of readers who might be trying to decide which experts to believe. Instead of clarifying the boundaries of accepted, credible science, textbooks often inhabited a fluid, unbounded world. Their hybrid nature was typified by Wheat and Fitzpatrick's *Advanced Biology*, which—like many others—straddled the worlds of journalism and scholarship. In an echo of Wharton's Professor Linyard (see p. 37), their introductory chapter noted that newspapers and periodicals "devote column after column to scientific matters, and even the writers of drama and fiction go to science for plots," with the result that many "discuss present-day science as they once discussed literature." Their book aimed to equip readers with basic scientific knowledge so that "this wealth of material may be easily interpreted and understood." Journalistic science was not to be scorned, but to be "interpreted"; laypeople would then be able to distinguish real from "pseudo-science," to base their decisions on "actual observable facts,"

79. Bailey 1906, 246, 243.
80. Bailey 1906, 242.
81. LaFollette 1990, 31–32.
82. Jordan and Kellogg 1907, 101–2.
83. Linville 1923, 3, 283–84.
84. Walter 1914, xii. For similar examples, see Conklin 1922, 273; Huxley 1926b, 13–14.

equipping them to play their part in shaping the future. Their readers would then have a solid basis "for the interpretation of future scientific readings."[85]

Nevertheless, as they tried to convey the excitement of biology, Wheat and Fitzpatrick's language revealed the biotopian mood's debts to journalistic styles. Under the subheading "To-morrow never comes," they noted that before the book had even left the printers, there was a chance that a new cure for cancer would be found, or "a new and unexpected theory may necessitate a check and revision of much of the work that has been set forth in these pages":

> You may be one who will add a chapter to the biology of to-morrow; you may learn and tell the farmers of to-morrow how to grow two blades of grass where one formerly grew. This *Advanced Biology* includes the story of the biology of yesterday and of to-day. If it has been well told, it should make your to-morrow a healthier, happier, and more complete day. (9)[86]

One reviewer described Wheat and Fitzpatrick's book as "quite different from that of the ordinary college introductory course."[87] Yet *Advanced Biology* was in many ways a typical textbook: it promised a better future based on biology (which the ordinary reader might help to create); it was a serious textbook with an accessible style (seemingly aimed at a wider readership); it was part of the civic-biology tradition (it was co-dedicated to George Hunter, whose textbook had shared the dock with Scopes) and thus focused on "problems relating to human welfare" (v); it discussed mutation (claiming de Vries's theory was "generally accepted by many scientists"—almost thirty years after its original announcement [377]); and, most of all it celebrated Burbank (who worked "like other scientists" to create "new species" [14, 344]). Biology textbooks (especially American ones) often promoted science, hard-headed commercialism, and utopianism side by side, conjuring a hybrid space in which the biotopian mood flourished.

The epitome of the new hybrid textbook was *The Science of Life* (1929–1930), written by H. G. Wells in collaboration with his son G. P. "Gip" Wells and Julian Huxley, both of whom held university posts.[88]

85. Wheat and Fitzpatrick 1929, 1, 6–7.
86. "Two blades of grass" was a reference to Swift's *Gulliver's Travels*; see p. 126.
87. Preston 1931.
88. Huxley's often amusing account of the collaboration was included in his autobiography, where he made it clear that Wells intended the book to be usable as

This massive, hugely popular textbook was directly modeled on Wells's earlier bestseller *The Outline of History* (1920); both initially appeared in serial form as fortnightly magazines with attractive color covers and were later republished as books. Despite being more detailed and technical than *The Outline, The Science of Life* was not written for specialists but for the general reader, "the ordinary man," defined by the authors as a lawyer, a schoolteacher, "or anyone of general intelligence but without specialised biological training, [who] might be expected to follow with interest and understand." Despite the book's size (over 1,500 pages), the authors hoped "not to confuse, weary and defeat" such readers, nor blunt their "natural widespread curiosity."[89] Like most textbooks, it emphasized the practical applications of biology, from pest control and curing disease to the improvement of both crops and people. The book's subtitle, "A Summary of Contemporary Knowledge about Life and its Possibilities," stressed the book's biotopian perspective: existing life was merely a starting point from which the human imagination would take flight.

The Science of Life never matched the commercial success of *The Outline*, but it sold well on both sides of the Atlantic and was widely reviewed. The leading British science weekly *Nature* commented that it showed its readers that "science is not only for illumination but also for 'the relief of man's estate,' as Bacon phrased it." (Just as Davenport had promised thirty years earlier, at the opening of the Cold Spring Harbor lab [see pp. 70–71], the vision of *New Atlantis* was, finally, to be realized.) The review continued, declaring that "mankind is at the dawn of a new era—the biological era," when the life sciences would become at least as important as the physical "for guidance in the control of human life." In this new era, *Nature* argued, "the big book of Wells, Huxley, and Wells will come to be regarded as an instalment of the relevant 'Law and Prophets.'"[90]

Wells and his co-authors manifestly relished their prophetic role, repeatedly emphasizing what they saw as science's most profound implications. Biology "throws new light upon our moral judgments; it suggests fresh methods of human co-operation, imposes novel conceptions of service, and opens new possibilities and freedoms to us" (1). The claim that science imposed "novel conceptions of service" echoed J. B. S. Haldane's claim that scientific and technological capabilities imposed new moral

---

formal school or college textbook, while remaining accessible to the general reader (1970 [1978], 148–63). See also Smith 1986, 262–63; LeMahieu 1992, 253–54; Sherborne 2013, 286.

89. Wells, Huxley, and Wells 1929–1930, 281.

90. Review of Wells, *The Science of Life* 1931, 478.

imperatives, one of several ways in which *The Science of Life* exhibited biotopianism's claim that ethical systems were as plastic as plants and that science would remake both.

Wells and his co-authors ranged widely over the whole of modern biology, bringing together an astonishing compendium of up-to-date knowledge. Naturally, evolution was a central concern and was presented in a more up-to-date and sophisticated form than Wells had previously offered. *The Outline of History* had begun with a chapter titled "Natural Selection and the Changes of Species," which presented evolution in the classic form that Wells had imbibed from Thomas Henry Huxley in the 1880s.[91] Despite being published in 1920, *The Outline* mentioned neither Mendel nor de Vries but embraced what Wells had previously dismissed as "Excelsior biology" (see p. 164). *The Outline* unfurled the great scroll of nature and flourished it as a banner of progress.[92]

The stadial version of evolution propounded in *The Outline of History* was vigorously attacked by the Catholic controversialist Hilaire Belloc in a lengthy series of articles for the Catholic *Universe*.[93] Perhaps Belloc's most telling criticism was that Wells's biology was out of date ("he knows nothing of all the modern work against Darwinism"), and the problem of the arrival of the fittest was at the heart of Belloc's argument. He noted that, thanks mainly to "the work of De Vries," it had been shown that "the killing off of the unfit is proved not to be the agent of change" (21). And he rubbed salt in Wells's wounds by citing recent scientific authorities (including Thomas Hunt Morgan) and concluded that if "you admit Mutation... poor old Natural Selection goes by the board" (40). Belloc's criticisms may explain why Wells started to collaborate with Gip and Julian Huxley soon afterward, relying on them to supply the updated biology that had been missing from *The Outline*.[94] (As Wells acknowledged in the introduction to *The Science of Life*, "The senior partner is the least well equipped scientifically" [2].)

In sharp contrast to *The Outline of History*, *The Science of Life* went into considerable detail about recent biology, including the controversies over the mechanism of evolution. Everything from neo-Lamarckianism to Henri Bergson's theory of creative evolution was analyzed, yet despite acknowledging the importance of the latest insights, the authors insisted on the centrality of Darwin's original theory ("Natural Selection has no

91. Wells 1921, 15.
92. Wells 1921, 1101, 1121.
93. Belloc's articles were later collected and published in book form (Belloc 1926).
94. Wells 1926.

more been 'exploded' by recent research than the rejection of underweight coins at the Mint has been exploded by the doctrine of relativity").[95] Darwinism needed to be considered in the light of recent experiments (particularly with *Drosophila*) and then restated in Mendelian terms of chromosomes, but it should not be abandoned (278–80).

When it came to mutation, *The Science of Life* (unlike many textbooks) tried to distinguish the original sense of the word from its more recent meaning, derived from work with *Drosophila* ("that most convenient of insects"). De Vries's work was briefly discussed (noting that he had been unlucky in selecting *Oenothera*, since it exhibited several rare genetic behaviors that had led him to misinterpret his findings). Although modern geneticists rejected his claim of a "special storm of mutation" (mutation periods), the discussion of mutation was clearly framed by the problem of the arrival of the fittest; the authors acknowledged that selection alone could not change a species unless the genetic material itself varied. Fortunately, "the germ-plasm is ever so slightly restless," thus ensuring that "the race alters a little." These alterations (each "a little jerk of evolutionary progress") are "the raw material on which Evolution works" and are what "biologists call *mutation*" (307–308). So far, so Mendelian, but mutation was explained with time-honored examples of sports, such as the Ancon sheep (which had also been discussed by Darwin in the *Origin*), which was used to illustrate how a species suddenly exhibits "some innovating change, and . . . an entirely new feature, a 'sport' appears." Sometimes genes were being reshuffled, but it was only "actual alteration in one of the genes" that created "a genuine innovation" and these "little gene-leaps that we call mutations are certainly one very important way in which the race changes." Mutations could be "so small as to be barely perceptible," but "every now and then the germ-plasm experiments and tries something new" (380–82). For their target audience (those with "general intelligence but without specialised biological training"), this description of mutation must have sounded similar to the many popular accounts of de Vries's original theory that had appeared over the previous thirty years.

The Science of Life emphasized the core promise of experimental evolution—control. However, for Wells and his co-authors, it was no longer de Vries's primroses that embodied this hope but Muller's experiments with X-raying *Drosophila*, which showed that artificially generated radiation provided geneticists with a powerful tool for inducing mutations

95. Henri Bergson (1859–1941) was a French philosopher of evolution who believed that a person's life-force (*élan vital*) could allow them to control their own evolution. Best known for *Time and Free Will* (1889) and *Creative Evolution* (1907).

at will.[96] This finding led to speculation that naturally occurring radiation was the ultimate motor of evolution (a claim that was often explored by science fiction writers; see chapter 8).

The book's optimistic mood was most evident in its novel focus on ecology (an obscure academic discipline at the time, still largely unknown to the public). When Julian Huxley had reviewed Wells's *Men Like Gods* in 1923, he had been particularly struck by its vision of ecological engineering, notably the wholesale elimination of supposedly noxious species.[97] As Peder Anker has argued, Huxley was influenced by the Oxford school of ecology, which grew directly out of the imperial tradition of preserving natural resources to ensure they remained available for continuing long-term exploitation.[98] *The Science of Life* also defined ecology as the management of resources for human needs; the colonization of one country by another was simply a form of efficient resource management (it had to be done carefully, to avoid introducing new pests, but the authors expressed no hint of any ethical concerns over the process). As Anker noted, Wells and his fellow authors saw nature as a part of the global economy, when they argued that "Man's chief need today is to look ahead. He must plan his food and energy circulation as carefully as a board of directors plans a business. He must do it as one community, on a world-wide basis; and as a species, on a continuing basis" (689).

Anker also notes that, in addition to being a successful mass-market text, *The Science of Life* influenced academic ecologists, providing them with a succinct summary of the idea that the Earth was a collection of exploitable resources. Thanks in part to the increasing use of aircraft for ecological research, Anker argues that "the master perspective from above" became central to the British ecological tradition—in which natural and human resources were equally available to be managed scientifically (116). (It served, also, as a reminder of the literal perspective of Wells's characters in *Men Like Gods*, as they are flown across Utopia and get an overview of the new world's perfected, garden-like landscape.) Thomas Huxley would surely have agreed with his grandson when the latter claimed that left to

96. Muller (see chapter 5) was the first to demonstrate that exposing the flies to radiation caused them to mutate, a discovery for which he was eventually awarded the Nobel Prize. See "Hermann J. Muller—Biographical," NobelPrize.org, accessed February 13, 2024, https://www.nobelprize.org/prizes/medicine/1946/muller/biographical/.

97. Wells "also imagines a purging of the organic world," all of which was simply "an extension of what has already been begun" (Huxley 1923, 592).

98. As Anker acknowledges, his analysis is indebted to Richard Groves's pioneering work on imperial resource management (Grove 1995).

herself, "nature's clumsy sequences" led only to "the sacrifice of the many for the few, broadcast waste to ensure the rare lucky survival, ruthless pruning, adaptation through strife and death" (687–88).[99] Both *The Science of Life* and *Men Like Gods* depicted Mother Nature as an old hag to be dispossessed—they were effectively part of a single utopian project.[100]

The Science of Life was also, as William Provine has argued, the product of the British national media market (particularly its national broadcaster, the BBC), which provided opportunities and encouragement for thinkers like Huxley, Wells, and Haldane to reach a genuinely national audience. Huxley's earlier book *The Stream of Life* (1926) was based on a series of BBC radio talks, which explained modern ideas about evolution, heredity, and eugenics to a broad, educated audience (he commented that "in no other country in the world has the Wireless gone so far in harmoniously combining its two great functions of popular entertainment and popular education as it has in Great Britain").[101] The earlier book hinted at claims that became more explicit in the later one (for example, that thanks to biology, "blind acquiescence . . . is giving place to the hope that destiny may in large measure come to be controlled" [1]). The skills Huxley refined in producing such talks contributed directly to the success of *The Science of Life*, and Provine argued that most of Huxley's important contributions to twentieth-century evolutionary thought can be partly attributed to his willingness to write for a general audience, for whom he pulled together a range of ideas from different specialists, explaining them in accessible language as parts of a grand vision of evolutionary progress.[102]

Like many textbooks, *The Science of Life* was a work of synthesis—comprehensive and packed with summaries of recent discoveries and theories—but its innovations (especially the use of ecology to bring order to what might otherwise have been a confusing compendium) gave the book its own distinctive character. The authors unapologetically embraced what they called "subjective" elements—biology's practical and ethical implications for human life. They had, they admitted, "departed from the atmosphere of clear, cold statement, proof and certainty," but "to shirk questions of feeling and will because they did not admit of the

99. Anker (2002, 112) notes that the ecology section of *The Science of Life* was written by Huxley, checked by Charles S. Elton at Oxford (to fit the Oxford school's view), and then rewritten to fit Wells's "science-for-all" ideal.

100. Krishan Kumar (1987, 190) argues that *The Outline of History* and *The Science of Life* were key texts in Wells's utopian project. John C. Greene (1990) argued that the book was also a product of Julian Huxley's utopian project.

101. Huxley 1926b, n.p.

102. Provine 1980a (1998).

hard precision of a purely objective treatment would be to rob our *Science of Life* of half its interest and two-thirds of its practical value" (918). This sentence introduced a section titled "Modern Ideas of Conduct" which took a biological approach to ethical guidance, reinforcing the book's claim that modern biology could offer guidance on morals and behavior as confidently as on human health or pest control.

The authors referred to Haldane's idea of "biological inventions" (citing examples from *Daedalus* and recommending the To-Day and To-Morrow series [973]). They noted that the pace of biological invention had quickened in recent years (for example, pest control, vitamins, and improved breeding on Mendelian principles), but acknowledged that their list was "not a long one"—it still fell far short of "the biological imaginations of mankind." They listed some of science's utopian dreams (many of which could have been found in *Daedalus*—and in *New Atlantis*):

> Man has dreamt of prolonging his life; of controlling the destinies of society as he can now control a business or a machine; of eliminating pain; of building a new race, all of whom should be strong and beautiful, clever and brave and good; of harnessing the forces of life to work for us as effectively as we have harnessed the forces of lifeless matter; of creating living matter anew; of getting rid of disease; of making synthetic food and drink and substances which should stimulate and enlarge this or that faculty without being followed by depression or injurious effects; of fashioning new kinds of animals and plants as easily as he fashions clay or wood or metal; of painless, quiet and happy dying; of the abolition of fear and worry, cruelty and injustice; of an intensification of human capacity for living—the abolishing of fatigue, the enhancement of vigour and enjoyment; of making life yield happiness, or if not happiness, then joy and divine discontent. (684)

The idea of "divine discontent" suggested an argument that Wells made in *A Modern Utopia* (1905), when he insisted that utopia "must be not static but kinetic, must shape not as a permanent state but as a hopeful stage, leading to a long ascent of stages"; a vision that hybridized stadial history with the Baconian ideal of unending scientific inquiry.[103] Nevertheless,

---

103. Wells 1905; Partington 2000. The phrase "divine discontent" occurred in D. H. Lawrence's *Sons and Lovers* (1913), when Paul embraces the term to express his loathing for a life of comfort and ease. Julian Huxley recalled that Lawrence visited him and Aldous while Julian was at work on *The Science of Life*. Whenever the Huxleys discussed science, Lawrence raged against its claims to be able to solve all the world's problems (Huxley 1970 [1978], 153).

Men Like Gods included a discontented minor character called Lychnis; when Barnstaple says he would like to remain in Utopia, she warns him that "there is no rest. Every day men and women awake and say: What new thing shall we do today? What shall we change?" In sharp contrast to the other inhabitants of her world, Lychnis complains about the divine discontent that drives her fellow citizens: "Research never rests," she observes; "curiosity and the desire for more power and still more power consumes all our world" (234). Barnstaple (who frequently expresses Wells's own views) is sympathetic but ultimately commits himself to the Utopians' kinetic vision—a future of ceaseless, unending experiment and change.

Unlike its companion novel, *The Science of Life* offered no dissident voice. The authors encouraged their readers to be dissatisfied with what nature provided, assuring them that utopian dreams could now be realized, thanks to "the sciences of life." Yet, they commented, "what a paltry beginning we have as yet made with their realization!" (684). The allure of the perverse was hinted at when the authors described artificial techniques such as grafting plants as a "thoroughly unnatural thing" by which something entirely new was "brought into existence." It was "perfectly possible" that biologists will "do these things with animal bodies in the near future." And what could be done to animals would eventually be done to humans, with—they speculate—artificially grown tissues being grown in incubators, allowing grafts to replace damaged skin and bones. Just five years earlier, Haldane had complained that the typical practitioner of the life science was "just a poor little scrubby underpaid man"; Wells and his fellow authors assumed that the biologist would indeed be recognized as "the most romantic figure on earth," once it became possible to directly manipulate human hereditary material, allowing "man" to master "a new art, with living protoplasm as his medium."[104]

The full vision of what the biologist would eventually accomplish was summarized toward the end of *The Science of Life* in a section with the telling title "Life Under Control." The authors optimistically asserted that "our species will survive and triumph over its present perplexities," as long as people were willing to "take control not only of [their] own destinies but of the whole of life." Wells's voice was unmistakable in the conclusion.[105] The garden world of *Men Like Gods*, full of perfect crops with every weed extirpated, lay within humanity's grasp: "Of every species of plant and animal man may judge, whether it is to be fostered, improved or

---

104. Haldane 1924, 77; Wells, Huxley, and Wells 1929–1930, 973, 976.
105. Wagar 2004, 176.

eliminated. No species is likely to remain unmodified." (Even the novel's tamed big cats are mentioned in the textbook: "The tiger may cease to be the enemy of man and his cattle; the wolf, bred and subdued, may crouch at his feet.") The end result will be that "the wilderness will become a world-garden" (976)—the defining image of biotopianism.

The Science of Life ended on a crescendo, proclaiming that "these might-ier experiences and joys of the race to come will be in a sense ours, they will be the consequence and fulfilment of our own joys and experiences, and a part, as we are a part, of the conscious growth of life, for which no man can certainly foretell either a limit or an end" (976).[106] For Wells and his co-authors, evolution was a story without an end, a progressive force that would propel humanity into a future where unimaginable possibilities awaited. The book exemplified both the power of biotopian vision and its hybrid nature. It merged journalism and scholarship, science and publicity, utopian fantasy and hard-headed commercialism (naturally, Luther Burbank was mentioned, and his spineless cactus was used to illustrate some of the possibilities of advanced plant-breeding).[107]

The book's hybridity helps explain why, despite the authors' attempt to clarify the modern meaning of mutation using up-to-date *Drosophila* experiments, their attempts to utilize a more accessible language ("sports," sudden innovations, and "gene-leaps") retained echoes of mutation's older, de Vriesian, sense. Popular understandings of mutation were also apparent in their section "Artificially-induced mutations," which "can be produced at will" (for example, by using X-rays), thus allowing modern biologists to "control the variation of animals and plants, not by the te-dious and indirect methods of selection, but directly and immediately."[108]

The book's audience was as mixed as its contents. It was one of the first textbooks to attempt an accurate survey of the whole of biology, one that tried to include the most up-to-date information while remaining accessible to the general reader. Reviews, particularly those in scientific journals and magazines, generally agreed that the authors had succeeded in their goals. In the USA, the *Science News-Letter* felt that it was "just the

106. Wells may have intended a deliberate echo of the closing words of the *Origin of Species*: "endless forms most beautiful and most wonderful have been, *and are being*, evolved" (Darwin 1859, 490, emphasis added).

107. Nevertheless, Burbank was treated with the skepticism so frequently applied by British writers, when they noted that he seemed unaware of the real nature of his achievements: Burbank was too unscientific to "realize" how he had created his plants and so "was Mendelian without admitting it" (Wells, Huxley, and Wells 1929–1930, 323).

108. Wells, Huxley, and Wells 1929–1930, 391–92.

sort of book needed by the general reader."[109] Britain's scientific weekly *Nature* welcomed the book's initial serial publication as a "new educational venture of great attractiveness" and commented that the "increased availability of science promises well for the future, for it is one of the most hopeful lines of human progress."[110] And, as we saw above, *Nature* welcomed the eventual appearance of the completed book with even more fulsome praise (and at much greater length), celebrating "the dawn of a new era—the biological era," which would offer "guidance in the control of human life."[111] The book sold well on both sides of the Atlantic, went through several editions, and was translated into French. It was widely used in US college classes (and some were still using it as late as 1960).[112]

The Science of Life's success partly reflects the fact that this period was one in which the "nonexpert" readership was growing to include scientists from other fields, policy makers, government officials, and the trustees of major funding bodies (for Warren Weaver, director of the Division of Natural Sciences at the Rockefeller Foundation, reading *The Science of Life* inspired a new direction for the foundation's policy).[113] The textbook's unique combination of reputability and accessibility entranced a wide variety of readers (it even seems to have inspired Walt Disney to launch his studio's True-Life Adventures series of natural-history films).[114]

Thanks largely to Wells's editorial direction, *The Science of Life* provides the most vivid example of just how complex the hybrid textbook genre became. The common themes that reverberated through both *Men Like Gods* and *The Science of Life* reveal that the textbook blended science fact and science fiction with utopian speculation almost as freely as the novel did.[115] As noted above, *The Science of Life* speculated that animal bodies would be reshaped and remade (in a utopian reimagining of Wells's *Island of Doctor Moreau*), and noted that cosmic rays might be driving evolution (a notion that provided the germ of Wells's 1937 novel *Star Begotten*, discussed in chapter 8). Yet alongside its more imaginative elements, *The Science of Life* was filled with pages of authoritative diagrams, photographs, equations,

109. "First Glances at New Books" 1929, 248.
110. "Biology for All" 1929, 442.
111. Review of Wells, *The Science of Life* 1931, 478.
112. Smith 1986, 262–63.
113. Abir-Am 1982, 348–49.
114. Bashford 2022, 156.
115. As Krishan Kumar 1987, 190, 28–29 has persuasively argued, *The Outline of History* (1921) and *The Science of Life* were key texts in Wells's utopian project.

and charts.[116] *The Science of Life*'s conclusion depicted the world-garden of *Men Like Gods*: the garden of Eden, remade by science, cleansed of original sin. Together, these two books were Wells's response to his mentor Thomas Huxley's melancholy conviction that humanity's struggle to clear an ethical space amid natural selection's howling wilderness was ultimately doomed to failure. The biotopian mood allowed the utopian romance and the biology textbook—not normally considered natural bedfellows—to be grafted onto one another.

116. Wells's main role was editorial, and he paid particular attention to smoothing out stylistic differences between the writings of his two, more junior contributors (Smith 1986, 262–63).

* 7 *

Counterfutures

On Christmas Eve 1919, a daily newspaper in Butte, Montana, informed its readers that "if not a new Heaven, at least, a new earth" was imminent. However, the source of their optimism was not the anniversary of the birth of Christianity's savior but de Vries's mutation theory, which meant that socialist revolution—a new "social mutation"—was coming. De Vries had finally explained "the arrival of the fittest," the missing piece of Darwin's puzzle, and as a result "it is no longer necessary to assume countless millions of years for the evolution of living forms. A plant enjoys a period of apparent stability, then it reaches a point where it 'explodes' and gives birth to new species. If a plant, why not a society?"[1]

Using mutation theory to justify revolutionary socialism might seem like an eccentric interpretation of experimental evolution, but—as we shall see—it was fairly widespread in socialist circles at this time.[2] Socialist writers were not the first to assume that what was true of plants might also be true of people, but the claim that evolution's mode had changed (from gradual to abrupt) inspired writers who wished to change the dominant social order to claim a scientific basis for a parallel transformation, from evolution to revolution. New biology and the new publishing conditions created opportunities for socialist and feminist writers to produce work that shared themes, ideas, and a common language with many other kinds of writing about recent biology, including early science fiction (see chapter 8).[3] In the hands of those who were usually excluded from elite scientific conversations, new ways of describing science and its imagined impacts

1. Lewis 1919.
2. Paul 1979, 120–26. Hofstadter (1944, 96–97) briefly touched on mutation's popularity among socialists earlier. See also Cotkin 1984, 201–5.
3. Williams 1978, 212–13.

were used to imagine alternative futures—where underdogs would finally have their day.

As earlier chapters have shown, following the new biological ideas wherever they appeared reveals that many supposedly distinct kinds of writing—from scientific reports to utopian fantasies—shared a quality of hybridity. Even that most apparently dull, well-defined genre—the scientific textbook—proved to be fluid and impure, blending sober science with optimistic speculation. This hybrid condition was partly enabled by new publishing genres, often created (like Burbank's new hybrid plants) by combining existing types to serve human needs (particularly the supposed need to make ever-larger profits). Textbooks shared features with cheap newspapers and pulp magazines, which reflected wider changes in the publishing world, as old formats were abandoned and new ones emerged. By the end of the nineteenth century, the triple-decker novel (a rock upon which more than one major Victorian publishing house had been built) was being overtaken by publications targeting the shallow pockets and short attention spans of newer readers. As Roger Luckhurst has argued, this audience was the one described by a cynical journalist in George Gissing's novel *New Grub Street* (1891) as "the quarter-educated," and new publications were devised to meet this new market's needs. The term *short-story* first appeared in 1884 to describe a form of fiction that soon became a staple of new periodicals on both sides of the Atlantic, where it appeared sandwiched between short articles, jokes, news, gossip, and competitions.[4]

Once again, Burbank featured regularly as a component of the underdog bricolage; his altruistic huckster image persuaded some socialist writers to co-opt him to their cause, as they tried to portray plausible futures based on current science. The *International Socialist Review* asserted that Burbank's work was "more prophetic of what may be accomplished when intelligence is applied to production . . . than that of any other man of this century."[5] And Charlotte Perkins Gilman's novel *Herland* (1915) described a women-only utopia whose inhabitants have perfected their world by careful scientific breeding—and are therefore described as "Lady Burbanks."[6] So, despite's Burbank's links to the world of profitable free enterprise, his promise of an end to scarcity through control of nature enabled some radicals to expand their sense of what was possible. The world of the underdogs also offered another kind of participatory culture;

4. Luckhurst 2005, 16–19.
5. [Simons] 1906, 507.
6. Gilman 2013, 76.

those who held advanced ideas about topics like socialism and feminism were often drawn to discussion groups, lectures, and radical publications, which were a haven for fans of both science and socialism. They were encouraged to read, write, and educate themselves, and to reach radical conclusions about what biology meant for the future, including the future of socialism itself.

The ways these writers appropriated early twentieth-century biology and put it to new uses made them part of what Kodwo Eshun has called "the histories of counter-futures," of attempts to imagine radical alternatives to dominant expectations.[7] Experimental evolution's rejection of naturalized ethics (and most other traditional guides to the regulation of human behavior) inspired some interpreters to offer egalitarianism as the human value that ought to shape the future. But the new biology's most important contribution was probably that it had apparently not only speeded up evolution but proved that it occurred through abrupt, revolutionary change. If the biologists were right, the oppressed need not wait centuries for an improvement to their lives—the future could be now.

Mutant Socialists

The Montana newspaper that hailed "a new earth" was the *Butte Daily Bulletin*, and its comments introduced extracts from the book *Evolution: Social and Organic* (1907) by popular Chicago-based socialist lecturer Arthur M. Lewis (figure 7.1). The *Bulletin* told its readers that "a thorough understanding of evolution is necessary for a true knowledge of life and labor" and that since Lewis's book "contains much information not found in the ordinary textbooks," they were serializing it to bring it to a wider readership.[8] The *Bulletin*'s readers were probably rather different from many of the audiences considered in previous chapters; it was a radical paper that had begun as a union bulletin published during a bitter strike in the copper-mining industry, which grew into a socialist daily with links to the Industrial Workers of the World (IWW). Its masthead carried the slogan "We preach the class struggle in the interests of the workers as a class," and, despite the fact that it never reached a large audience, it was well-known for its uncompromising politics (Montana's Democrat governor, Samuel Vernon Stewart, called it the most "radical or revolutionary

7. Eshun 2003, 301.
8. Editor's note, above: Lewis 1919.

PRICE 10C

Lewis - Harriman Debate

ARTHUR MORROW LEWIS

Socialist Party vs. Union Labor Party

·SIMPSON AUDITORIUM

LOS ANGELES, CALIFORNIA

Common Sense Pub. Co., 211 New High St.

FIGURE 7.1. Popular Chicago-based socialist lecturer Arthur Morrow Lewis regularly staged public debates where he promoted his evolutionary ideas.

sheet in the United States . . . Why it is allowed to circulate through the mails is more than I can understand").[9]

Arthur Lewis and the *Butte Daily Bulletin* were part of a working-class autodidact culture that was particularly strong in socialist and left-leaning communities; Lewis's debates with conservative opponents at Chicago's Garrick Theatre were especially popular, and his contemporary, the radical pacifist and socialist Jessie Wallace Hughan, claimed that his "popularizations of science are widely influential among workingmen."[10] He regularly turned his lectures into books, most of which went through several editions and were often recommended or advertised in socialist periodicals. He frequently appropriated the latest biology to support socialist ideals, as in *Evolution: Social and Organic*, where he used the claim that de Vries had speeded up evolution to argue that the mutation theory had also made revolutionary change scientifically respectable.[11]

A similar claim had been made the previous year by *Wilshire's Magazine* (one of the most popular socialist publications in the USA), which noted that mutation's link to socialism was vital because "to be called unscientific is about the greatest insult that can be hurled at a Socialist" (figure 7.2).[12] Another Montana socialist paper argued that the recent abrupt granting of universal suffrage in Russia was an example of "what Hugo de Vries calls 'evolution by mutation,'" while one in Utah attacked the argument that "a social revolution would be against the scientific laws of development."[13] The Utah writer Jaime Anguelo noted that conservatives routinely claimed that "evolution, it is said, proceeds slowly"—both societies and organisms needed to adapt slowly to changing conditions, which implied that "a social revolution, a sudden change . . . would be premature, unscientific." However, he assured readers that such arguments ignored the most recent developments in "the science of evolution"; the mutation theory proved that "evolution has proceeded . . . by leaps, by bounds," so social evolution could do the same.

---

9. Montana Historical Society 2013; Swibold 2006, 191–92, 204–5.

10. Hughan 1912, 58. See also Cotkin 1981, 274; Pittenger 1993, 144–45.

11. Lewis 1908, 94–95.

12. Wilshire 1905b (1907), 250. For *Wilshire's*, including estimates of circulation (c. 250,000), see Mott 1957, 207; Hillquit 1965, 353; and Salvatore 2007, 220. Nelson (2014, 78–79) argues that the magazine's circulation has been exaggerated, and it may never have reached two hundred thousand. Nevertheless, it appears to have sold better than any other socialist periodical, apart from the *Appeal to Reason* (Quint and Wilshire 1969).

13. "The Woman Vote in Russia" 1906; Anguelo 1909.

FIGURE 7.2. The "millionaire socialist" Gaylord Wilshire was one of several US socialists who interpreted the mutation theory as providing a scientific foundation for revolution.

For a brief period, American socialists were so interested in mutation that an article in the *International Socialist Review* chided "many of our comrades" for exaggerating the importance of "the theory of evolution by mutation which we owe to de Vries."[14] By contrast, British socialists (like British biologists) were less interested. Nevertheless, even in Britain a few joined the argument: the Independent Labour Party's *Socialist Review* argued that "what De Vries has called 'Mutations' . . . are distinct, sudden changes . . . and lead to the formation of new species." As a result, socialists could reject the claim that evolution only occurs via "small, slow elaborations."[15] And the *New Age*, which was linked to the Fabian Society (to which H. G. Wells briefly belonged), argued that many were "inclined to contend that the mutation theory is a biological justification for revolution" (but their writer added that "we revolutionists require no biological backing"—present-day injustices were "sufficient justification").[16] These early twentieth-century revolutionary interpretations of mutation built on a long-standing relationship between evolutionary and socialist theory; a brief overview of that history explains the initially surprising fact that John Edwin Peterson, a Swedish migrant railway worker, and a member of both the IWW and the Socialist Party of America (SPA), owned a copy of de Vries's *Species and Varieties*, which sat alongside the *Communist Manifesto* and the works of Emma Goldman on his bookshelves.[17] Socialist faith in evolution ensured that de Vries's theory would have a political impact its founder probably never anticipated as it became embroiled in debates over the future of American socialism.[18]

According to Chicago-based socialist publisher Charles H. Kerr and Company, "to understand modern Socialism, you must understand Evolution." With that goal in mind, they published a library of ten essential books (including Lewis's *Evolution: Social and Organic*), at the accessible price of 60 cents each, postpaid.[19] As Mark Pittenger has shown, the assumption that socialism and evolution were inseparable had first been established in the mid-nineteenth century and became particularly widespread in the

14. Spargo 1909b, 912.

15. Herbert 1910b, 33.

16. Eder 1908. For Wells, see Hyde 1956, 217–18.

17. Peterson 1986, 166–68.

18. De Vries himself was a political liberal who hoped science would play a role in the solution of social problems (Theunissen 1998, 476).

19. "Library of Science for the Workers," full page ad inserted in Pannekoek (1912) (and numerous other Kerr publications). The SPA had no party journal or official line, so publications like those from Kerr and Co. and *Wilshire's Magazine* were crucial to the developing socialist culture (Pittenger 1993, 121).

USA.[20] Danish-born American Laurence Gronlund wrote *The Coöperative Commonwealth* (1884), the first major American popularization of what he called modern "German socialism." The book emphasized that "modern Socialism . . . teaches that *the Coming Revolution is strictly an Evolution.*"[21] Gronlund's evolutionary vision of socialism sold sixty thousand copies in the USA and another forty thousand in the UK. The phrase "cooperative commonwealth" was soon being used by some socialists to distance themselves from the image of violent revolution and conflict traditionally associated with Marxism.[22] The claim that socialism could be achieved slowly and peacefully underpinned Bellamy's *Looking Backward*, which inspired a whole generation of readers to identify themselves as socialists (see chapter 3). When the novel's tiresome cicerone, Dr. Leete, told the time-traveling protagonist, Julian West, how America had become a socialist utopia, he explained that it was "the result of a process of industrial evolution which could not have terminated otherwise. All that society had to do was to recognize and cooperate with that evolution."[23] Bellamy and his followers rejected most of the core tenets of Marxism (including class struggle and revolution), but still claimed that their views were based on scientific extrapolation from present-day tendencies—socialism would be the inevitable result of history's ineluctable forces.

Despite using the word *evolution* regularly, neither Bellamy nor Gronlund even mentioned Darwin; they believed in stadial evolution—the nineteenth-century vision of history as an inevitable, gradual ascent up the ladder of progress (see p. 16). Unsurprisingly, Gronlund cited Herbert Spencer, but no biologists, to support his account of evolution (another reminder that Spencer was not invariably regarded as a conservative figure at this time). Underpinning this commitment to evolutionary change was a common organicist metaphor built on analogies (for example, individuals were cells in a social body whose parts should function harmoniously), which implied that socialism was natural.[24] Again, Spencer was an important influence. His interpretation of biological evolution as progress toward increasing perfection relied on just such an analogy; evolutionary

20. Pittenger 1993, 1–11.

21. Gronlund 1884.

22. William Dean Howells's *A Traveller from Alturia* (1894) was directly influenced by Gronlund's ideas. William Dwight Porter Bliss, who found the American Fabian Society, was also directly influenced by Bellamy and Gronlund. And Eugene Debs also directly acknowledged Gronlund's influence (Pittenger 1993, 43–44, 60–62).

23. Bellamy 1888 (1917), 49. See also Williams 1978, 206.

24. Pittenger 1993, 43–44. Britain's Fabian Society offered similar gradualist arguments; see Ogilvie 2002.

history showed organisms becoming increasingly complex and specialized. Spencer assumed that human society would evolve in a similar fashion, with the division of labor leading to each group in society assuming
increasingly specialist but complementary functions. Tension between
different groups would thus fade away, leading to a perfect, conflict-free
society.[25]

Partly because of Spencer's influence, many socialists interpreted
Marxism as a science that had features in common with evolution; Kerr
and Company's ad (see p. 235) argued that "socialists predict the speedy
end of the capitalist system as a result of irresistible NATURAL LAWS." Nor
were Marxists alone in describing biological and social evolution as parallel forms of inevitable progress: Pyotr Kropotkin offered a similar vision
in books such as *Anarchism and Modern Science* (1923), which concluded
that his version of communist anarchism was the "inevitable result of the
intellectual movement in natural sciences."[26] The claim that socialism was
inevitable—because it was based on evolutionary science—was also widespread in the USA, whose socialists were equally eager to underpin their
arguments with science's prestige.[27] In addition to offering beleaguered
revolutionists the comforting thought that (evolutionary) history was on
their side, a stadial interpretation of socialism made revolution unnecessary, which appealed to those who rejected violence.

Although many socialists' vision of evolution did not draw directly
on Darwin's theory of natural selection, his work was acknowledged as
having provided substantial evidence for common descent and was regularly interpreted as having fatally undermined all forms of religious belief
(which appealed to secularist socialists). Like many of their contemporaries, socialists often used the terms *evolution* and *Darwinism* interchange-
ably, and, of course, Marx and Darwin had long been linked (when Engels
spoke at Marx's funeral, he made what would become a well-known claim:
that "just as Darwin discovered the law of development of organic nature, so Marx discovered the law of development of human history").[28]
Marx had been more diffident about the connection than his collaborator;
nevertheless, the ideas of the two bearded sages soon became so firmly
linked that Darwin's most prominent German supporter, Ernst Haeckel,

25. Richards 1987, 260–61.
26. Kropotkin 1923, 91. See also Shpayer-Makov 1987, 384; and Nicolosi 2020,
147–49.
27. Cotkin 1981.
28. Quoted in Colp 1974, 337. For Engels's importance to the American evolutionary socialists, see Pittenger 1993, 22.

felt compelled to separate them, arguing that "Darwinism, or the theory of selection, is thoroughly aristocratic; it is based on the survival of the best." By contrast, "Communism and the demands put up by the Socialists in demanding an equality of conditions and activity is synonymous with going back to the primitive stages of barbarism."[29]

Despite the efforts of conservatives like Haeckel, many socialists continued to claim Darwin's support. The popular *Socialism for Students* (1910) stressed that for "proletarian science, evolution and revolution are twin forces," and so "science and socialism belong together."[30] Dutch astronomer and socialist theorist Anton Pannekoek wrote a brief introduction titled *Marxism and Darwinism* (which was translated into English and published by Kerr and Company as a 10-cent pamphlet).[31] It set out the basics of each man's ideas in an accessible format (and used Charlotte Perkins Gilman's popular poem "Survival of the Fittest" as its epigraph). Pannekoek's pamphlet illustrates how early twentieth-century socialists connected the two sets of ideas, beginning by describing both Marx and Darwin as "scientists" who had "revolutionized" people's conception of the world: "The scientific importance of Marxism as well as of Darwinism consists in their following out the theory of evolution, the one upon the domain of the organic world, of things animate; the other, upon the domain of society" (7).

Darwinism was explained in then-conventional terms, for example by arguing that unlike earlier theorists, Darwin had produced proper evidence to support his theory (as a result, the *Origin* "struck like a thunderbolt," and Darwin's "theory of evolution was immediately accepted as a strongly proved truth" [10]). And when "we turn to Marxism we immediately see a great conformity with Darwinism"; the earlier utopian socialists had been daydreamers, mere speculators (like the pre-Darwinian evolutionists), but Marx had "discovered the propelling force, the cause of social development" (16).[32] In Pannekoek's argument, the human world was propelled forward by class struggle; the nonhuman world by natural selection.

Nevertheless, Pannekoek argued that "Darwinism and Marxism are two distinct theories." They complemented each other, but—as the writings of Spencer and Haeckel showed—"false conclusions" would be

29. Ernst Haeckel, *Freie Wissenschaft und freie Lehre* (Stuttgart, 1878): 73–74; see Weikart 1993. Translation quoted from Pannekoek 1912, 28–30.

30. Cohen 1910, 89; Cotkin 1984, 205.

31. Pannekoek 1912. First published in Dutch and German in 1909 (Steen 2019, 148).

32. Arthur M. Lewis (1908, 49–51) compared the utopian socialists to Lamarckians.

reached by anyone who tried "to carry the theory of one domain into that of the other, where different laws are applicable" (33). Like Marx himself, Pannekoek had no sympathy for social Darwinism in any form (both arguments that used Darwin "in favor of capitalism" and those "who base their Socialism on Darwin, are falsely rooted").[33] Pannekoek accepted that separating the social and biological was not straightforward, since humans had evolved from other animals and Darwin had shown (in the *Descent*) that human ethical instincts had their roots in the evolved social behaviors we share with other animals. He cited books such as Kropotkin's to illustrate that animals exhibited behaviors that favored the group over the individual—and argued that the same instincts had evolved further in humans.[34] These facts refuted the views of "bourgeois Darwinists" who believed that "the extermination of the weak is natural" in order to "prevent the corruption of the race" (37–39). Yet Pannekoek appeared willing to commit the naturalistic fallacy when it suited him (for example, to prove that human altruism was merely a survival tactic and could be explained without spiritual factors). Similar arguments were found in many other socialist interpretations of Darwinism, such as Enrico Ferri's influential *Socialism and Modern Science* (1894) and Kropotkin's *Mutual Aid* (1902).[35]

Nevertheless, the fact that "bourgeois Darwinists" regarded natural selection as "the scientific proof of inequality" explains why some on the left distanced themselves from Darwinism. Writers like Pannekoek tried to dispel such fears (by arguing that the conservatives misread Darwin), but Marx himself had sometimes implied that Darwinism was a bourgeois ideology.[36] In an effort to deny that Darwinism made capitalism natural, well-known American socialist John Spargo quoted Marx as saying: "Nothing ever gives me greater pleasure than to have my name thus linked onto Darwin's. His wonderful work makes my own absolutely impregnable. Darwin may not know it, but he belonged to the socialist

33. Marx regularly expressed contempt for left-wing Social Darwinists, e.g., after reading Ludwig Büchner's book *Darwinism and Socialism* in 1868 he described it in a letter as "superficial nonsense" (Colp 1974, 332).

34. Kropotkin 1902 (1904); 1923. See also Avrich 1980; Shpayer-Makov 1987; Stack 2003, 37–41; Nicolosi 2020.

35. Ferri 1905 (1909); Kropotkin 1902 (1904); Pittenger 1993, 124–25; Nicolosi 2020, 149.

36. According to Marx, it was "remarkable how Darwin recognises among beasts and plants his English society with its division of labour, competition, opening up of new markets, 'inventions,' and the Malthusian 'struggle for existence,'" K. Marx to F. Engels, June 18, 1862, quoted in Radick 2003 (147). See also Young 1985.

revolution."[37] The source of this implausible quotation was a conversation Marx supposedly had with W. Harrison Riley (editor of the *International Herald*) in the 1860s; despite its dubious historicity, it was too useful to be overlooked by those who wished to claim Darwin's imprimatur for their views.[38] Socialists and biologists were both part of a larger battle over who was entitled to call themselves a Darwinian, and the questions of tempo and mode that beset the biologists were equally important to the socialists.

Pittenger has demonstrated that early twentieth-century American socialists often built their arguments on what now appears like an incoherent synthesis of Darwinian and Spencerian evolutionism. Yet as we saw (chapter 5), Spencer's publications included utopian socialist elements as well as aspects of laissez-faire economics that most socialists rejected. Many American socialists simply assumed that "evolution" (whether Darwinian or Spencerian) implied "progress," as Spargo did in a short pamphlet titled *Where We Stand*. He noted: "We believe in evolution," but he acknowledged that "everybody believes in evolution nowadays"; what distinguished the socialists was their belief that "there is a law of social evolution which is but the counterpart of the law which pervades the organic world." It might seem contradictory to be both an evolutionist and a revolutionary, but there was "no necessary antagonism between Evolution and Revolution, as any scientist will tell you."[39]

However, Spargo's conviction that a belief in evolution was entirely compatible with a commitment to revolution was not shared by those on the right of the socialist movement, such as Victor Berger (one of the founders of the SPA) and his supporters, who used gradual evolution to argue for reform rather than revolution.[40] The American debate reflected a much wider one within the international socialist movement between reformists (who believed the road to socialism must be traveled slowly in small steps), and revolutionaries (committed to rapid change). The debate first surfaced within the German Social Democratic Party (SPD) around 1900, with Eduard Bernstein (revisionist) and Karl Kautsky (orthodox) representing the two sides. Because the SPD was the world's largest, most successful socialist party, their internal quarrel assumed international significance. Bernstein argued that increasing misery and the hardening of class divisions would not lead inevitably to socialism; the workers would become too dispirited to take part in politics, hence the need for reforms

37. Spargo 1912b, 200. For Spargo, see Pittenger 1993, 139–40.
38. Colp 1982, 472. The same quote can be found in Runkle 1961, 108.
39. Spargo n.d. [c. 1908], 5–7.
40. Pittenger 1993, 17, 129–31; Nelson 2014, 58.

and gradual change.[41] His gradualist, evolutionary reformism was widely influential in the USA (especially on Berger, who was proud to be known as the "American Bernstein").[42] And it found echoes in Britain; when the Labour member of parliament James Ramsay MacDonald wrote the preface to the 1905 edition of Ferri's *Socialism and Positive Science*, he declared that "socialism is naught but Darwinism!" (And when he helped translate his friend Bernstein's book into English, he suggested that its English title be *Evolutionary Socialism*.)[43]

Meanwhile, Kautsky and his allies argued that truly revolutionary classes always produced scientific theories based on abrupt change (when the bourgeoisie were struggling to replace the feudal order, they had accepted geological theories based on catastrophic change, but once in power they embraced Charles Lyell's and Darwin's theories of "gradual imperceptible development"). A bourgeois scientist might not be a conscious apologist for capitalism, but his thinking would inevitably be shaped by the "mental attitude of the class amid which he lives." So conservatives regularly asserted that "nature makes no leaps, revolutionary change was impossible," and life could only be improved through the "accumulation of little changes and slight improvements, called social reforms."[44] However, these bourgeois arguments were being destroyed because evolutionary science increasingly accepted the role of "sudden profound transformations." As an example, Kautsky cited de Vries's arguments that species are normally stable, and thereby prone to extinction, but "other species are more fortunate; they suddenly 'explode,' as [de Vries] has himself expressed it," giving birth to successful new forms. Kautsky did not argue that de Vries had proved that social revolutions were natural, merely that they were not impossible (thus nobody could "appeal to nature for proof that a social revolution is something unnecessary, unreasonable, and unnatural" [16, 20]).

As the Kautsky–Bernstein debates reveal, socialists were disputing who best represented the legacy of Marx and Engels, at the same time (and often in similar terms) that the biological evolutionists were fighting over who held the rights to Darwin's name. The two debates were so entangled because Marx and Engels claimed to have founded the first genuinely *scientific* form of socialist thinking, dismissing their predecessors as fanciful utopians and condemning those who wasted the workers'

41. Weikart 1998, 157–64; Stack 2003, 76–77.
42. Pittenger 1993, 23.
43. Stack 2003, 53–54, 78–79.
44. Kautsky 1902, 12–14.

time on trying to build utopian communities in the midst of a capitalist system. By contrast, Marxism was supposedly based on scientific laws that explained history and allowed future developments to be predicted and controlled (much as experimental evolution was supposed to function in the biological realm).[45]

However, some socialists interpreted Marxist science as meaning socialism was inevitable and must thus evolve in a conventional Darwinian fashion—slowly and gradually. As a result, some socialists wondered if it was possible to be both a good revolutionary and a good evolutionist—until the mutation theory came along. Arthur Lewis was typical of the period's largely self-taught propagandists, who saw evolution and socialism as inseparable. And, as we saw above, he used de Vries's theory to argue that it was "no longer true that species require thousands of years for the simplest change." Biology no longer contradicted "the Socialist position that a new society may be born of a sudden revolution."[46] Lewis even edited a short-lived socialist magazine simply called the *Evolutionist*.[47]

The most detailed exposition of Lewis's views was the book that the *Butte Daily Bulletin* serialized: *Evolution: Social and Organic*. It described the mutation theory as a step in what he called the "Battle of the Darwinians"—over whether or not natural selection was sufficient to explain evolution—and even quoted Harris's claim that natural selection "cannot explain the arrival of the fittest."[48] Lewis told his readers that Darwin himself had struggled to explain the origin of new variations, as had many later evolutionists until de Vries's "modest mutating primrose" helped solve this "riddle of the universe."[49] Socialists should be particularly interested because the mutation theory exploded the argument that "it took as long for one class in society to displace another . . . as it does for one species to develop from another," a proposition that left the ruling class safe "for a thousand generations" (92–93). As he put it in a later book, critics of socialism liked to "speak of slow, gradual evolution as against sudden revolution. But even in science this apparent conflict has been resolved"—by de Vries, who showed that apparently stable species

45. Pittenger 1993, 6–8; Paden 2002.

46. Lewis 1908, 94; Cotkin 1984, 209–10.

47. The *Evolutionist* is available online at Hathi Trust, https://catalog.hathitrust.org/Record/000058139.

48. He also noted that the anti-evolutionary German botanist Albert Julius Wilhelm Wigand (1821–1886) had argued that "selection does not do more than determine the survival of what is offered to it, and does not create anything new" (Lewis 1908, 35, 83).

49. Lewis 1908, 86–87.

regularly enter "what might be called a revolutionary period, when these plant forms 'explode' and new species appear."[50]

Lewis advised socialist readers who wanted to educate themselves to read de Vries's *Species and Varieties*.[51] The same advice was offered by several socialist periodicals; *Wilshire's Magazine* described the book as "the greatest step forward we have had since the publication of Darwin's epoch-making 'Origin of Species'" (a claim that mirrored the politically conservative Charles Davenport's assessment; see p. 40).[52] And Spargo put de Vries's book at the head of his list of recommended readings about evolution in his *Elements of Socialism* (1912).[53] It therefore seems more predictable than surprising that someone like the IWW railway worker Peterson owned the book; not only was it the first full-length exposition of de Vries's ideas in English, but it had originally been a series of lectures and was thus more accessible than *Die Mutationstheorie* (which did not appear in English until four years later). *Species and Varieties* was considered so important that the *International Socialist Review*'s editor, Algie M. Simons, analyzed it in detail to argue for mutation's importance "to the socialist." For years the "pseudo-scientist has always been quoting at us, that 'there are no sudden leaps in nature,'" but de Vries had proved them wrong.[54] *Oenothera*'s sudden changes were similar to the "sudden change of social character (called revolution) brought about by the accession of a new social class to power"; this similarity was not a coincidence but an example of the "universality of natural law" that increasingly characterized modern science.[55]

Not only were socialists as eager as biologists to associate their ideas with Darwin, many of them (like the battling biologists) claimed to be better Darwinians than any of their opponents. Just as slow, gradual evolution had provided comfort for those who resisted revolution, the new world of rapid experimental evolution provided analogies that could be used to support one kind of socialism over another. The Utah paper's writer, Anguelo, asserted that in the distant past, random mutations among fish would have produced many useless variations, but a few created basic lungs and "naturally they soon overcrowded the land." A parallel process of social evolution was underway, "with the new conditions of production . . .

50. Lewis 1911, 26–27.
51. E.g., Anguelo 1909.
52. Wilshire 1905b (1907), 252.
53. Spargo and Arner 1912, 75.
54. Simons 1905; Pittenger 1993, 133–36.
55. Simons 1905, 175.

lungs are needed: in this case an adequate social system." Naturally, the new conditions were stimulating the production of mutations, but most were useless ("sports without the much demanded lungs, parliamentary Socialists, opportunists, vote-catching parties, craft unions, etc., in the struggle for life they are destined to be outlived by the original bourgeois stock"). However, Anguelo told his readers that "when the true sport with lungs, the uncompromising Socialist, will have bred a sufficient progeny, he will survive the others by a social revolution, and evolution will have advanced another step."[56] While some of his contemporaries simply used mutation theory to attack the revisionists, Anguelo went further: the idea that the uncompromising socialist was a "true sport," a leap into a new evolutionary stage, hinted that the mutation theory might be more than just a handy stick with which to beat one's fellow socialists. For a relatively brief period, a few socialists interpreted mutation as showing that when the fittest arrived they would prove to be a new species of socialist.

Biotopian Socialists?

Although some socialists used the mutation theory to attack revisionists, socialist appropriations of mutation and experimental evolution do not map straightforwardly onto pro- and anti-revolutionary factions. Well-known socialist writer William Ghent noted that radicals had an answer to conservative arguments for slow change even "before De Vries and Burbank came to our aid with their proof of mutations"; history showed (as in the case of the French and American revolutions) that "social evolution has other movements than those of gradual and uniformitarian transformations."[57] This claim might sound as if Ghent were another Marxist revolutionary, but he was not: he had originally been a Bellamyite and was associated with American Fabianism before finally joining the SPA (where he was aligned with the reformists, serving as secretary to Victor Berger during his term as a Congressman).[58] Meanwhile in Britain, the Labour parliamentarian Ramsay MacDonald wove the work of de Vries and Burbank into his distinctly *non*revolutionary vision of evolutionary socialism; he argued that the apparently abrupt changes revealed by recent biology were the final result of slowly accumulated changes, and that

56. Anguelo 1909.
57. Ghent 1910, 48.
58. Smith 1957, 122–23, 203–8.

society must follow a similar path.[59] Scrutinizing the different uses that socialists made of de Vries, Burbank, and Mendel shows how experimental evolution changed the terms in which socialists debated the future.

For socialists, as for most of the period's other science fans, an interest in experimental evolution included more than the mutation theory. Reginald Punnett's introductory textbook *Mendelism* (see p. 31) was not of obvious interest to socialist readers, yet its first US edition was published by the socialist publisher Wilshire Books, whose eponymous founder—Henry Gaylord Wilshire—provided it with a preface. He explained that he believed socialism must come about in a single leap, but his "scientific friends" said his view was "against the Darwinian theory," which proved that evolution "necessarily proceeded slowly, step-by-step." It was in response to this "so-called Darwinian objection" that he had produced a cheap (50 cent), popular edition of Punnett's book.[60] As was typical at the time, Wilshire saw Mendelism as an aspect of mutation theory (he argued that "the law of Mendel is of intense interest" because it provided "a final clincher to the argument for mutation in biology"). And he quoted the university of Chicago's Jacques Loeb, who had written to him to argue that the logical outcome of Mendelism "is that evolution takes place by mutation only." And if Loeb was right, Wilshire commented, any "change in society must proceed by mutation."[61]

There was little in Punnett's book to suggest that the new science had revolutionary implications. Its author was a respected Cambridge scientist, a fellow of the conservative Gonville and Caius College, who worked alongside Britain's leading Mendelian scientists, notably William Bateson. The book described genetics with an attractive blend of brevity and clarity (so much so that it even made a brief appearance on a British bestseller list).[62] De Vries's work was mentioned in passing (noting that his claims were still being evaluated), but the closest Punnett came to a radical statement was in making the claim that it seemed "reasonable to regard the mutation as the main, if not the only basis of evolution."[63] His

59. MacDonald 1909, 8–9. MacDonald's pamphlet featured a full-page advertisement for Bernstein's *Evolutionary Socialism*, describing it as "a critical examination of Marxist Socialism." MacDonald was chair of the Independent Labour Party at this stage in his career.

60. Wilshire, preface to Punnett 1909 (3–4). See also Pittenger 1993, 140–41, 160.

61. Wilshire, preface to Punnett 1909 (5).

62. Punnett's book appeared alongside the latest novel by the enormously popular Mary Corelli as the two best-selling books of the week in the *Westminster Gazette* (Crew 1967, 315).

63. Punnett 1905, 51.

book promoted the experimental approach: its expanded second edition (reviewed in the *Atlantic Monthly*; see p. 31) had included the claim that "experiment would give us the solution" to some of evolution's missing pieces.[64] As a result, Punnett anticipated considerable progress in future, especially as human genetics became understood. Yet, beyond evoking the "tremendous powers of control" heredity implied, he offered little hint of future revolutions—in biology or anywhere else.

Nevertheless, Punnett's lack of obvious political messages did not stop socialists taking an interest in his book's implications. As John Spargo commented (when he reviewed the Wilshire edition of Punnett) the words "Mendelian Law" had recently been occurring more frequently among socialists (despite the fact that few socialists understood them) and were "bound to loom large in Socialist discussions during the next few years." Given the level of interest in experimental evolution, he recommended Punnett's book because it explained Mendelism "with admirable lucidity and conciseness." Spargo, despite being on the right wing of the SPA, agreed with Wilshire that Mendelism "links on to the mutation theory of De Vries" and proved that neither biological or social evolution had to be "slow and almost imperceptible."[65] As noted earlier (p. 66), mutation and Mendelism were frequently conflated at this time (for example, the popular nonsocialist weekly the *Living Age* reviewed Punnett's book and referred to the "Mendelian theory of evolution by mutation" that was replacing natural selection).[66]

Wilshire himself was as complex and contradictory as his interpretations of biology. He was one of the group known as the "millionaire socialists" and a relentless self-publicist (one of his supposed comrades waspishly commented that "*Wilshire's magazine* is entirely devoted to advertising Wilshire's van dyke whiskers, his check pants and tan shoes—no other possible reason for its existence being disclosed anywhere in its pages").[67] His substantial personal wealth had been derived from such not-obviously-socialist enterprises as gold mining, billboard advertising and property speculation (Los Angeles's Wilshire Boulevard is named after him, having been built on land he donated—mainly to increase the value of his own properties in the area). Nevertheless, he regarded himself

64. Punnett 1907, v–vi.

65. Spargo 1909a, 80–81. Pittenger (1993, 164–65) notes that Spargo and Wilshire were moving in opposite political directions by 1909, when they each discovered Mendelism. See also Cotkin 1981, 277.

66. Review of Punnett, *Mendelism* 1909, 383–84.

67. Elbert Hubbard, *Philistine* 14 (December 1901): 18–21. Quoted in Mott 1957 (207). For millionaire socialists, see Reynolds 1974.

as a Marxist (and hung a large portrait of Marx over his desk).[68] His magazine's slogan, "Let the nation own the trusts," encapsulated his idiosyncratic interpretation of Marxism: socialism could not be a quasi-religion built on the hope that people were innately good but must be based on a scientific understanding of history's laws. Wilshire's view owed at least as much to Bellamy and Spencer as it did to Marx, in that he forecast that socialism would emerge inevitably from the economic forces unleashed by capitalism, including the growth of monopolies (especially America's trusts).

For Wilshire, part of the attraction of Mendelism was its mathematical precision; Mendel's ratios and formulae allowed breeding experiments to be predicted, which meshed well with Wilshire's vision of social evolution—"You can count up the number of machines and count up the number of men, and can prophecy the time almost exactly when Socialism must come" (Wilshire published his collected editorials under the title *Socialism Inevitable*).[69] Such views were obviously compatible with the slow inevitability of stadial evolution, but Wilshire rejected it. Despite its title, *Socialism Inevitable* included his article "The Mutation Theory Applied to Society," which hailed de Vries's work as "epoch making" because it offered a riposte to those who accused revolutionary socialists of being unscientific (see p. 233). He made some familiar analogies between biology and politics (for example, that society had "mutated from feudalism to capitalism" in the past and would surely "mutate from capitalism to Socialism" in the future). The change would be both rapid and abrupt—"Society must jump from capitalism to Socialism"—because monopoly capitalism (exemplified by the growth of trusts) would result in overproduction and mass unemployment, "and then like a bolt from the blue, will come the crash, leaving society to mutate into Socialism or die."[70] In response to readers who asked why they should struggle for socialism if its arrival was inevitable, Wilshire responded that "Socialism will not come without our working for it any more than the egg would be hatched unless the chick worked itself out of its shell." The effort was essential, not least so the workers would realize that socialism was possible. He acknowledged that the economic preconditions for socialism had to evolve over time, but once humanity was "ready to be hatched from the shell of capitalism" it would take a revolutionary fight before the workers could be born "into

68. Nelson 2014, 43.
69. Wilshire, in Punnett 1909 (3–6); Wilshire 1907.
70. Wilshire 1905b (1907), 250, 259, 257.

the new life of Socialism."[71] Organicist metaphors were nothing new in socialist thinking, but mutation and Mendelism gave a distinctive twentieth-
century flavor to Wilshire's idiosyncratic interpretations.

Wilshire's innovative appropriation of new biology reveals some of
the complexities of analyzing how socialists used the mutation theory.
Experimental evolution provided new ammunition for one of the socialists' favorite pastimes—attacking other socialists—but it also produced
unexpected links between thinkers as different as Wilshire and Ramsay
MacDonald, both of whom interpreted mutation as the dramatic outcome
of small accumulated changes.[72] The mutation theory was even used by
Spargo to try and effect a reconciliation between the two wings of the
US socialist movement. He described his pragmatic approach as "applied
socialism" and described the way he thought socialists should bring the
"irresistible evolutionary process" under conscious control by the "exercise of human directive energies" (an argument that seems indebted to the
optimism around experimental evolution).[73] Spargo expounded his view
by discussing recent biology, "particularly the 'mutation' theory of Hugo
de Vries and the rediscovery of the Mendelian law," which made rapid evolution plausible. Socialists who supported revolution had used the "work
of Mendel and De Vries . . . to add to the strength of their position," by
arguing for the inevitability of violent revolution. Despite not being a
revolutionary, Spargo echoed Kautsky's claims that such arguments were
"a too literal application of the laws of biology" to society (even if "the
entire primrose family had been transformed by a single mutation," that
would not prove that "society can likewise be transformed by a sudden
revolution, or mutation"). He argued that the phrase *social revolution* only
described how *complete* the transformation would be, not how swiftly or
violently it would come about. Socialists should "reject the catastrophic
theory" of violent revolution and realize that the work of de Vries and
Mendel was a continuation of Darwin's; species evolved by accumulating
small changes that sometimes led to dramatic ones. It was therefore more
realistic to see socialism arising through "a *series* of mutations" (original
emphasis). There was "nothing in the theory of evolution by mutations

71. "A psychological problem," in Wilshire 1907 (33–34).

72. The tipping-point analogy was also used to explain the emergence of new species, initially by Francis Galton. It became a familiar analogy for mutation theory;
Morgan (1903b, 289), e.g., argued that de Vries recalled Galton's "apt comparison
between variability and a polyhedron which can roll from one face to another." It
resisted smaller disturbance, but larger ones "may cause the polyhedron to roll over
on to a new face" and that "new position corresponds to a mutation."

73. Spargo 1912a, 19–20.

that is incompatible with the Marxian theory of social evolution"—and hence no reason for socialists to see abrupt and gradual change as incompatible (28–29, 30).

However, reconciliation between revolutionary and evolutionary socialists was not easily achieved. Spargo had made similar arguments when he reviewed Jack London's novel *The Iron Heel* in the *International Socialist Review*, arguing that the book represented a return to "the old and generally discredited cataclysmic theory," and would thus weaken the socialist's electoral efforts.[74] This view was energetically disputed by another socialist writer, Robert Rives La Monte, who took Spargo to task for his seeming ignorance of current science. La Monte argued that "Nature has always utilized both methods—evolution and revolution," but in recent years, scientific interest in slow, gradual change was fading to the point that "our most advanced scientists" were increasingly sure that "nature makes nothing but leaps." De Vries (whose work was described as "cataclysmic biology") and Mendel were both cited as evidence of this trend (and La Monte recommended Lewis's *Evolution: Social and Organic* and the Wilshire edition of Punnett's *Mendelism* to those who wished to know more).[75]

La Monte elaborated his points by explaining the shift to "cataclysmic biology" as evidence of Marx's dictum that a society's ideology (including its science) was determined by its socioeconomic system. This claim was proved by the fact that Mendel had published his work in the nineteenth century, but it had been ignored. There was "as it were, no market for the discovery that the raw material for natural selection to work upon must consist of 'leaps.'" By contrast, the twentieth century had seen an increasing "'demand' for cataclysmic theories, so science began to furnish the 'supply.'" Why the change? The importance of Mendel's work had to wait

until the German Social Democracy had polled more than 1,000,000 votes and the British Labor Party had returned more than a quarter of a hundred members of Parliament, that a leading British biologist [i.e., Punnett] began teaching that mutations were "the main, if not the only, basis of evolution."[76]

74. Spargo 1908, 629.
75. An editorial note appended to La Monte's article informed readers that Punnett's book had been recently published by the Wilshire Book Company. It showed that the "theories of Mendel regarding organic evolution are in striking accord with the Marxian views of the development of capitalist society" and so the book should be read by "all Socialist students" (La Monte 1909, 8–9).
76. La Monte 1909, 8.

This optimistic (if rather implausible) history asserted that the growth of socialism, mutationism, and Mendelism were all aspects of the same revolutionary shift: a new class, the proletariat, was coming to power and its ideologies were challenging those of the bourgeoisie.[77]

Experimental evolution offered socialists both fresh ammunition for their internecine disputes and new ways to imagine a socialist future. Some focused on the same things that had excited other fans of the new science: new plants and the promise of an end to hunger, which also seemed to bring socialism a little nearer. American socialists (like so many of their fellow citizens) occasionally evoked Burbank to illustrate that promise. The *International Socialist Review* examined Harwood's *New Creations in Plant Life* shortly after it appeared, and while they criticized Harwood for his "bombastic, sensational" writing and lack of expertise (such as claiming that Burbank had overthrown "all the laws of science"), the magazine's reviewer saw marvelous possibilities in Burbank's achievements, notably the spineless cactus. They welcomed Harwood's observation that Burbank had disproved those "pessimistic theorists" who believed humanity would starve as its population expanded (which completed "the grave of already long dead Malthusianism").[78] Meanwhile, in Britain, the Social Democratic Federation's newspaper *Justice* reviewed Punnett and commented that Mendelism was of particular interest to socialists because of the "the possibilities it conjures up . . . [and] the wonderful workings of nature that it opens to our astonished gaze." Their reviewer also conflated Mendelism with mutation, to argue that gradualism was being overthrown by revolutionary change (Mendel's conclusions "support the mutation or 'jump' theory . . . the production all of a sudden of a distinct species").[79]

Whether represented by Burbank or Punnett, de Vries or Mendel, a key part of experimental evolution's appeal to socialists was that it finally laid Malthusianism to rest—scientific ingenuity would disprove his dismal anti-utopian forecasts.[80] New plants seemed to provide new support for Marx's arguments about technological and social change creating material abundance. Conventionally, such claims rested on the revolutionary potential of the physical sciences and the technologies they produced,

77. La Monte's argument echoed those of the influential Soviet theorist Georgi Plekhanov. As Diane Paul (1979, 125–26) has shown, the idea that mutation theory was somehow more "dialectical" than orthodox Darwinism briefly held sway in Soviet Marxist circles. See Plekhanov 1907 (1976); 1929, 104.

78. Harwood 1905b, 217; [Simons] 1906, 508.

79. Reprinted as Quelch 1909.

80. [Simons] 1906, 507. The review was unsigned, but Pittenger (1993, 132–35) credits it to Algie Simons, who edited the *Review* at this time.

but experimental evolution promised a parallel biological revolution (as Haldane would suggest). The *International Socialist Review* quoted Burbank's claim that his methods had opened up enough possibilities "for the government to put at work a thousand experts," who would change "the whole face of nature" through "intelligent, patient and systematic" plant-breeding.[81] Their writer seized on this point to argue that Burbank's methods were evidently "ill suited to capitalist methods of business, but could best be carried on by governmental departments."[82] A socialist government could direct research so that it served the needs of the many, not just the greed of the few; Harwood's book about Burbank and modern American scientific plant-breeding, *The New Earth*, directly inspired Vladimir Lenin to inaugurate scientific agricultural research in the infant Soviet Union.[83] And, as Luis Campos has shown, the claims about remaking nature being made by American biotopians like Burbank were strikingly similar to those of the Stalinist biologist Trofim Lysenko and his mentor, I. V. Michurin (for example, "We must not wait for favours from nature; our task is to wrest them from her," since "Man can and must create new breeds of plants better than Nature").[84]

Among the many American writers who used Burbank to connect the promises of experimental evolution to specifically socialist goals was the extraordinary twelve-year-old socialist orator Queen Silver. She attacked the anti-evolutionary opinions of William Jennings Bryan, arguing "that *Bryan does not know what he is talking about*" (original emphasis) because Burbank had proved that new species could arise from existing ones.[85] Silver offered a typically progressive vision of evolution (today's human resembled "the ape" they had once been, rather than "the man he will become after a few million more years of Evolution"), but she was not willing to sit back and wait for evolution to make the change. Silver argued that human control would bring the exciting socialist future closer: all that was

81. In Harwood 1905b (59).

82. [Simons] 1906, 509.

83. Vavilov 1992, 284–85. See also Campos 2017. For the biology and science fiction in the USSR, see also Krementsov 2014; 2015.

84. Lysenko quoting I. V. Michurin, in *The Situation in Biological Science* (Lenin Academy of Agricultural Sciences of the USSR; Moscow: Foreign Languages Publishing House, 1949): 34, 86. Quoted in Campos 2017 (163–64).

85. Silver 1923 (2000), 179. See McElroy 2000. Among the books on evolution that Silver recommended was Marshall Gauvin's *Illustrated Story of Evolution* (1921), which offered the "plant creations of Luther Burbank" (118) as one of its proofs of evolution. Silver's book list appeared in *Queen Silver's Magazine* 1, no. 3 (2nd quarter, 1923): 8, where "Monkey to Bryan" first appeared.

required was "putting man's brains behind natural laws, and forcing them to operate intelligently, toward a certain goal instead of blindly." Burbank proved that evolution could be accelerated; his corn-breeding, for example, had accomplished in seventeen years what nature "required thousands of years to do."[86] Wilshire made similar efforts to recruit Burbank's work to the socialist cause. He characterized the Californian breeder's well-known bonfires as "selection of the fittest according to what man thinks best," as opposed to leaving the matter to the struggle for existence. Only human standards could produce plants, like the thornless cactus, that served human needs but could never have been the result of natural selection. This claim led Wilshire to a distinctly socialist interpretation of Burbank's achievement; under capitalism—as in nature—survival does not go to the "soundest, sweetest and most beautiful fruit on the tree of humanity, but the one with the most thorns, the thickest skin and hardest heart." But, if capitalism's remorseless competition were to be abolished, hard hearts would be as redundant as thorns. Wilshire interpreted Burbank's *Training of the Human Plant* as showing that "wonderful changes might be made in man" if only children were as "carefully reared and protected" as Burbank's plants. Burbank's Lamarckian view was that better environments created both better plants and better people, and Wilshire seemed to be agreeing with it when he quoted a recent report on the appalling condition of children's health in New York. However, Wilshire argued that by itself Burbank-style Lamarckianism would be too slow; experimental evolution would speed things up—and so, he concluded, "our children should not fear a mutation into a different life from the present."[87] This eclectic version of biology (which had echoes in the work of Charlotte Perkins Gilman, discussed below), illustrated how incongruous mixtures of incompatible biological ideas might be held together by nothing but biotopian optimism.

Herland

In 1915, three men discovered a lost world. On a high plateau only accessible by aircraft, they found a land inhabited only by women, who had nevertheless been giving birth to successive generations of daughters for almost two thousand years. When the explorers learned that the whole population was descended from just one woman, they naturally asked

86. Silver 1923 (2000), 176, 179–80.
87. Wilshire 1905b (1907), 258–59.

FIGURE 7.3. Charlotte Perkins Gilman mixed various elements of recent biology to add plausibility to her vision of a utopia without men in *Herland*.

how, then, the women had become so different from one another; they were told it was due "to the law of mutation."

These mutant women were the inhabitants of *Herland* (1915), a utopian novel by Charlotte Perkins Gilman, which utilized various aspects of then-current biology to explore the possibilities of a world without men (figure 7.3).[88] It first appeared in serial form in the magazine the *Forerunner* (which Gilman published, edited, and wrote), whose circulation never exceeded 1,500; as a result, Gilman's story remained largely unknown for

88. Gilman 2013. All subsequent page references are to this edition.

half a century.[89] Nevertheless, the novel is evidence of yet another reader making imaginative use of the possibilities unleashed by early twentieth-century biology, and is particularly interesting because Gilman (in contrast with almost all the male writers discussed in earlier chapters), used biology to attempt a radical reimagining of gender roles and relations.

In Gilman's story the three explorers—Van, Terry, and Jeff—named the country they discover Herland. After making contact with its inhabitants and being taught their language, they learned its history. More than two thousand years earlier, while most of the men were away at war, a volcano and earthquake blocked the only accessible pass. Most of the few remaining men were slaves, who revolted and killed the remaining masters, hoping to enslave the women, but they revolted in turn and slew their would-be oppressors. As a result—in the narrator, Van's, words—there "was literally no-one left on this beautiful high garden land but a bunch of hysterical girls and some older slave women." The women assumed that extinction must await them, but then "the miracle happened" when one of the surviving women bore a child. They assumed at least one man had survived, but none could be found. Nevertheless, their "wonder-woman" bore a series of daughters, each of whom inherited their mother's miraculous ability to reproduce asexually, eventually giving rise to a wholly female nation. The fact that the women evolved into a new species overnight suggested that the new ability was a mutation, which had appeared in response to "the stress of a new necessity" (83). (The explanation echoed de Vries's speculation about why *Oenothera* had begun to mutate).[90] Imaginary futures often relied on some form of time travel: Wells was the first to imagine making such journeys in a machine; Bellamy achieved the same effect by having his protagonist sleep for over a century (creating the time needed for slow social evolution); but Gilman used a parthenogenetic mutation to hurl her women abruptly into an alternative future.

Gilman used up-to-date biology to add a degree of credibility to her writing (which was to become a common use for mutation in the next few decades; see chapter 8). She addressed the story's scientific basis in the

89. The *Forerunner* got a few reviews, in the *Chicago Evening Post*, the Chicago *Dial*, and the *Vegetarian*. Both Eugene Debs and Upton Sinclair quoted from it in public (Scharnhorst 1985, 85).

90. De Vries suggested that some combination of external conditions, possibly connected to the plant having been transplanted to new growing conditions, could have jolted *Oenothera* into a mutation period. He noted (1905c, 706) that the similarities between different mutants suggested "their origin to common external agencies." Such suggestions led to various attempts to induce mutations; see Endersby 2013, 480–81; Campos 2015; Endersby 2018, 454–55; Curry 2016.

Forerunner, when she noted that "several subscribers have asked if there is any foundation in biology for the condition of parthenogenesis—virgin birth—alleged in Herland." She admitted it was not possible in humans but said she had used the device "to bring out as clearly as possible the essential qualities of a purely feminine culture."[91] Nevertheless, her choice was not arbitrary; parthenogenesis was causing excitement in the biological community at the time, primarily because Jacques Loeb of the University of Chicago had apparently been able to induce it artificially. Loeb's achievement appeared to be the first step toward creating completely artificial life, and it was widely discussed in the popular press (Garrett Serviss, who also wrote extensively on Burbank [see p. 110], told the readers of *Cosmopolitan* that science might not yet be able to make an egg, "but it has done something hardly less surprising—it *has discovered a way to fertilize the egg without sperm!*").[92]

Gilman demonstrated her biological expertise by telling her readers that asexual reproduction was widespread among insects, such as bees and aphids, but was a primitive form of reproduction, which reduced the variability of a species and would thus be undesirable in humans, even if it were possible.[93] (In the novel, Terry, who embodies the most traditional kind of masculinity, made exactly this point when he argued that there must be men somewhere, since if the Herlanders "were parthenogenetic they'd be as alike as so many ants or aphides [*sic*].") The harmful effects of asexual reproduction led the visitors to ask the Herlanders how they had achieved "so much divergence without cross fertilization," and the women's explanation revealed more of Gilman's biological thinking. They attributed their diversity "partly to the law of mutation. This they had found in their work with plants, and fully proven in their own case" (103). The reference to plant-breeding would have reminded readers of work like Burbank's, while the "law of mutation" invoked de Vries's work, perhaps hinting at the possibility of inducing further mutations.[94] However, Gilman's biological thinking was a complex form of bricolage (with some similarities to Wilshire's ideas, discussed above). In explaining their diversity, the Herlanders invoked "the careful education, which followed each slight tendency to differ." By exercising their minds and bodies they had

91. Gilman 1916a, 83. She may have derived the information from Geddes 1901, 192–93, a book she owned and regarded as important (Ceplair 1991, 88–89).

92. Serviss 1905a, 463. See also Pauly 1987 and Turney 1998, 67–72.

93. Gilman 1916a, 83.

94. Some commentators have assumed that Gilman was using mutation in the post-*Drosophila* Mendelian sense, e.g., Hausman (1998, 498). However, given *Herland*'s date this usage is most unlikely, as earlier chapters have shown.

gradually been able to produce heritable changes, allowing variations to be consciously developed in specific directions.[95] This Lamarckian belief in the inheritance of acquired characteristics was losing ground among some biologists, as Gilman acknowledged by having Terry remind the women that "acquired traits are not transmissible . . . Weissman [*sic*] has proved that." (The German biologist August Weismann's experiments were widely interpreted as disproving Lamarck's view.)[96] In response, the Herlander (Zava) simply comments, "If that is so, then our improvement must be due either to mutation, or solely to education. . . . We certainly have improved" (103).

The possibility that mutation and Lamarckianism might offer complementary sources of evolutionary novelty was explicit in Gilman's sequel, *With Her in Ourland* (1916). One of Herland's main characters, Elladar, accompanied her new husband Van on a journey into the wider world and was deeply critical of the oppression of women she discovered there. She argued that if—as seemed the case—men refused to marry innovative and rebellious women, the trait for innovation would be bred out of the female population. (Restoring the ability of women to choose their mates was another important strand in Gilman's biologically infused feminism, analyzed below.) Leaving men to do the selecting "means extinction—the end of that variety of woman," because the men have "successfully checked mutation in women," eliminating a major source of novelty. At the same time, they reduced the other source of novelty, Lamarckianism, by limiting women's education.[97] Gilman arrived at her own, idiosyncratic view of biology's implications, but her synthesis of mutation and Lamarckianism shared qualities with the work of her contemporaries (as noted, Bellamy, Wilshire, and Burbank had similar views, as did the sociologist Lester Ward, whose work influenced Gilman). Although experimental evolution was prompted by mutation and Mendelism, the excitement it created spilled over into other kinds of biology; Lamarck's ideas were over a century old, but writers like Gilman reimagined them as part of speculative heredity's toolkit.

Whatever their precise theoretical bases, the Herlander's achievements were primarily the result of their expertise in scientific agriculture.[98] As

95. For Lamarckianism, see p. 12. The women's choice of words in the novel suggests Gilman may also have been interested in Bergson's creative evolution, which was widely discussed in the US at this time; see Pauly 2000, 199.

96. For Weismann's work and its apparent political implications, see Hale 2014, 252–300.

97. Gilman 1916c (1997), 71–72.

98. Christensen 2017, 286.

the men commented when they first arrived, it was "a land in a state of perfect cultivation, where even the forests looked as if they were cared for; a land that looked like an enormous park, only it was even more evidently an enormous garden."[99] Van commented, "This is a *civilized* country! . . . There must be men," but the landscape provided the first of many challenges to the men's preconceptions. The women had used plant-breeding to subordinate nature to human standards by working out "a system of intensive agriculture" that exceeded anything any society of men had ever accomplished.[100] Early in their history, the women had decided that the best use of their small country was to grow only trees, so "the very forests [were] all reset with fruit or nutbearing trees" (94). However, this policy created a dilemma over one beautiful tree—whose nuts were inedible. The women's aesthetic judgment led them to spend nine hundred years carefully breeding the tree until it produced the edible crop that justified its continued cultivation (104). As Van summarized, the women were as "proud of their record of ever increasing efficiency; they had made a pleasant garden of [their country], a very practical little heaven" (118), in which beauty was as essential as food. Untamed nature had not been allowed to dominate.

The Herlanders shared a single driving passion—motherhood. As one of them, Moadine, explains to the men, the "children in this country are the one center and focus of all our thoughts" and repeated "we are Mothers"—and Van mused "as if in that she had said it all" (92). Herland's children were brought up in ways Burbank would surely have approved: they "grew up as naturally as young trees" and were raised largely out of doors, "naked darlings playing on short velvet grass, clean-swept." The country had been engineered to provide the perfect environment for its children, whose whole education took place "in an environment which met their needs; just as young fawns might grow up in dewy forest glades and brook-fed meadows. And they enjoyed it as frankly and utterly as the fawns would" (125, 122–23).[101] Despite this Edenic vision, Gilman clearly treated "nature" as a cultural category (a carefully cultivated, artificial landscape is the best place to raise children "naturally"). Her approach allowed Herland's unnaturalness to be deployed to challenge supposedly natural gender roles, as when the visiting men discover that children were usually raised by specialized professionals, not by their birth mothers. Van

99. Park-cities were common in late nineteenth-century American utopias (Burt 1981, 179).
100. Christensen 2017, 287–88.
101. Ward 1913, 741, 753–54.

admitted to a feeling of "cold horror," but the Herlanders believe raising children is too important to be left to anyone without a special aptitude (one of them compared their childcare workers to dentists in the outside world—a mother's love, however strong, would not qualify her to fill her own children's cavities). This argument did not impress Terry, who observed, "The whole thing's deuced unnatural. . . . And an unnatural condition's sure to have unnatural results" (107–8, 106). By contrast, the Herlanders seem to have accepted that both disease and decaying teeth were entirely natural—avoiding them required overcoming nature and achieving "unnatural results."

The Herlanders' divergence from supposedly natural gender roles leads the men to compare them to the insects whose asexual reproductive system they share. When Terry insisted (despite all the obvious evidence to the contrary) that women could never cooperate with one another, Jeff countered with the evidence of social insects, whose worker castes are all female (and products of parthenogenesis); "Don't they co-operate pretty well?" Jeff asked, adding, Herland is "just like an enormous ant-hill—you know an ant-hill is nothing but a nursery" (93). All the women of Herland are literally sisters, which made it easier for some to voluntarily forgo childbearing in order to protect the country from overpopulation (as one of them explained, they "each have a million children to love and serve—*our* children"). Jeff could see the utopian aspects of this state of affairs, but Van struggled to imagine how the women felt, eventually admitting, "I suppose that is the way the ants and bees would talk" (97). (Insect societies have a rich history as metaphors for human ones. Beehives, for example, have often been used as a model of well-ordered industriousness. Utopian ants are rarer—ant-heaps have more often been models for dystopias—but of course Herland is a dystopia—for a man like Terry.)[102]

The unnaturalness of Herland was vividly portrayed when Ellador tells Van how she first became interested in her specialization, forestry. As a child she had caught a strange butterfly and learned it was an Obernut moth, potentially very destructive of one of Herland's most cherished trees. The teacher who identified it added, "We have been trying to exterminate them for centuries," and Ellador and the other children were

102. For examples of utopian bees, see Bernard Mandeville, *The Fable of the Bees* (1724). See "The Beehive as a Model for Colonial Design," in Kupperman 1995. Gilman herself hinted at the analogy in her story "Bee Wise," in Gilman 1999. For ants, see Sleigh 2007 and Chew 2019.

promptly set to work to find and kill any others(123).[103] Given that Herland's children had been raised in the most supposedly natural way possible, their aggression is rather startling; Gilman plainly did not assume that
their upbringing would create a mystical affinity with nature. Gardening
conflates nature and culture; the inhabitants of a biotopia do not adapt
themselves to nature but constantly force nature to adapt to them.

Faith in Evolution

Herland's male visitors discover the women have no domestic animals
other than cats, who are retained to control mice and similar "enemies
of the food supply." However, the Herlanders find the mating cries of
tomcats irritating, and Van asks the reader, "What do you think these
lady Burbanks had done with their cats?"—"By the most prolonged and
careful selection and exclusion they had developed a race of cats that did
not sing!" (and although they still purred, they did not kill birds [76]).
Even the deepest instincts could be remade to suit human convenience. By
imagining the women doing rapidly what "Mother Nature" did slowly (or
could not do at all), Gilman dramatized her resistance to the long-standing
assumption that women should be considered part of nature, while only
men create culture (including science and technology).[104]

Andrew Christensen has argued that *Herland* cannot be considered
an ecological or ecofeminist utopia because the women have liberated
themselves by dominating nature. They have made nature serve their
needs—just as Bacon prescribed.[105] Western science has long been
characterized—in highly gendered terms (the very title of Wells's *Men
Like Gods* and its "adolescent sons" calling Mother Nature an "old Hag"
epitomized science imagined as indelibly masculine). The Baconian scientific tradition envisaged men seeking power over nature, partly in an effort
to usurp the female power to create life. Given this history, it is initially disturbing to read about Herland's children eagerly exterminating unwanted
species, but perhaps the unease is the product of an unexamined assumption that women ought to be "naturally" in tune with nature.[106] Gilman's
Herlanders exhibit an instrumentalist view of nature—a collection of raw

103. Some scholars (e.g., Shishin 1998, 111) have described Herland as an ecological utopia, but recent criticism has challenged this interpretation; see Knittel 2006;
Christensen 2017.
104. Ortner 1972.
105. Christensen 2017, 293–94.
106. Some critics nevertheless seem determined to read Gilman's book as promoting harmony with nature or suggesting that women are somehow natural ecologists;

materials to be used. Jennifer Hudak has linked Herland's artificiality to Thomas Huxley's argument that nature offers no ethical guidance (chapter 5). Just as Huxley argued that natural selection's hostility to human values was best resisted by gardening, Gilman compared Herland's landscape to a park: a landscape that rejects nature's weeds, its pests—and its gender roles.[107] The Herlanders apply expert scientific skills to managing everything—including their own "natures," because they are as committed as Wells's Utopians to the "revision and editing" of what nature provided.

Nevertheless, despite the similarities between Gilman's and Wells's visions, she would surely have disagreed with Wells's Utopians when they told their Earthling visitors to renounce their faith in "Nature" and "Evolution," since these were simply names for "a Power beyond your own which excuses you from your duty," implying that humans can only end their childhood when they take full responsibility for their decisions (see p. 186).[108] For Gilman, nature's ends had to be critically re-examined but could not be simply ignored. She expressed her view as having a "faith in evolution," especially since a "great many people nowadays" have lost their faith in providence. They could instead trust in "the force called Evolution," which was always "pushing, pushing, upward and onward."[109] Gilman's exhortation, "Let us have faith in Evolution," would seem to restore nature's normative role, which sits incongruously alongside the denaturalizing elements of her argument. To understand this apparent paradox, we need to look more closely at what kind of evolution she believed in.

One of Gilman's earliest publishing successes was the poem "Similar Cases" (in the Bellamyite *Nationalist* magazine, 1890), an evolutionary satire in which various creatures are mocked for their desire to evolve into something higher.[110] The Neolithic man's vision of humanity's future is dismissed as "Utopian! Absurd!" and he is told that before such things can happen, "*You must alter Human Nature!*" This refrain recurs throughout the poem, which concludes that it "was a clinching argument" but only to "the Neolithic Mind"—for twentieth-century thinkers, experimental evolution had made altering human nature a practical possibility.

Gilman rejected the idea that stadial evolution meant simply letting nature take its course; like her fellow socialists (including Wilshire and

---

see Keller 1984, 264; Gilbert and Gubar 1989, 75; Graham 1998, 118–19; Shishin 1998, 111. This view has been disputed by Knittel (2006, 49) and Christensen (2017).

107. Hudak 2003, 466–68.

108. Gilman 1910, 28.

109. "Having Faith in Evolution," *Forerunner* 6, no. 11 (November 1915): 299–300, in Gilman 2013 (234–35).

110. "Similar Cases," in Gilman 1895 (72–76). See Scharnhorst 1985, 22–23.

Queen Silver), she argued that nature needed human help and so (as in the case of Herland's Obernut moth), people should "discuss the merits" of each organism "according to its 'use' to us"—and act accordingly. She explained the idea more fully in an article tellingly titled "Assisted Evolution" (1916), which noted that Luther Burbank "in part of one lifetime" had done more than Mother Nature had managed in "long, long ages." Humans need not accept evolution's conventionally slow tempo. It had "taken Mother Nature long, long ages to turn fierce greedy hairy ape-like beasts into such people as we are," but if humans were to emulate Burbank and take charge of evolution, it would only require "but two or three close-linked generations to make human beings far more superior to us than we are to the apes."[111] Part of her argument for trusting evolution was that such trust had "already done wonders" in the hands of plant and animal breeders who had nevertheless intervened to push evolution into creating new organisms. The same principles could just as easily be applied to humans. "We must provide right conditions," she asserted—guiding nature into the channels we chose—and then "that great pushing life-force of Evolution does the rest" (a view that echoed those of Burbank and of Orator Cook).[112]

Gilman had faith in evolution as a progressive force that could be harnessed and directed. Yet she rejected the core assumptions of naturalized ethics, primarily because she understood gender as more cultural than natural. The women of Herland possess no traditional feminine attributes (much to Terry's disgust), but the more enlightened Van realizes that "those 'feminine charms' we are so fond of are not feminine at all, but mere reflected masculinity—developed to please us because they had to please us" (85). In the absence of men, they would obviously disappear. Gilman's views contrasted with those of most male social Darwinists (who assumed that supposedly natural laws should govern society). A cultural feminist, she argued that men had usurped women's natural biological power and so what she called the "androcentric" world was anything but natural.[113] Gilman explained her views on gender and its relation to evolution in her best-known book, *Women and Economics* (1898), which she claimed had no bibliography because it used only two sources: Patrick Geddes

---

111. Gilman 1916b, 5.

112. "Having Faith in Evolution," *Forerunner* 6, no. 11 (November 1915): 299–300, in Gilman 2013 (234–35). Gilman's mentor, Lester Ward (1913, 746), also argued that there was "universal upward tendency in nature," an energy that was both "constructive and creative." See also Lears 2009, 239–42.

113. Deegan 1997, 31–32.

and J. A. Thomson's *Evolution of Sex* (which applied Darwin's theory of
sexual selection to human society) and Lester Ward's gynocentric theory
of evolution (which argued that the female was the primary sex throughout nature).[114] Gilman's book was hugely successful, going through nine
editions in the USA and UK and being translated into seven languages.
It helped to make her one of the USA's best-known women intellectuals
(when H. G. Wells visited America in 1906, he mentioned Gilman as one
of the American "curiosities" he was anxious to see).[115]

Lester Ward had first published his ideas in an article for a nonspecialist
magazine in 1888, in which he argued that the ability to reproduce must
have evolved before separate sexes did; an asexual organism was thus "in
all essential respects a mother," and so the female was the essential "type of
life," with "the insignificant male appearing to be a mere afterthought"—
evolving later and dependent for his survival on the female's reproductive power. As a result, Ward believed it was "from the steady advance of
woman rather than from the uncertain fluctuations of man that the sure
and solid progress of the future is to come."[116] In *Women and Economics*,
Gilman summarized Ward's argument as having "clearly shown the biological supremacy of the female sex." Conservatives claimed feminism
would halt human progress, which supposedly depended on women fulfilling their reproductive role. Gilman responded by citing Ward's claim
that it was only males for whom biology was destiny; the female was the
essential sex. And yet in human societies, economic inequality meant that
women were being "sacrificed not to reproductive necessities, but to a
most unnecessary and injurious degree of sex-indulgence under economic
necessity"—prostitution was merely the least-camouflaged version of the
general relationship between men and women.[117] Ward's ideas clearly lay
behind the Herlanders' discussion of the animals with two sexes with
which they are familiar. They tell the men that among birds "the father
is as useful as the mother," but when it comes to mammals like their cats
(and, by implication, humans), the "father is not very useful" (75). The
parthenogenetic women have rediscovered the original primacy of the
asexually reproducing female.

114. Gilman made the claim in her autobiography; however, Mark Van Wienen
(2003, 622–24) has argued that she was slightly oversimplifying her intellectual
influences.
115. Ceplair 1991, 90; Hamlin 2014a, 94–96.
116. Ward 1888 (2013), 264, 266–67.
117. Gilman 1915, 171–72.

Gilman's use of Ward's work was a distinctive feature of *Women and Economics*, but in other respects the book was typical of late nineteenth-century feminist interpretations of Darwinism. As Erika Milam has shown, several Victorian feminists used the human reversal of sexual selection to support their claim that women's place in Victorian society was unnatural—and would be much improved by returning to a more natural condition.[118] Gilman built on the pioneering work of Antoinette Brown Blackwell (*The Sexes throughout Nature*, 1875), who argued that because Darwin had strengthened the link between humans and other animals, the relations between the sexes in nonhuman animals provided a natural model with which to reform those between humans (for example, Blackwell noted that nonhuman species did not prevent females from contributing to its survival when they became mothers, and concluded that human pregnancy and childbirth should be regarded as unproblematic natural processes, not as a kind of illness).[119]

Arguments like Blackwell's led Gilman to think deeply about the wider significance of evolution. Like many Victorian women, she was particularly interested in sexual selection, which Darwin used to explain some otherwise inexplicable features of organisms.[120] Darwin once admitted that "the sight of a feather in a peacock's tail, whenever I gaze at it, makes me sick!"—because natural selection seemed unable to explain the evolution of such an extravagant—but apparently useless—feature.[121] Gilman agreed, noting that the ornaments of male animals were of no use "in self-preservation" (a male's "mane or crest or tail-feathers . . . do not help him get his dinner or kill his enemies," and might even make him more vulnerable).[122]

In the *Descent of Man* (1871), Darwin had noted that "the long train of the peacock . . . must render them a more easy prey to any prowling tiger-cat than would otherwise be the case." How could natural selection explain such features? Darwin's proposed solution relied on the fact that peacocks "display their attractions with elaborate care in the presence of the females," almost always "during the season of love," and it seemed impossible, he wrote, "that all this display should be purposeless."[123] He

118. E.g., Eliza Burt Gamble, *The Evolution of Woman: An Inquiry into the Dogma of Her Inferiority to Man* (1894), cited in Milam 2010, 24–26.

119. Hamlin 2014a, 97, 102.

120. For women's responses to sexual selection, see Richards 1983; 1997.

121. Darwin to Asa Gray, April 3, 1860, Darwin Correspondence Project letter no. 2743, https://www.darwinproject.ac.uk/letter/DCP-LETT-2743.xml.

122. Gilman 1915, 32–33.

123. Darwin 1871, vol. 1, 97, 399.

therefore assumed that the peacock's tail must be a sexual ornament, which had evolved because there were random variations in both the tails of the peacock's ancestors and in the proto-peahen's preferences. Perhaps the females were attracted to big tails because they were usually attached to large and vigorous males, but their preference could have been completely arbitrary—the effect would have been the same. The big-tailed males would have a greater chance of mating with the big-tail-fancying females—and would produce numerous offspring who would inherit either their father's large tail or their mother's preference for large tails. As long as the big-tailed males mated early and often enough, it would not matter that prowling tiger cats might devour them before they reached a ripe old age. The rules of sexual selection's game were simple: live fast, die young (if you have to), but leave a good-looking body of big-tailed sons and big-tail-fancying daughters. If Darwin was right, the result would be a kind of runaway selection that explained the extravagant structure of the modern peacock.[124]

As Evelleen Richards has noted, Darwin modeled male choice on the practices of animal breeders; sexual selection was analogous to the work of an expert, male pigeon-breeder, while human wives were pigeons—passive objects of male scrutiny.[125] Like many of Darwin's metaphors, this one reflected prevailing Victorian cultural standards. Nineteenth-century fashions, for example, encouraged women to display themselves advantageously in order that men could choose (fashionable hats, bedecked with plumes, were a key aspect of those "feminine charms" that the Herlanders lacked [77]). Gilman argued that these displays were a corruption of nature: in contrast to most species, men's economic dominance gave them the power to select while forcing women to "compete in ornament," often by bedecking themselves in the feathers of *male* birds. This reversal of sexual roles was, she argued, "another sign of excessive sex-distinction," and she condemned the way that forcing "elaborate ornamentation" on girls "interferes with their physical activity . . . and fosters a premature sex-consciousness." It was a "menacing" foretaste of the utter dependency that awaited them in adulthood, which forced exaggerated sexuality on

---

124. For more on sexual selection, see Deutscher 2004; Cronin 1991; Endersby 2003; Milam 2010; Richards 2017.

125. Richards 2017, xxvi. Richards (2017, 162–63, 184) and Erika Milam (2010, 14–15) have shown how prevailing gender roles made it difficult for male naturalists to accept female mate choice as a significant evolutionary factor in evolution. Darwin believed it was important in nonhuman animals, but less so among people, while many of his contemporaries and later followers found imputing intellectual abilities to females of any species implausible.

women, creating "a race with one sex a million years behind the other."
As Sandra Gilbert and Susan Gubar comment, Gilman believed that
economic inequality turned every woman into a parasite, or as Gilman
summarized it: the male of the species "is her food supply."[126] Women's
"arrested development" meant she was "imprisoned" in an early period
of human evolution. Having been prevented from evolving as far as men,
women were stuck at what Gilman and her contemporaries would have
called a "savage" level of development.[127]

The assumption that different groups of people were at different evolutionary levels was the source of Gilman's racism. The women of Herland
are racially homogeneous (as Van notes, "there is no doubt in my mind
that these people were of Aryan stock, and were once in contact with
the best civilization of the old world" [80–81]), and Gilman's prejudices
against immigrants were expressed repeatedly in the novel's sequel. A few
years before these novels appeared, Gilman had published a "suggestion
on the negro problem," in which she proposed to conscript all unem-
ployed African Americans into a labor army, to be trained, educated, and
put to work improving the South ("an undeveloped country" that could
be improved by this "undeveloped race"). She acknowledged that African
Americans had been "forcibly extradited" from Africa, and that whites
were the "original offender" whose "injuries" to black people "greatly out-
numbered" those blacks had supposedly inflicted on whites, but she nevertheless insisted that the inferiority of African Americans was indisputable.
Her faith in evolution led her to argue that some African Americans were
already "self-supporting and well behaved," proving they could evolve rapidly, thanks to the "advantage of contact with our more advanced stage
of evolution." With this evidence in mind, the state should treat its black
population as it did "its poor, its defectives, or . . . its children," and she
offered the existing achievements of some African Americans as "proof
that social evolution works more rapidly than the previous processes of
natural selection."[128] This nightmarish racist proposal was predicated on
the assumption that there was a single scale of evolutionary progress (and
Gilman assumed not only that white people were at the top but that her
fellow New Englanders were the best white people).[129] Darwin had used
sexual selection to explain the diversity of the human species (local standards of beauty led to the evolution of unmistakable differences in the

126. Gilbert and Gubar 1989, 74.
127. Gilman 1915, 55, 70, 330, 22.
128. Gilman 1908, 83, 78–80.
129. Ganobcsik-Williams 1999, 28–29.

appearance of each variety of people), so the lack of sexual selection in Herland (and its isolation from all other human groups), would explain its racial homogeneity (something Gilman evidently thought was desirable).

As a result of the absence of men (and thus of sexual selection), the Herlanders had not evolved the "exaggerated" and "excessive" sexual characteristics that the three male visitors have come to expect. They are even more disconcerted to discover that the Herlanders they marry have no interest in sex other than for reproduction (Ellador asks Van whether married people in his world really "go right on doing this in season and out of season, with no thought of children at all?"—and makes it clear that she is not going to do so). Van is forced to acknowledge that what he had assumed was "physiological necessity" was no more than a habit, which he discovers he can unlearn. He eventually agrees with Ellador that excessive sexual desire simply fades when true human companionship is available (146–47).

Gilman asserted that—even without parthenogenesis—the distorted sexual relations created by economic inequality would disappear in a fully human world built on gender equality. One source for her argument was *The Evolution of Sex*, in which Geddes and Thomson argued that there had been an important shift in human evolution from emotional to intellectual motivations. As a result, human mate choice had become more rational and less emotional than in other animals, allowing humans to exercise a degree of control over their own evolution.[130] A similar argument was made by Alfred Russel Wallace, who shared many of Gilman's socialist and feminist commitments; he argued that empowering women, allowing them to become educated and economically independent, would let them exercise free mate choice. And once women were "free to choose," it was inevitable that "the worst men among all classes . . . will be almost universally rejected" (the argument Wells's Utopians used to explain why they had no need to punish antisocial behavior). Wallace averred that unrestricted female choice would "improve the character, as well as the strength and the beauty, of our race," which would make eugenic legislation unnecessary.[131] In effect, Wallace disconnected Darwinism from its Malthusian roots; socialism would replace the remorseless struggle for existence, while sexual selection would protect humanity from degeneration. (Richards has argued that Wallace's views were the unacknowledged

130. Geddes 1901, 285–86.

131. *Social Environment and Moral Progress* (New York: Cassell and Co., 1913), 151–53; Wallace 1890, 337. Both quoted in Milam 2010 (22). See also Stack 2003, 26–29; Hamlin 2014b.

target of Huxley's *Evolution and Ethics*; Huxley brought the "Malthusian Serpent" back into Wallace's socialist, feminist garden of Eden, so that the specter of overpopulation could again serve as the barrier to utopia.)[132]

Herland went beyond arguments like Wallace's. Gilman agreed about the need to restore the natural relations between the sexes but refused to let nature dictate utopia's values (by arguing for a reengineering of nature). The Herlanders remade everything from trees to cats according to their own design, and they enthusiastically exterminated species that got in their way. As Van comments, the women cared "for their country as a florist cares for his costliest orchids" (48)—and of course the costliest orchids are tropical ones (prized precisely because they look almost artificial and require expensive hothouses to thrive).[133] Herland is the quintessential Baconian garden, scientifically created to satisfy human standards, and its gardeners applied the same standards to themselves. The parthenogenetic mutation gave the women complete conscious control over their own reproduction, and "very early they recognized the need of improvement" so "devoted their combined intelligence to that problem—how to make the best kind of people" (86). Through unique imaginative forms of both positive and negative eugenics, the women have become beautiful, strong, athletic, fearless, and intelligent. There have been no criminals in Herland for almost six hundred years because women with antisocial characteristics have not been allowed to reproduce. The women's choices have even succeeded in changing the most fundamental human instincts (as easily as they modified those of their cats); as Van explains, the women had "no sex-feeling . . . or practically none" since two thousand years "of disuse had left very little of the instinct." This seemingly natural fading had been reinforced by conscious choice to eliminate the sex drive: "those who had at times manifested it as atavistic exceptions were often, by that very fact, denied motherhood" (107, 115).

At times, the Herlanders' willingness to embrace the unnatural echoes ideas like Haldane's: if even purple seas can come to seem natural, then so can a world without sex, where children are cared for by professionals, hunt moths to extinction, and play with silent cats. However, Gilman insisted that what she called "excessive sex-distinction" was a "perversion" or "distortion" of natural human instincts by which women "have been injured in body and in mind."[134] (And those Herlanders who were prevented from passing on their "atavistic" interest in sex could only have

132. Richards 2017, 509–13.
133. See "Sexy Orchids," in Endersby 2016 (157–83).
134. Gilman 1915, 339.

been expressing their desire for other women—behavior that Gilman was unwilling to endorse—in public, at least.[135]) Reinstating female mate choice would restore nature, yet her utopia argued that women should match or exceed men's ability to master nature.[136] Biotopianism could be defined as the acceptance that once you have decided that teeth should not be left to decay (nor cats be allowed to yowl), you have forfeited the right to condemn anything as "unnatural." Yet, Gilman would surely have disagreed with Haldane's claim that "every biological invention is a per-version" to be celebrated until accepted. He illustrated his argument with the example of the "radical indecency" of the dairy industry (see p. 172), and when Herland's visitors discover the country has no dairy industry they ask how the women manage "without milk?" The response is "*Milk? We have milk in abundance—our own*" (75–76). The women are appalled when they learn how calves are killed for food and to make the cow's milk available for human consumption—the natural use of milk (each species drinking its own) is preferred to the cultural perversions Haldane em-braced. Vegans will surely applaud, but other readers might find *Herland*'s relentless focus on motherhood as the most (almost the only) natural use of women's intelligence and energy uncomfortably reactionary. It is worth noting that the novel first appeared just as many men were arguing for a renewed "maternalism" (which became increasingly strident after World War I), so it is discomforting to find similar-sounding arguments in a feminist utopia that repeatedly challenged women's supposedly natural roles.[137]

Herland was as much a socialist utopia as a feminist one (the two were inseparable in Gilman's eyes), utilizing the latest biology to create the ease and abundance that socialism promised.[138] It was also a reminder that the biotopian garden was always about more than simply feeding its inhabitants—*human* nature was also going to be remade. This prospect created the possibility of a new kind of socialism, as Wilshire noted when he argued that recent science had provided socialists with an indisputable riposte to the common assertion that "you can't change human nature"

135. Gilman's biographer, Anne Lane, is one of several scholars who believe that some of Gilman's intense friendships with other women were almost certainly sexual (Lane 1997, 166).

136. Gilbert and Gubar 1999, 75; Christensen 2017, 291.

137. Donawerth 2009, 216; Weeks 2017, 139–41. Similar points could be made about Gilman's much-debated racism and opposition to immigration; she regularly seemed uncomfortable with what she saw as unnatural mixing of different human stocks. See Gilman 1923 (1991); Pittenger 1993, 73–74.

138. Van Wienen 2003, 603–8.

("a song we have heard many a time" sung by some and that "horrified some old conservative" every time socialists promise to improve the world and its people). According to Wilshire, the "human nature" objection was obsolete: "Burbank has shown us what can be done with plants," while de Vries proved that the "most astonishing leaps" were part of plant evolution. He argued that animal—and eventually human—nature would soon be subject to the same improvements, thanks to recent work by Swiss entomologist Max Standfuss (another of "our leading biologists" who disproved Darwin's claim that creating new species "required millions of years").

Standfuss had been able to transform butterflies as dramatically as Burbank had done plants, and he had also used a method "so very simple that any one can repeat Standfuss's experiments in his own home" (socialists were, understandably, drawn to the claim that even nonexperts could help shape evolution). By simply varying the temperature at which their cocoons were incubated, Standfuss produced butterflies that "are without a counterpart on this earth and which would normally have made their appearance thousands of years hence."[139] If experimental evolution could change evolution's tempo and mode and accelerate butterflies into the future, it could not be long before the same could be done for humans.

Wilshire concluded that Standfuss, Burbank, and de Vries were all part of a wider scientific rejection of slow, gradual change, which should "help convince certain minds of the possibility of Socialism in our own day."[140] As we have seen, Gilman also mocked the oft-repeated argument that socialism was impossible because selfishness and greed were ineradicable aspects of human nature, and she, too, used experimental evolution's potential to imagine a future in which both class and gender relations had been thoroughly reinvented. By contrast, few of the male socialist activists (or the more elite socialist biologists) even attempted to reimagine gender to any significant degree. Gilman made that leap, yet she seemed to have been unable to avoid the naturalistic fallacy when she argued that women were natural mothers, that sex was only for reproduction, and that a desire for continuous sexual indulgence was an unnatural aberration. Her "faith in evolution" was partly inspired by the idea of restoring the human sex instinct to its supposedly natural level (and she even argued that a supposedly excessive sex drive should be bred out—in the real world, not

139. Wilshire 1905a, 8. These words were part of several lengthy quotations from an unidentified source, which was Kaempffert 1905.
140. Wilshire 1905a, 8.

just in utopia).[141] Meanwhile, her condemnation of some expressions of sexuality as a "perversion" suggests that perhaps she found it difficult to completely shake what Frederick Jameson called "the mesmerizing (and crippling) prestige of the natural."[142] However, Gilman's views were more complex than such an analysis would suggest; men controlled science in Gilman's day and no feminist could be indifferent to the impact that male science could have on women's and children's lives. Gilman's utopia exemplifies Michelle Murphy's argument that feminist utopias dream "*with and through* (not just about) technoscience."[143]

New biologies and the technologies they offered made it possible to dream about new things (from speculative investment in Burbank's spineless cacti to a world without men). Such dreams involved new facts but also changed the ways people felt about the future. It is surely no coincidence that those who expressed unqualified confidence in the promise of biotopia were all white men (often wealthy and highly educated), confident that they would be the future masters of nature, not the victims of misguided scientific arrogance. As the next chapter shows, Gilman was not the only woman to imagine the future from a more critical perspective. Ruha Benjamin has argued that social change "requires novel fictions that reimagine and rework all that is taken for granted about the current structure of society. Such narratives are not meant to convince others of what is, but to expand our own visions of what is possible."[144] The underdogs' uses of biotopianism can be used to extend Benjamin's argument beyond the issues of race that she considers; much of the writing considered in this chapter—fiction and nonfiction—illustrates science being used to redraw the horizon of expectation.

141. Gilman, "Birth Control" (1915), in Gilman 2013 (230).
142. "Science Fiction as a Spatial Genre," in Jameson 2005 (307).
143. Murphy 2015.
144. Benjamin 2016, 2.

✳ 8 ✳

(Science) Fictional Futures

When young Americans wanted to know what the future would be like (and how soon it might arrive), one of the places they looked was the new, cheap "scientifiction" magazines that began appearing in the late 1920s. Such cheaply printed and shoddily written publications would probably have been largely ignored by fans of Edith Wharton or Charlotte Perkins Gilman, yet both writers shared common sources and ideas with science fiction and many other kinds of paraliterature.

For example, browsers at America's corner newsstands who picked up the Autumn 1928 issue of *Amazing Stories Quarterly* would have discovered "Stenographer's Hands," a new story by David H. Keller (figure 8.1). It told how a large industrial firm struggled to keep female stenographers or prevent those they hired from making endless errors. They hired an "eminent biologist and sociologist" to solve the problem. After months of research, he announced his solution: "We will secure better stenographers by breeding them!"—explaining that "Burbank bred a spineless cactus—we will breed errorless stenographers!"[1] By shortening the breeding cycle of the female stenographer and selecting for the most useful traits, the company used experimental evolution to fling its typists into the future, just as Gilman had done with the women she wrote about. Keller's fiction took "Burbanking" into fantastic new realms, yet lab-bred errorless secretaries were not that distant from some of Burbank's more uncanny creations (spineless cacti being only slightly queerer than palm/oak hybrids that grew apples and breadfruit; see p. 112). And the pulp writers regularly used the same sources as Wharton—such as textbooks and media reports—as the starting points for their interpretations of how science would shape the future.

---

1. Keller 1928, 524.

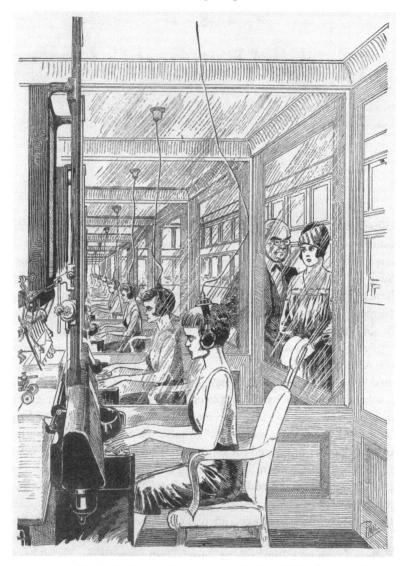

FIGURE 8.1. "Burbanked" secretaries, as imagined by David Keller (*Amazing Stories,* 1928).

The links between stories like Keller's, novels like *Herland,* and many of the socialist speculations discussed in the previous chapter demonstrate how Burbank's name had come to signify much more than the man who actually lived and worked in Santa Rosa. Pioneering SF writer Jack Williamson described Burbank (alongside Einstein, Edison, and Henry Ford; see figure 9.1) as one of the "actual public heroes who were scientists and technologists," whose work inspired him and other early SF writers—while helping them keep their feet on the scientific ground.[2] Yet the boundaries between these "public heroes" and their fictional personae were porous. Garrett Serviss (who also wrote popular scientific articles about Burbank; see chapter 4), wrote an unauthorized sequel to *The War of the Worlds* in which a fictionalized version of Thomas Edison leads a human force to conquer the red planet and exact humanity's revenge.[3] Edison's name brought a kind of plausibility to Serviss's tale, just as Burbank's did in Keller's ("Burbank" was to biology what Edison was to more conventional technologies).[4]

The use of real scientists' names (as Haldane did in *Daedalus*) helped establish the plausibility that is part of SF's appeal. Some hint of scientific explanation helps readers suspend their disbelief (and for some, is what separates SF from its close cousins, fantasy, horror, and the gothic—despite their overlaps). Some writers used science to make their fictions more believable, but the reverse also happened—SF tropes lent a different kind of persuasive quality to many nonfiction sources. Haldane used the time-traveling device of a report from the future to frame his speculations in *Daedalus*; Jaime Anguelo described the future socialist as a mutant, a "true sport with lungs"; and, when Harwood wanted to express the uniqueness of Burbank's newly created plumcot, he argued that even "a dweller upon some other planet" could not "come down through space bearing a fruit as yet untasted by the world-men" which would be more distinctive or delicious than Burbank's latest creation.[5]

Harwood was not alone in using an SF trope to enrich the imaginative possibilities that could be attached to the Californian breeder's work. In 1901, the Californian magazine *Land of Sunshine* described work on scientific plant-breeding as "a new literature" that was as "fresh, bright, helpful, more fascinating than any novel."[6] Their writer was Charles Shinn (the University of California's Inspector of Experiment Stations), who also reviewed Wells's *Men Like Gods* (see p. 24). He described the latter as one of the "Wellsian masterpieces" because the questions it raised "are living

2. McCaffery and Williamson 1991, 233.

3. Serviss 1905c, 63, 65; Serviss 1898 (1947). The first part of Serviss's sequel appeared just six weeks after the last part of Wells's original (Serviss 1898 [1947], xxxi).

4. A full account of Edison's fictional afterlives would require a volume in their own right, since they straddled everything from the "Edisonade" genre of boy-inventor adventure stories (from which Serviss's tale is descended) to Villiers de L'Isle Adam's *L'Eve Future* (1886). See Luckhurst 2005, 56–58; Villiers de L'Isle-Adam 1886 (2001).

5. Harwood 1905b, 207.

6. Shinn 1901, 3–5.

problems which we must solve if we would salvage our civilization."[7] He was discussing *Men Like Gods* but could just as easily have been reviewing a book about experimental evolution. A similar blurring of genre boundaries occurred when *Hearst's International* serialized *Men Like Gods* (giving the story its widest circulation to date). As the tale was drawing to a close, the magazine included an article by Burbank (described as the "world's foremost botanist") in which he asserted that "what can be done with plants can be done with human beings," a comment that prompted the editor to ask whether botany "will make us what Wells pictures in *Men Like Gods?*"[8]

Each of these different kinds of writing made use of real-world science, either to add credibility to speculations or an imaginative touch to factual accounts (sometimes both at once). Yet while Keller's and Wells's stories are unambiguously SF, many would hesitate to apply the same label to *Herland*, and nobody would use it of Edith Wharton's "The Debt" or "The Descent of Man." Prescriptive definitions of "science fiction" tend to be as sterile as they are tedious, not least because they seem intended to cleanse SF of the impurity that—particularly for historians—makes it so fascinating.[9] It is more productive, as Roger Luckhurst has argued, to use SF to analyze important connections between histories of science and more general, cultural history.[10] And Samuel Delany has argued that "science fiction is nothing *more* than a way of reading; it is nothing *less*; it is nothing *other*." For Delany, SF is the result of reading texts in a particular way, not a property of those texts.[11] Adopting a descriptive and historical approach reveals the kinship between SF and the many other kinds of writing analyzed in this book, not least because the insights that have emerged from analyzing science fiction prove useful when considering science fact.

Despite SF's ill-defined (perhaps undefinable) quality, it emerged as a distinct genre bearing the name "science fiction" (originally, but briefly, "scientifiction") in the American pulp magazines of the late 1920s. As with most genres originally created by publishers, SF began as a marketing strategy: its distinctive iconography, for example, helped readers find what they were looking for, while stylistic features (such as using objects and technologies as central characters, rather than human emotions and relationships) helped writers who wished to flourish learn how to fulfil

7. Shinn 1923, 44–45.
8. Burbank 1923, 114.
9. Luckhurst 2005, 6–11.
10. See, for example Luckhurst 2006; 2010.
11. Delany 1994, 273–76.

those readers' expectations. SF's horizon of expectations was unusually unstable because readers, writers, publishers, and editors were part of a conversation that was (and still is) always redrawing the boundaries of the common territory they were carving out.[12]

SF can and has been interpreted in many different ways, but this book seeks to take what might be called a cultural anthropological approach: to ask "What can this artefact tell us about the time, place, and society in which it was made?" Among other things, SF is "a literature of technologically saturated societies"—a set of imaginative responses to concerns about the impact of the growing physical and cultural power of science (it shared some of these concerns with literary modernism, a genre that emerged at the same time, partly in response to similar concerns).[13] Such things as paper quality, pricing, and the products being advertised tell us that SF (like most of the other writing considered in this book) was the product of a well-developed industrial, capitalist society, in which technology was transforming the production, circulation, and reception of all sorts of texts. Many of those texts reflected explicitly on the world that made them, for example by speculating about the significance of new scientific discoveries. However, science fact and fiction were most often explicitly and intimately connected by SF, a genre in which, as Jameson has argued, "some nascent sense of the future" took over the space that the past occupied in the earlier genre of the historical novel. SF explored the implications of the future-oriented understanding of heredity ushered in by the new biologies, because SF's "multiple mock futures" transform "our own present into the determinate past of something yet to come" (as with, for example, the essay from the future section of *Daedalus*).[14]

Paul Kincaid has argued that SF evolved out of other kinds of writing by a process not unlike natural selection, as its creators—writers, editors, and publishers—competed for new audiences by borrowing and adapting from diverse sources (and, as we have seen, they were far from unique in doing so). Competing creators picked what they believed would succeed—their goals being as diverse as their source materials—yet despite that diversity, the surviving examples exhibited a pattern of "family resemblance" (like a group of species whose common features establishes their relationship). No particular common feature could be considered

12. My thinking on these points has been influenced by the work of Mikhail Bakhtin, e.g., Bakhtin 2010; Holquist 2002; Tattersdill 2016, 8–13.

13. Luckhurst 2005, 3. See also Trotter 1992, 1–5; Cheng 2012, 22–23; Tattersdill 2016, 12–15.

14. Jameson 1982, 150–52.

essential to identifying the resemblance between the SF family of texts, nor can we specify the precise degree of likeness that allows recognition, yet somehow large communities came to agree on what to point to when they said "science fiction."[15]

Among the strengths of Kincaid's account is that the resemblance works backward as well as forward; later texts borrow from earlier genres (such as utopias or the gothic), but earlier texts may also "become" SF (in that they can be read as if they were SF, Mary Shelley's *Frankenstein* being the most famous example), so there can be no definitive historical starting point for the genre. Nevertheless, Luckhurst argues that SF depended on specific "conditions of emergence," which included: the industrialization of publishing; the spread of compulsory, state-funded education; and, the resulting huge expansion of the reading public (which led to a proliferation of markets, formats, and genres).[16] He also regards the popularization of Darwinism as vital, and the Darwinian theory of evolution by natural selection was the product of many of the same conditions (the world's first industrialized, capitalist economy provided Darwin with the crucial metaphor that allowed him to see nature as a perfect free market). Richard Gerber's pioneering study of English utopian fiction made a similar point: that the vastly expanded time which geology and evolution made available to the imagination allowed narratives of infinite future progress—and ultimate perfection—plausible for the first time (see also Koselleck on modernity; p. 6). As a result, utopia assumed "the nature of a genuine myth where 'the line between the barely conceivable and the flatly impossible' cannot be drawn at all."[17] Such arguments allow us to define SF descriptively rather than prescriptively, while retaining a clear sense of how the genre is firmly embedded in history, even if no definite founding date can be determined.[18]

For most readers, science fiction is most easily identified by recurring themes and images (such as time machines, rocket ships, or extraterrestrials). *Mutants* (a term that, as we have seen, was first used in English to describe de Vries's new primroses) have become one of these tropes. Following the mutants into SF allows us both to trace the long-standing cultural influence of early twentieth-century experimental evolution and

---

15. Kincaid 2005, 46–49. Frederik Pohl defined science fiction as "that thing that people who understand science fiction point to, when they point to something and say 'that's science fiction!'" (originally in Pohl's *Science Fiction: Studies in Film*). Quoted in Hollinger 2014 (141).

16. Luckhurst 2005, 15–24.

17. Gerber (1955, 11) was quoting Johan Huizinga's *Homo Ludens* (1949, 129).

18. Young 1985; Radick 2003; Endersby 2009.

to uncover the origins of images and assumptions that still influence the public's understandings of what evolution might mean for the future.

The Prophetic Mutant

In 1911, a curious figure greeted British readers: Victor Stott, just four and a half years old, but almost bald, with a "broad cliff of forehead" and a very small nose. He generally ignored other people, but when he did make eye contact, "the dominating power of his brain" made those who met him shrink "into insignificance." Victor proved to be an astonishing prodigy, able to read and apparently memorize the entire *Encyclopaedia Britannica* before he reached school age. Many regarded the boy as a "freak," and the exasperated local vicar was so disconcerted by the infant's sophisticated denial of God's existence that he talked of "maleficent possession" (and even tried to have Victor confined in an asylum before he could unleash "revolutionary tendencies towards socialism" among "the uneducated, agricultural population").[19] Meanwhile, others speculated that this strange child might one day "revolutionise our conceptions of time and space" (125–26).

Victor Stott was the central character in *The Hampdenshire Wonder* (1911), by the British writer John Davys Beresford. The book belonged to the British tradition of "scientific romance" and bears numerous marks of Beresford's admiration for H. G. Wells (the book even contained a nice in-joke: Victor's father was a cricketing prodigy—just like Wells's own father).[20] Wellsian themes in the book include Victor's "bloated, white globe of a skull," which made his body looked small and puny by contrast (73–74, 55–56), a description that recalls the big-brained creatures who populated many of Wells's stories (from "Man in the Year Million," to the Martian invaders in *War of the Worlds*, and the Grand Lunar of *First Men*

19. Beresford 1911, 127–28, 122–23, 155–56. Subsequent page references are to this edition.

20. Both fathers were bowlers: Victor's took a double hat-trick at his first first-class match, while Wells's took four wickets with four balls (Beresford 1911, 46–47; 1914, 13). Note for baffled US readers: a bowler is a pitcher; taking a wicket means dismissing a batter; a hat trick refers to dismissing three batters with three consecutive deliveries (pitches); and a first-class match is one that takes place over two or more days between recognized teams, usually representing a county (as opposed to village or club teams). If you would like the leg before wicket (LBW) rule explained, please send a large amount of cash (in unmarked used bills) to the author.

in the Moon).[21] The story also contained various scientific characters, including the anthropologist Henry Challis whose record of Victor's "case" adds plausibility to the account. His investigations led Challis to wonder whether their science was simply "raking up all kinds of unsavoury rubbish to prove that we are born out of the dirt," while Victor's superhuman intellect suggested that they should instead be thinking about the human future, "a future that may be glorious" (113–14). Abandoning the past, the "dirt" of human history (a phrase suggestive of both original sin and ape origins) to work toward a "glorious" future is typical of both Wells and wider early twentieth-century optimism.

The Hampdenshire Wonder is now widely considered a pioneering work of SF; the first in a long series of rival imaginings of future human evolution, but when it was published it could not be assigned to any recognizable genre. The book received numerous positive reviews, which praised its uncanny qualities, but almost none connected it to science of any kind (and only the *Academy and Literature* mentioned Wells as a likely influence).[22] The reviewers' puzzlement may have stemmed from the fact that in the first (British) edition of his book, Beresford had offered no real hint as to *why* Victor had been born with his strange characteristics. *The Hampdenshire Wonder* made vague gestures toward the idea that willpower directed evolution, as well as to the folk myth of maternal impressions, mixed in with hints of Lamarckianism and of Bergson's creative evolutionism, but the result was reminiscent of Victor's description of all human science—"Inchoate ... a disjunctive ... patchwork" (169–70).

The fact that reviewers failed to recognize Beresford's book as a scientific romance may have prompted him to add a new chapter to the book's first US edition (1917, when it was retitled *The Wonder*). He added an account of Victor meeting a "representative of twentieth century science"— Hugo Grossmann, a leading German biologist. Grossmann was engaged in "investigations and experiments on the lines first indicated by Mendel," and had used the arrival-of-the-fittest problem to undermine "the whole principle of 'Natural Selection.'" One of Challis's friends, Sir Deane Elmer, arranged the meeting between Victor and Grossmann because the

21. See Wells 1893; Wells, Hughes, and Geduld 1898 (1993); Wells 1901. Beresford 1914 (20) knew "Man in the Year Million" and described it as being—like *The Time Machine*—an "essay in evolution."

22. British reviews included *Academy and Literature* 81, no. 2052 (September 23, 1911): 385; *Athenaeum*, no. 4371 (August 5, 1911): 153; *Bookman* 40, no. 240 (September 1911): 263–64; *Bystander* 32, no. 415 (November 15, 1911): 358; *Country Life* 30, no. 763 (August 19, 1911): 286; *Nation* 9, no. 21 (August 19, 1911): 749; *Observer*, September 10, 1911: 4; *Spectator* 107, no. 4341 (September 9, 1911): 389.

German's theories had undermined Elmer's "somewhat too optimistic prophecies" about eugenics. In Grossmann's view, there was never enough "progressive variation from the normal"—no really new types upon which natural selection could act. By contrast, Elmer believed that Victor "demonstrates the fact of an immense progressive variation."[23]

By adding such details, Beresford changed Victor from an inexplicable "freak of nature" into something more reminiscent of a de Vriesian mutation, a sudden leap into a new type. The reader learns that Victor represents immense evolutionary progress (Challis believes humanity cannot hope to learn from Victor, because he is "too many thousands of years ahead of us"). As a result of his intellectual progress, Victor is able to develop his own "theory of origin, evolution, and final adjustment," which so horrifies Challis that he "dare[s] not allow himself to be convinced by Victor Stott's appalling synthesis" and decides neither to share nor record it (171–72, 266). Much of the power of Beresford's story stems from the fact that Victor is both uniquely gifted—and cursed; when he asks Challis "Is there none of my kind?" Challis has to confirm that the child is indeed, entirely alone (173–74). Victor has no future (the novel ends with his death, apparently a murder by the fearful, conservative vicar).[24] Like a rare sport, Victor could never find another like him with whom he could have reproduced, but he was to give birth to numerous fictional progeny—by establishing what might be called the prophetic mutant as a recurring trope in SF.[25]

A quarter of a century after Beresford's novel appeared, another British writer, Olaf Stapledon, produced a kind of sequel—(Stapledon acknowledged his debt to Beresford in the book's preface). Stapledon's eponymous hero also had a huge head and prodigious mental powers, atop a rather feeble and unattractive body. However, he developed Beresford's hints and explicitly made his central character a mutant. John discovers he has telepathic powers that allow him (unlike Victor) to find others of his kind, whom he calls "wide-awakes." And the book's narrator refers to John's kind as *Homo superior,* implying he and his kind are the next stage in human evolution.[26] It becomes explicit that the wide-awakes are the result of a de Vriesian mutation when John finds others: "When a species mutates,

23. Beresford (1917, 222–24).

24. Vint and Bould 2011, 32.

25. Beresford's book is generally credited with founding the mutant trope in SF, but it had one or two obscure precursors, e.g., *Another World* (1895) (Aîné 2012). See Bleiler 1948; and "Supermen and Other Mutations," in Pierce 1987 (25–37).

26. Stapledon 1935 (1972), 46.

it often produces a large crop of characters so fantastic that many of the new types are not even viable" (193). The new types are found all over the world but seem more common in Asia, so John speculates that they might have all "sprung from a single 'sporting point' centuries ago, probably in Central Asia. From that original mutation, or perhaps from a number of similar mutations" (227).

Stapledon seems to have been the first to imagine humanity undergoing a de Vriesian mutation period. The idea of a new variety of humans—perhaps a successor to *Homo sapiens*—appearing as a crop of major mutations has remained a staple of SF ever since (as early as 1944, the fan-produced *Fancyclopedia* was defining mutants as "an entire new field of sf stories").[27] Yet by the time *Odd John* appeared—1935—de Vries's ideas were almost forgotten in the scientific community. A few textbooks and popular writers were still using *mutation* in ways that carried echoes of de Vries's original meaning, but it was more common to find references to the post-*Drosophila* sense of mutations (small and random; occurring continuously, not periodically; then sifted by natural selection). It was largely because the prophetic mutant became an SF trope that de Vries's all-but-forgotten theory has continued to haunt biology's visions of the future right into the twenty-first century.[28] A century after de Vries first published, the first of the internationally successful *X-Men* movies came out.[29] As the opening credits rolled, audiences heard the authoritative tones of Professor Charles Xavier (Patrick Stewart) pronouncing: "Mutation: it is the key to our evolution. . . . This process is slow, normally taking thousands and thousands of years, but every few hundred millennia, evolution leaps forward," and in the film series, humanity has entered a "mutation period"—an era in which new mutant types appear, in large numbers, at a leap, with no intermediate forms.

The *X-Men* mark the most prominent recent cultural reappearance of the prophetic mutant; long after its origins have been forgotten, de Vries's theory has been smuggled into the future by SF, bringing with it faint echoes of the biotopian dream (the "good" mutants, like Xavier, offer humanity hope for the future). The first major crop of mutant stories appeared after World War II and often used the radiation unleashed by atomic weapons as the source of mutations. Nevertheless, they remained

---

27. Speer 1944.

28. To the best of my knowledge, Robert Silverberg is the only person to connect SF mutants with de Vries. See the introduction to Silverberg 1974. However, he did not discuss the theory itself, nor its historical context.

29. *X-Men* (2000), directed by Bryan Singer.

distinctly de Vriesian, often featuring rapid and optimistic transformations into a new, improved type. An early example was the bleak "Tomorrow's Children" (Poul Anderson and F. N. Waldrop, 1947), which imagined the aftermath of a nuclear holocaust in which radiation is mutating humanity into unrecognizable, often monstrous forms. Nevertheless, the story concluded with the hope that some of the mutants would prove well-adapted to this melancholy new world.[30] Many later stories were even more optimistic and imagined mutation creating various kinds of *Homo superior*, harbingers of a better future. They can be found in A. E. van Vogt's "Slan" (1940); Lewis Padgett's "Baldy" stories (collected as *Mutant*, 1953);[31] Philip K. Dick's "The Golden Man" (1954); Wilmar Shiras's *Children of the Atom* (1954); John Wyndham's *The Chrysalids* (1955); the original Marvel *X-Men* comics (f. 1963), which were apparently inspired by Shiras; and, many, many other tales.

However, mutants were to prove as inherently paradoxical as utopias: one person's good mutant is another's nightmare. Many of these stories were ambiguous, while in others explicitly sinister mutants represented a threat to humanity (for example, Edmond Hamilton's "The Man Who Evolved" [1931]; John Taine's "Seeds of Life" [1931]; and the "Mule" character in Isaac Asimov's *Foundation and Empire* [1945]). The idea that a hidden group of mutant humans (either mentally or physically more powerful than nonmutants) was waiting to take over could fuel paranoid fears (as it does in the *X-Men* movies), but it also clearly appealed to some SF readers (who perhaps imagined discovering hidden powers within themselves). After the atomic bombings of Hiroshima and Nagasaki, mutants became commonplace in postwar SF. They were less common in the pre-war period but were similarly tied to then-current science (such as the induction of mutations using X-rays). The stories not only reveal the intimate connection between science and its fictions but also highlight the role of SF writers and readers as interpreters of science, who created another route by which biology became public culture.

The appearance of new pulp magazines (for example, *Amazing Stories* [f. 1926]; *Science Wonder Stories* [f. 1929]; and *Astounding Stories* [f. 1930]) mark SF's emergence as a distinct genre in the years just before *Odd John* appeared. Of course, science-based stories had appeared in the earlier pulps (it was their popularity that persuaded Hugo Gernsback, the editor who founded *Amazing*, that there might be an audience for

30. See Silverberg 1974, 13–44.
31. Lewis Padgett was the pseudonym of the husband-and-wife team of Henry Kuttner and Catherine Lucille (C. L.) Moore.

his new magazine).[32] Nevertheless, the new magazines, with their distinctive visual and literary style, helped to create communities of readers who identified with the genre and were encouraged by early editors to participate as both writers and critics.[33] These communities were similar to the other kinds of interpretive communities discussed earlier, not least because SF readers usually identified themselves as fans of science, not just of its fictions.[34] The magazine's editors generally shared this perspective and regularly reported scientific news (not least because new stories were often scarce—recent science news might encourage readers to transform themselves into writers). The intimate connections between the participatory culture of various fans and the history of science are particularly well exemplified by the rise of the mutant story, which demonstrates how the latest science news was made use of by SF writers, helping science fact and fiction develop an increasingly shared language.

Although the earliest mutant stories bore traces of de Vries's ideas, they were seldom directly inspired by the mutation theory; the catalyst was Muller's 1927 announcement that he had succeeded in inducing mutations in *Drosophila* using X-rays (see pp. 95–96). The media reported the facts in terms that sounded distinctly like science fiction.[35] The reputable *Science News-Letter* (which provided news to many other publications) announced, "X-Rays Speed Up Evolution over 1,000 per Cent," and claimed that evolution and crop-breeding were "under the spell of a new magic" that "holds out the prospects of producing new forms of life a hundred times faster than has hitherto been possible." This prospect was made possible by "what the modern scientific breeder calls 'mutations,' which are the same things that the old-fashioned gardener called 'sports.'" Before Muller, no one had known what caused these sudden changes. Although "Luther Burbank's reputation was built largely on mutations," he had to "wait years for the variant he wanted." Despite planting "millions of seeds" he could only "trust to luck to produce mutations." The same was true for evolution (since "nature acts the part of a super-Burbank"); natural selection weeds out undesirable mutants ("as Charles Darwin pointed out long ago"). However, while natural selection had all the time in the world, "man is an impatient creature" who "wants what he wants when he wants it—and that includes mutations. Assisting or forcing nature in some way,

---

32. Ashley and Lowndes 2004, 63–69; Tattersdill 2016.

33. Carter 1977; Ashley and Lowndes 2004; Cheng 2012; Sleigh 2018.

34. Moskowitz 1994.

35. The media attention on Muller's work led to earlier studies of induced mutations being largely ignored by subsequent histories (Campos 2015, 225).

so that new things will be produced faster than at the old, poky rate, has for centuries been the breeders' dream."[36] Muller's work finally realized a key goal of the original mutation theory: control, which allowed both evolution's tempo and mode to be changed.

The *Science News-Letter* also noted that there were "such things as human mutations," offering the possibility of improving people that was of particular interest to newspapers. The *Los Angeles Times* speculated that "startling results may be accomplished in improving both physical and mental characteristics of human beings" (although "latent bad or undesirable heredity particles" might also appear).[37] And the *New York Times* commented that although it was still too "early in the history of experimental evolution" to start breeding a race of supermen, we were already "on the road toward controlling human evolution."[38] Press interest intensified the following year when Robert Millikan announced the discovery of cosmic rays.[39] The *New York Times* was one of many publications which linked Millikan's and Muller's discoveries; under the headline "Cosmic Rays and Evolution," it noted that a "new field of research" had been opened, in which "astrophysics and biology are strangely united." Perhaps chromosomes and genes were "the playthings of the terrific forces which tear down and build up atoms in stars millions of light-years distant," a possibility that suggested that perhaps the earth "came to be peopled with species of plants and animals . . . because our biological destiny is controlled by millions of incandescent suns?" Their writer concluded with words that seemed almost intended to spark an SF writer's imagination: "If man's feeble laboratory X-rays can switch evolution from one track to another, what may not be expected of the more powerful gamma rays of radium or those cosmic X-rays?"[40] If naturally occurring cosmic rays proved to be the ultimate source of the mutations that were natural selection's raw material, there was no telling what their laboratory equivalents might accomplish.

The link between Muller's work and cosmic rays took on a more permanent (and widely read) form when H. G. Wells and his co-authors described it in *The Science of Life* (1929–1930). They commented on the "disturbing idea, that life has evolved and is still evolving under the spur of those strange rays, shot casually into the world from unknown corners

36. Thone 1927, 243–44.
37. "Speeds Breeding Types" 1927.
38. Kaempffert 1928, 73.
39. Millikan and Cameron 1928. See also Rouyan 2017, 136–37.
40. "Cosmic Rays and Evolution" 1928.

of the universe!"[41] Wells himself used this idea as a basis for fiction in *Star Begotten: A Biological Fantasia* (1937). The novel reimagined his own *War of the Worlds* (1897) by supposing that the Martians have realized that the Earth is not suited to their biology and so are using cosmic rays to mutate humans into a new "Martian" type, who will eventually inherit the Earth. While some "Earthlings" are panic-stricken at the prospect, the story's protagonist gradually comes to realize that the "Martianised" humans are more intelligent, calmer, and more rational than *Homo sapiens* and ends up welcoming the "invasion."[42] As in *Men Like Gods*, *Star Begotten* saw Wells using state-of-the-art biology to create a utopia (as the *Times Literary Supplement* commented, "the 'Martians' are his 'Utopians' in another guise").[43]

The Science of Life also predicted that the human species would both survive and thrive once "man" had taken "control not only of his own destinies but of the whole of life." Wells and his co-authors told readers who wished to know more to consult "many of the little volumes in the To-day and To-morrow series" (which included *Daedalus* and Bernal's *The World . . .*), which would provide interested readers "with plentiful food for his imagination in these matters."[44] Among those "little volumes" was *Metanthropos; or, The Body of the Future* (1928), which suggested that cosmic, or "Millikan," rays had probably pushed the "obstinate" and "conservative" genetic material into changing in the past. Yet *Metanthropos* retained a faint echo of de Vries when it noted that "evolution has been rather rhythmical and explosive in its workings, hinting at cosmical or astronomical occasional causes and that there have been places and periods of passivity"—perhaps sporadic bursts of cosmic rays were the ultimate cause of mutation periods?[45] Given this possibility, *Metanthropos* was dubious about being able to induce useful mutations in humans using X-rays, but nevertheless agreed that humanity taking control of the "*machinery* of evolution" was "perhaps the most interesting and important [idea] in all biology" (51, 57)—and one that was steadily becoming more plausible.[46]

Writers in the newly founded SF pulps showed a similar fascination with the possibilities of radiation-induced mutations. The year after Muller's announcement (and just months after Millikan's) *Amazing*

41. Wells, Huxley, and Wells 1929–1930, 392.
42. Wells 1937 (2006).
43. Murray 1937.
44. Wells, Huxley, and Wells 1929–1930, 973.
45. MacFie 1928, 11–12.
46. See Wood 2009, 22–23; Bowler 2017b, 185; Saunders 2019, 18–19.

Stories published "The Metal Man" by Jack Williamson (who acknowledged Burbank as one of his inspirations; see p. 271), which seems to
have been the first story to imagine an entirely new life-form evolving as
a result of exposure to radiation. A Professor Thomas Kelvin, who was
prospecting for radium, discovered a mysterious lake hidden in an isolated
crater somewhere in the Andes (and—like Herland—unreachable except
by plane). The crater contained strange, seemingly intelligent crystalline
entities who had evolved as a result of the intense concentration of naturally occurring radiation. (Sadly, the radiation turns all living things into
metal—including Professor Kelvin—so we never learn more about the
mysterious crystal creatures).[47]

Williamson's story was directly based on the latest science, and links
like these became even stronger when Gernsback's new magazine, *Science
Wonder Stories*, decided to include a "Science News of the Month" column
to keep readers up to date. In 1929 the column included items based on
stories from *Science*, about two University of California experimenters
(E. B. Babcock and J. L. Collins), who had found that naturally occurring
radiation from rocks within the earth (as opposed to outer space) induced
mutations in fruit flies.[48] The magazine explained that "earth radiation has
played and is playing an important role in the great drama of organic evolution" and quoted Babcock and Collins's conclusion that their discovery
had "important practical aspects for agriculture," since most plants were
more "easily treated experimentally with X-rays or radium" than animals.[49]

As the pulps developed into a new home for bricolage, the idea that
radiation might accelerate evolution became so common that it was
soon a cliché. Among its early exponents was Edmond Moore Hamilton,
who produced a series of stories that explored the idea. He was a prolific
writer (perhaps because he seems not to have wasted time polishing his
prose—or coming up with new plots) and the themes he explored in his
mutation stories rapidly became as overfamiliar as those of space opera
(a subgenre he also helped create).

Hamilton's "The Man Who Evolved" (1931) appears to have been
the first story that portrayed the use of "cosmic rays, as discovered by
Millikan" to create superhuman mutants (figure 8.2). Hamilton's scientist, John Pollard, supposedly discovers a way to dramatically accelerate evolution. The process normally proceeds slowly by "successive

47. Williamson 1928.
48. Babcock heard de Vries lecture in 1903, which shaped his lifelong interest in
mutation (Smocovitis 2009, 305).
49. "Science News" 1929.

evolutionary mutations," but focusing cosmic rays into a powerful beam allows him to increase their frequency. He tests the idea on himself in order to discover what "the future course of man's evolution [is] going to be," assuming that "through the ages life has been raised from the first protoplasm to man, and is still being raised higher."[50] This view was an obviously stadial interpretation of evolution as predictable progress (see p. 16), which was revived and dramatized repeatedly in the pulps, allowing SF authors to turn accelerated evolution into a form of time travel; organisms—including humans—could be flung into futurity to see how they would turn out (much as Gilman had done in *Herland*). Pollard asserts that by stepping into the rays he will "be changed millions of times faster than ordinarily" (just like Muller's flies in the *Science News-Letter*'s story), and so will go "in hours or minutes through the evolutionary mutations that all mankind will go forward through in eons to come!" (22). Each fifteen minutes under the ray is supposedly equivalent to fifty million years of evolution. Pollard emerges from his first exposure as "a great figure of such physical power and beauty" that the narrator is stunned—the future looks bright indeed. But, predictably, things soon go wrong; as Pollard rapidly mounts the ladder of progress he develops a "weak body" supporting "an immense, bulging balloon" of a head, which (like Victor Stott's) is almost hairless, "its great mass balanc[ing] precariously upon his slender shoulders and neck" while a "great bulging forehead dominate[s] the face" (27).

The ever-expanding head became another familiar image in mutation stories, one that was also built on earlier assumptions about the pattern and predictability of evolution. The claim that the human intellect had been the key to our species' apparent evolutionary success had a long pedigree, beginning in the nineteenth century when the German evolutionist Ernst Haeckel was among the first to predict that the human evolutionary sequence would be marked by increasing skull size—an expectation sometimes referred to as "big brain first" (which some saw as evidence of the key mutation that had created humans; see p. 214).[51] The expectation of ever-larger heads being needed to house ever-larger brains was a common feature of linear models of evolutionary progress, and it became a standard cliché of SF iconography (figure 8.2).

50. Hamilton 1977, 21.
51. The idea had its roots in the early nineteenth-century science of phrenology and later developments such as craniology, which used skull size to estimate intelligence; see Cooter 1984; Gould 1997; Tomlinson 2005.

FIGURE 8.2. Edmond Hamilton's "The Man Who Evolved" was one of the earliest examples of the "evil mutant" trope that still haunts the science fiction imagination.

In Hamilton's story, Pollard develops almost godlike intellectual powers but seems oblivious to the dangers of his experiment and insists on pushing forward until he becomes nothing but a giant, pulsating brain, devoid of both a human body and human emotions but able to communicate telepathically (the ultimate logical outcome of the big-brain-first assumption). Pollard's growing mental power finally strips him of all empathy with unevolved humans, and he becomes an arrogant, inhuman superman, able to dominate or exterminate the rest of humanity. He forces his visitors to turn on the rays one last time to see "the last mutation," convinced that the fittest had yet to arrive. However, the result is just a mass of formless protoplasm; evolution has gone full circle, returning humanity back to its starting point (33–35).[52] One of Hamilton's earlier stories, "Evolution Island" (1927), also featured a mutating ray, "the Garner ray" (but this ray is derived from "vast masses of radio-active substances in the earth's interior," rather than from outer space). However, it pushes the eponymous test island's animals up the ladder so rapidly that they become extinct (cyclical evolution again), while the plants climb up to become intelligent and mobile, almost destroying their human creators (luckily the Garner ray has a fabulously implausible reverse mode, which allows the humans to put the plants back in their place).[53]

Hamilton's view of how evolution might be manipulated was far from unique. In the same year that "The Man Who Evolved" appeared, another pulp author, John Taine, also used the idea that accelerated evolution could enhance human intelligence.[54] His story "Seeds of Life" (1931) certainly gave its readers their money's worth (among other things, it included mutant, giant Black Widow spiders; dinosaurs created by reverse-evolving chickens; a sinister corporation threatening to blast humanity backward through evolutionary time; and inter-species sex—resulting in humans giving birth to a mutant baby which "is not a mammal"). Insofar as these diverse plot elements were held together, it was by the idea that evolution follows a predictable path; at the story's heart was a giant X-ray tube that could move species both up and down the ladder of progress. Taine's mutant central character, de Soto, had initially been a biotopian; he had aimed at the "creation of life and the remaking of it to my will, in spite of chance and blundering evolution. This was my dream" (490). But as he regressed (thanks to a radiation overdose) he despaired at being able to make anything worthwhile out of humanity.

52. Cyclical evolution—a descendant of earlier ideas about cyclical human history—was another way in which SF authors regularly tried to find predictable patterns in evolutionary change; see Gould 1988.

53. Hamilton 1927. Intelligent plants, who have evolved far enough to enslave humans, also feature in Hamilton 1931 (304).

54. Taine 1931.

Despite its more fantastic elements, "Seeds of Life" was firmly grounded in experimental evolution: the protagonists discuss inducing heritable mutations in flies using X-rays; the story's main biologist explains accelerated evolution in a paper to the Biological Society called "New Light on Evolution" (the same title under which the mutation theory had first been announced in the USA; see chapter 1)—it even features a distinguished elderly scientist called de Vries (and the real de Vries was still alive, aged eighty-three, when the story first appeared). Gernsback's original definition of *scientifiction* included the demand that stories must have a basis in real life science; the laws of nature might be stretched (often to breaking point), but authors were expected to offer an explanation for the phenomena they described. When Gernsback introduced "Seeds of Life" to the readers of *Amazing Stories*, he emphasized that Taine was the pseudonym of "one of the very few younger members of the National Academy of Sciences," who had "had opportunities for first-hand information on the subject" of radiation's effects, including some "which has not yet been published in scientific journals."[55] Similarly, Gernsback insisted that the story's theme of "the control of evolution" was not a "wild dream"—Muller had already shown how much is possible.[56]

As noted, many SF readers regarded themselves as primarily *science* enthusiasts. They shared Gernsback's concern with the scientific details, and the magazines' participatory culture encouraged them to acquire and use their own expertise. One of *Amazing*'s readers, Charles Campbell, wrote in to complain about the materialistic assumptions of "Seeds of Life." He argued that Taine's story was wrong because evolution was not purely mechanical but the result of a divinely implanted instinct for self-preservation (he supported his view by quoting *The Divine Pedigree of Man*, by Thomson Jay Hudson—which used the arrival-of-the-fittest problem to argue that while natural, selection "is preservative of species,—not creative").[57]

55. "Taine" was the pseudonym of Scottish-born mathematician and author Eric Temple Bell, who lived in the US from 1902 and published at least 250 papers and several studies in mathematical history and theory (Brian M. Stableford and John Clute, "Taine, John," *The Encyclopedia of Science Fiction*, eds. John Clute, David Langford, Peter Nicholls, and Graham Sleight [London: Gollancz], updated May 3, 2021, https://sf-encyclopedia.com/entry/taine_john). The details of "Seeds of Life" may also owe something to the fact that Bell had been on the faculty of Columbia University at the same time as Muller and, at the time he wrote the novel, was a faculty member at the California Institute of Technology—where Thomas Hunt Morgan then worked (Reid 2001, 394–95).

56. Gernsback [editor's introduction to] Taine 1931.

57. Hudson 1900, 34.

However, the letters pages were a forum for debate, and Campbell's claims were immediately disputed by another reader, William Kober, who acknowledged that the question of how variations arose had long been "the great objection to Darwinism," suggesting that evolution must involve some "factor independent of Natural Selection." However, there was no need to invoke divine purpose, because recent researches on a "certain type of fly" had revealed that the "explanation is 'Mutation.'" He went on to explain that although mutations were random, if "the mutated type is better suited to the environment than the normal, it will survive, and eventually completely replace the old type. If not suited, it will die out without a trace." Hence, if the number of mutations could be increased "evolution is speeded up!" (precisely echoing the *Science News-Letter's* report of Muller's work). Kober noted that X-rays were among the ways in which an increased mutation had been produced, so he concluded that, despite taking some imaginative artistic liberties, "Taine has stuck very close to actual, present-day scientific knowledge, merely exaggerating the controllability and the extent of difference from normal of the mutations which can undeniably be produced by X-rays."[58]

The creative ways in which *Amazing's* readers made use of science are a further reminder of the inadequacy of terms such as *diffusion* or *popularization*. Half a century before fan cultures like those around TV's *Star Trek* existed, the pulp's letters pages were already creating an eclectic culture based as much on being fans of science as of its fictions, which allowed its enthusiasts to play a role in defining both.[59] Campbell demanded both scientific credibility and divine design, while Kober blended an understanding of recent research with the idea of the ladder of progress (and was untroubled by the absence of natural selection in Taine's story, without which an increased mutation rate would have no effect on evolution).

Such letters (and there were hundreds like them in the pages of the SF pulps) illustrate important aspects of the culture of SF fandom: readers took the stories seriously enough to care about their scientific accuracy but also offered their own perspectives on what science could do and why it mattered. There were similarities between the SF community and that of the socialist autodidacts, and the two overlapped at times (as for example when a Detroit labor organizer complained about *Amazing's* "concealed slaps at labor" and "your vicious attacks on communists").[60] A few years

58. Kober 1932.

59. Jenkins 1992 (2013); Bacon-Smith 1992; Penley 1997.

60. Letter from W. Mollenhauer Jr. of the Detroit Federation of Musicians ("The Reader Speaks," *Science Wonder Stories* 1, no. 1 [June 1929]: 91). Mollenhauer's

later, in 1937, John B. Michel argued at the Third Eastern Science Fiction Convention that science fiction had to be more overtly political if it was to avoid becoming irrelevant. Like his friends Donald A. Wollheim, Frederik Pohl, and Robert W. Lowndes, Michel was a member of the Communist Party of the USA and the group became known as "Michelists" as they tried to enlist the SF community in the CPUSA's project of building a broad popular and cultural front against the threat of fascism.[61] Like many SF fans, the Michelists believed science could create a better future, and they saw the goals of communism as inseparable from the kinds of techno-scientific utopianism that Gernsback's magazines embodied (even though they generally loathed Gernsback, who was an uncompromising capitalist). The overlap between these communities was evident when Michel chose to call his speech "Mutation or Death"; for some Marxists and SF fans, the term *mutation* remained synonymous with positive, revolutionary change.[62]

SF readers were encouraged to be more than passive consumers, and their central role in shaping the genre was evidenced by the fact that SF magazines carried far more letters from readers than other pulps did. As John Cheng argues, neither authors nor publishers determined the shape of the genre; the readers and writers (often the same people) brought their own interests to bear, drawing on what they learned about science from newspapers, nonfiction magazines, and textbooks.[63] SF's culture had much in common with the diverse readerships of other science writing, providing further evidence of the lack of stable boundaries between the makers and consumers of science. In some respects, SF fans like Campbell and Kober were doing the same kind of interpretative work as the journalists who wrote about Burbank, or novelists like Wharton and Gilman; they were all exploring—and extending—biology's imaginative possibilities.

While SF had numerous similarities to other kinds of scientific writing, it also provided particularly vivid examples of the ways genres hybridized—gleefully erasing the boundaries between fact and fiction.

complaint concerned a story by Frank Gates ("The Man who Died by Proxy," *Amazing Stories* 2, no. 2 [May 1927]: 145–47, 179), which referred to Felix Dzerzhinsky (head of both the Cheka and the OGPU) as the "Soviet Monster." In 1938, arguments like these led to left-wing fans in New York forming the Committee for the Political Advancement of Science Fiction (CPASF) and campaigning for explicitly socialist SF; see Cheng 2012, 236.

61. Cashbaugh 2016. See also Ross 1991, 114–16; Cheng 2012, 236–37.

62. Michel 2017.

63. Cheng 2012, 52–55, 8.

Even Julian Huxley, despite being a sober, formally qualified scientist, contributed to these processes when he reviewed Wells's *Men Like Gods* in *Nature* and took its scientific forecasts seriously (see p. 169). Huxley (always keen to reach nonspecialists) seems to have shared Gernsback's view that stories might communicate scientific ideas to readers who would never read a nonfiction text.

Huxley even wrote his own scientifically based short story, "The Tissue Culture King," which gave a scientific twist to the Rider Haggard style of African adventure story. His narrator discovered a kind of lost world, an African kingdom carefully protected from encroachment, whose inhabitants included gigantic humans and two-headed toads. These turn out to be the work of a British medical researcher, Hascombe, an expert in the then-new science of tissue culture (growing living cells in the lab), who has mischievously taken advantage of various indigenous beliefs to make himself into an all-powerful, scientific witch doctor. Hascombe utilizes the power of the modern laboratory ("the mass-production methods of Mr. Ford," as he puts it) to create various biological novelties with which to bamboozle the credulous Africans.[64] Apart from inventing the trope of tinfoil hats as protection against telepathy, Huxley's dated and racist tale is of little interest, but its publishing history epitomizes the fluid genre boundaries of the period. The story initially carried the subtitle "a biological fantasy" when it appeared in Britain's rather highbrow *Cornhill Magazine* (April 1926, then edited by Julian and Aldous's father, Leonard). It crossed the Atlantic to reach a similarly educated audience in the *Yale Review*, but reached a very different (and much wider) readership when it was republished in *Amazing Stories* (August 1927, a time when the magazine claimed a print run of 150,000 copies a month).[65] During these years, Gernsback struggled to fill the pages of his new magazine, so he regularly reprinted older stories; as a result, Huxley's story appeared in the same issue as the serialization of his future collaborator H. G. Wells's *War of the Worlds*, with a cover that exemplified standard SF iconography (Martian tripods marching remorselessly across the earth).

Huxley's story was adapted to its new pulp environment by having its "biological fantasy" subtitle removed. It not only looked very different in its new setting, it *became* pulp fiction (as opposed to, for example, the kind of highbrow, cynical literary jeu d'esprit Julian's brother Aldous was

---

64. J. S. Huxley 1927, 456.

65. See "The Tissue-Culture King," ISFDB, accessed February 13, 2024, http://www.isfdb.org/cgi-bin/title.cgi?57847. Circulation figures are from Gernsback's editorial in the same issue.

then writing). Huxley's story might also have persuaded the readers of *Amazing Stories* to read some of his nonfiction, such as *The Science of Life* (especially as the book also carried Wells's name and was regularly advertised in the pulps; see figure 6.3). Gernsback introduced Huxley's story to American SF readers by reminding them that its author was "grandson of Thomas Henry Huxley famous English scientist, and himself Professor of Zoology in King's College, London." As with the claims about Taine noted above, such details helped readers suspend their disbelief. Similar claims were made to establish the plausibility of more highbrow books. When left-wing publisher Victor Gollancz (also a pioneering SF publisher) produced the UK edition of Muller's *Out of the Night*, he probably anticipated that the American author's name would be unfamiliar to British readers, so the book's cover consisted of a quote from Haldane explaining that Muller was "one of the world's leading biologists," whose forecasts were "entirely practicable" ("whether or not they are desirable") and promised a transformation as important as that created by the industrial revolution.[66]

Science fiction offers a particularly vivid illustration of how entangled science and its various fictions were. Readers of fiction might have learned about experimental evolution's potential and do-it-yourself ethos from a wide range of authors (including Gilman, Haldane, Hamilton, Huxley, Keller, Muller, Stapledon, Taine, and Wharton), any one of whom might have been judged as persuasive, depending on their audiences. All readers of scientifically inflected fiction took on the role of deciding which writer's interpretations were most convincing, but such judgments were explicitly expected of SF fans. The SF magazine editors used the word *backyard* to refer to the letters column and, by extension, the community of readers and contributors, a term that suggested a particularly American vision of local, domesticated democracy—a space, above all, where amateurs could flourish.[67]

The backyard was where an SF reader could dream of becoming an author and perhaps start writing their first story, while the adjacent home could be a place for science experiments (such as testing the radio you had built yourself—following instructions from a magazine like Gernsback's *Electrical Experimenter*). Or perhaps the scientifically minded youngster could turn instead to the Luther Burbank Society's pamphlet *Start the Boy Right* (see p. 104) and become a backyard evolutionist. The SF pulps

66. Haldane, front cover of Muller 1936.

67. Cheng 2012, 52. As Charlotte Sleigh (2018) has shown, British fans—partly inspired by the US pulp magazines—created their own subculture, largely through publishing their own magazines.

encouraged their readers to join the world of science in much the same way as the hype around Burbank welcomed readers into his scientific world; as noted above, Muller clipped and kept a newspaper story headed "How to Do What Burbank Does" (see p. 108), and the story's claim that he might one day help make new species may have set him on the path to becoming a geneticist. *Start the Boy Right* promised its readers easy access to the world of experimental evolution by providing simple instructions for creating new plants ("Any boy who is handy with his jackknife should be able to graft a seedling or a scion on a tree with success").[68] William Harwood's article "Every Man His Own Burbank" shared the frequently made promise that anyone and everyone could have a go at scientific plant-breeding.[69] Nor were journalists the only ones who used Burbank's example to encourage readers to get involved in experiments; de Vries himself stressed that Burbank's methods were "simplicity itself. Every one can do the same things in his garden" using only "the most ordinary garden tools."[70]

Such claims inspired the Luther Burbank Society to publish *Give the Boy His Chance* (mainly intended to promote the society and its publications; see figure 8.3), which encouraged parents to get their sons involved in practical plant-breeding ("If the boy can have ten feet in the backyard for his experiments, well and good," the pamphlet suggested, but if not "perhaps he can have five").[71] Youthful gardeners who were encouraged into science by Burbank's example would doubtless have been inspired by a story in the *Washington Post* (1902) headed "Scientific Investigation Pursued in a Washington Back Yard," which described experiments done with ordinary American tomatoes that apparently proved the correctness of de Vries's recently announced mutation theory (as Sharon Kingsland has noted, many Americans interpreted the mutation theory as a democratizing idea that seemed to make science more accessible).[72] The experimenter in this case was Charles A. White of the Smithsonian Institution; his scientific resources and expertise doubtless dwarfed those of the typical readers of pulp fiction or gardening pamphlets. Nevertheless, the newspaper explained that Smith's work had been done "in a little patch of ground hardly six feet by three," yet had been hailed by de Vries himself

68. *Start the Boy Right* 1914, 26–31.
69. Harwood 1905a. See also J. S. Smith 2009, 9.
70. De Vries 1906b, 1136.
71. *Give the Boy His Chance* 1913, 15.
72. Kingsland 1991, 493–94.

FIGURE 8.3. Traditional rural imagery met the promise of backyard biological experimentation in the Luther Burbank Society's promotional materials. A biotopian future was on offer to young American men, but it was one that would not deprive them of the traditional virtues that Burbank himself embodied.

"as an important and startling confirmation of his theory."[73] In contrast to many of the physical sciences, no grand laboratory full of expensive equipment was needed.

73. "Variation in Species" 1902. See also White 1902b; 1905. White was among those who first introduced the mutation theory to the USA; see White 1902a.

Had the Luther Burbank Society's planned network of male-dominated clubs of experimental evolutionists been established, they would doubtless have resembled the clubs of wireless enthusiasts that evolved into the early SF fan clubs and encouraged consumers to become creators. Like many other publications, the *New York Times* emphasized the accessibility of experimental evolution when it described Daniel MacDougal's mutation experiments at the city's botanic gardens. He was referred to as a "gardener-scientist" who tended the plants himself, and the story included some practical details of how each plant was "potted, ticketed, and preserved for the purpose of continued observation and experiment."[74] And, of course, textbooks on biology often included detailed instructions so that any of their readers—not just students—could try various experiments for themselves.[75]

The idea of verifying a scientific claim by repeating the original experiment had, of course, long been considered part of the scientific method. In the nineteenth and twentieth centuries, such verification increasingly became the work of experts, but experimental evolution offered a return to the backyard simplicity that Francis Bacon (a keen gardener himself) had originally identified as the key to his new experimental philosophy—and the utopia it would create.[76] The *San Francisco Chronicle* marked de Vries's Californian visit with a full-page, illustrated article (see p. 52), which stressed the revolutionary implications of the mutation theory and reminded readers that *Oenothera lamarckiana* was "very common in California and it is *open to all* to test the conclusions of Professor Hugo de Vries."[77] The celebrated Dutchman made the same point when he told his large audience at the University of California that experiments like his did not require "costly laboratory equipment" and so anyone "who has a small garden at his disposal" could try them.[78]

In the USA, Burbank and de Vries were the key figures used to inspire direct participation, while endless British texts delighted in describing Mendel in such terms as "a simple monk who spent part of his day on his knees, and part of his day examining the common or garden pea."[79] Readers of introductory works on Mendel were told, "You may be able

74. Harding 1905, 1.

75. E.g., Osterhout 1905, 452; Hunter 1907, 81; Punnett 1911, 187–89; Trafton 1923, 50–51.

76. Davis 1984, 31.

77. Wilson 1904, emphasis added.

78. De Vries 1905c, 10.

79. Lunn 1934, 175.

to contribute some useful facts."[80] But the same point was often made about more specialist texts; a review in the highbrow *Dublin Review* (of such serious tomes as Bateson's *Mendel's Principles of Heredity*, Punnett's *Mendelism*, and Lock's *Recent Progress in the Study of Variation, Heredity, and Evolution*), concluded by noting that "it is in the power of anyone who has a modest greenhouse, or even an ordinary garden, to carry out such observations for him or herself" (implying that the work was so straightforward that—on rare occasions—even *female* experimenters might be allowed to join in).[81]

Evolving Women

Early twentieth-century science often seemed exclusionary, dominated by highly trained experts—invariably men working in large labs full of expensive, incomprehensible machines. As Peter Bowler has argued, in Britain there was a widespread perception that modern science (relativity being the most obvious example) was just too complicated for the public to understand. They should content themselves with marveling at science's wonders and then cheerfully pay their taxes to fund future research. Such assumptions persuaded numerous publishers to started producing cheap, popular books that offered the public simplified accounts of what they were paying for.[82] The British public often seemed content with the role the elite assigned them (or perhaps they struggled to voice their discontent), but in the USA things developed slightly differently. American publishers produced the same kinds of cheap, accessible textbooks and magazines as their British counterparts but, as we have seen, Americans seemed less willing to be merely passive recipients of new knowledge. The enormous prestige of self-taught American "wizards" such as Edison and Burbank encouraged many people to take an active interest in invention and experiment and even to join in (especially if profits might be made).

Meanwhile, on both sides of the Atlantic, more attention was being paid to formal scientific education in schools. In Britain, such education was often dominated by the physical sciences, which led proponents of the life sciences to organize a National Conference on the Place of Biology in Education (1932), at which Sir Walter Morley Fletcher, secretary of the Medical Research Council, berated the government for their ignorance of biology. He told the delegates about "an important Government

80. Drinkwater 1910, 2.
81. Windle 1907, 356.
82. Bowler 2006, 165–69; 2009.

official" who had recently asked him, "What are genetics?" This question ought to be as funny as "What are Keats?"—"yet we laugh at the latter's illiteracy, even though his ignorance of poetry only affects him, and don't worry about scientific illiteracy of government officials," which affected the whole country.[83]

When the conference's proceedings were reviewed in *Nature*, the reviewer argued that change was urgently needed because everyone needed to understand the part "played by biology in the changing order of civilisation."[84] Despite such demands, change happened slowly in Britain, partly because its elite schools and universities valued the traditional classical education over more practical (and supposedly lower-class) subjects. By contrast, Americans were often more enthusiastic about the democratic possibilities of what Katherine Pandora calls "science in the vernacular," which helps explain why science teaching changed faster in the USA; the now-common phrase "do-it-yourself" was first used in 1910 by the *Popular Science Monthly*, to describe Boston Tech's new hands-on teaching approach.[85] Backyard science and its fans contribute to this emerging do-it-yourself scientific culture, which was in part a reaction to elite science's increasing inaccessibility.

The trend toward the democratization of knowledge was celebrated (or bemoaned) across many kinds of early twentieth-century publications. However, early SF probably included and celebrated its audience more than any other kinds of writing. Gernsback in particular constantly ran competitions that encouraged his readers to become writers (partly, of course, to fill future issues), and among those who responded were numerous women, who began to read, edit, and write SF—contributing new voices to the endlessly unfinished dialogue about its boundaries.[86] One of SF's earliest women writers was Clare Winger Harris, who won a prize in one of Gernsback's competitions and thus became the first woman to publish in his magazines, at least, under her own name (she also produced one of the first published definitions of the genre).[87]

Harris addressed the dystopian side of biology regularly, as in her prizewinning story, "The Miracle of the Lily," which imagined what might

83. Fletcher 1933, 129–30.

84. Review of *Biology in Education* 1933.

85. Pandora 2001, 487; MacLaurin 1910, 493. This is the first citation under "do-it-yourself" in the *OED*.

86. Attebery 2003; Ashley and Lowndes 2004, 22–32; Cheng 2012, 22–23, 57–59; Sleigh 2018.

87. Harris 1931; Yaszek and Sharp 2016, 9–25; Matzke 2017.

happen if humans successfully exterminated insects.[88] Both *Men Like Gods* and *Herland* incorporated pest control into their visions of biotopia, but Harris imagined a world where the insects' struggle for survival has become so intense that they've evolved sufficient intelligence to fight back, threatening humanity's survival. Humans try to cut off the insects' food supply, exterminating all the earth's plant life in the process, so they have to rely on synthetic food and oxygen-manufacturing factories. The humans win but at the cost of creating a sterile planet, devoid of plant life and entirely covered in artificial habitats. For writers like Wells and Gilman, a human-created world without noxious insects was a utopia, but Harris offered a darker reading of where human arrogance and its technology might take us.

Harris also wrote "The Ape Cycle" (1930), which depicted a future that included "the great American desert blossoming as a rose." Burbank had hoped his spineless cactus would achieve a similar miracle, but in Harris's story the key is training apes to work for humans. The apes' limited intelligence restricts their usefulness, so the story's scientist decides that the "evolutionary development of apes must be hastened," which he does using "the vital substance" that controls mental growth, extracted from human glands (a nice variation on the then-popular interest in using "monkey glands" to rejuvenate men by restoring their sexual prowess).[89] The apes evolve rapidly to do more and more work and make life increasingly comfortable for humans. Burbank's great dream, "reclamation of the desert wastes," is accomplished by "artificial rainfall and ape-slavery," and the immorality of slavery is explicitly debated in the story to dramatize the horrifying cost of this apparent utopia. Harris seemed to question the idea that nonhuman animals could simply be treated as a means to human ends (in contrast to the way the Herlanders treat their cats). Her tale appeared thirty-three years before Pierre Boulle's *La Planète des singes* (Planet of the Apes), which inspired the long-running movie series, but foreshadowed its plot: the increasingly intelligent apes rise up against their human overlords, briefly conquering them.[90]

Harris offered another variant on the accelerated evolution trope in "The Evolutionary Monstrosity" (1929), which (like the stories her male contemporaries were writing) made use of recent real-world science but

88. Harris 1928.
89. Rémy 2014.
90. Harris 1930.

substituted bacteria for mysterious rays (figure 8.4).[91] A mad lone scientist, Ted Marsden, injects himself with ever-increasing doses of the transforming bacteria and (predictably) evolves into a giant brain with immense psychic powers, becomes highly rational, loses all human emotions, and (of course) threatens to take over the world. However, the story adds an interesting twist; Ted injects his fiancée, Dorothy, with the bacteria so that she can catch up with him. The initial result is positive: she changes from a stereotypical empty-headed bimbo into a perceptive, articulate, and intelligent women (which allows the hero, Frank, to fall in love with her, inspiring him to save both her and the world). Despite the story's trite romantic ending, Harris seemed to imply that a little advanced evolution is a good thing for women, whereas unchecked male arrogance can only lead to disaster. The moral that (male) scientific ambition has to be controlled reveals her story's debt to *Frankenstein*. Mary Shelley's potent myth has shaped the modern world's sense of the threat of science and, of course, Harris was neither the first nor the last SF writer to utilize it.[92] However, as Patrick Sharp has noted, all Harris's stories are more critical of scientists and their power than most of the evolutionary SF written by men at the same time.[93]

Haldane's high-tech vision of reproduction's future—ectogenesis—was one that attracted particular interest from women writers, many of whom used it to dramatize the potential impacts of unlimited scientific power in male hands. As Susan Squier has shown, there was widespread interest in Haldane's idea of ectogenesis after World War I, as part of a wider debate over population, maternalism, and eugenics.[94] Among those who reacted critically to *Daedalus* was the British writer Vera Brittain, who wrote *Halcyon; or, The Future of Monogamy* (1929, also part of the To-Day and To-Morrow series). *Halcyon* borrowed Haldane's "report from the future" format to allow Professor Minerva Huxtawin to report from the mid-twenty-first century on how the relations between the sexes had been improved (SF writers also invented fictitious distinguished women scientists, often to critique their almost complete absence in the real world).

91. Harris drew on Ivan Wallin's theory of speciation by symbiosis (*Symbionticism and the Origin of Species*, 1927). *Amazing* summarized the theory as: bacteria growing in all cells changed their size and shape, which caused all other organisms, including humans, "to evolve from earlier and simpler forms." See Gernsback, editorial note to Harris 1929.

92. Turney 1998, 6–7.

93. Sharp 2018, 105–15.

94. Squier 1994, 66–95.

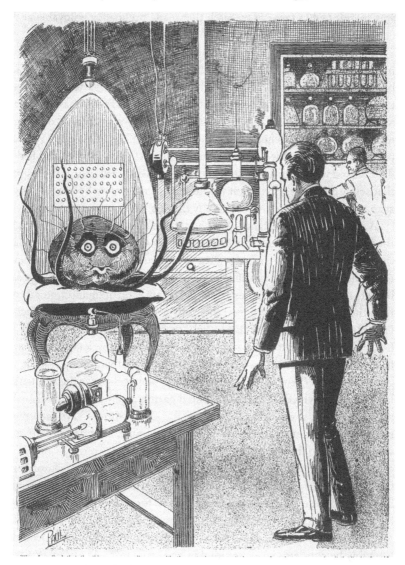

FIGURE 8.4. The evil effects of accelerating evolution, as imagined by Clare Winger Harris.

Brittain's professor discussed "biological inventions" (including those that had been forecast in a "remarkable little volume of predictions called *Daedalus*") that separated sex and procreation. However, in Huxtawin's future, ectogenesis had proved a disaster: children born this way suffered psychological injuries "from lack of parental affection" that outweighed any gains "through being selected from the best stock." The experiment was abandoned in favor of research that made pregnancy enjoyable and childbirth painless.[95]

As Squier argues, Brittain's opposition to ectogenesis was founded on her conviction that children's relationships with their mothers (which began in the womb, where the mother's health directly affected the fetus) were essential to their healthy development. In re-emphasizing the nurture of children (in explicit opposition to those who focused on their nature, that is, their genetic make-up), Brittain offered a vision of moral progress to counter the emphasis on merely perfecting bodies that Haldane shared with Bernal. At the heart of Brittain's alternative was the claim that scientific progress was not inevitable but needed to be consciously shaped by human needs—particularly those of women and children.[96]

A similar hint of resistance to the inevitable march of science appeared the year after Brittain's book in *Science Wonder Quarterly*. The story ("Via the Hewitt Ray," by a woman called M. F. Rupert), imagined another world dominated by scientifically advanced women whose biological breakthroughs include domesticating giant insects, bred and trained to raise the women's food and tend their offspring. Like the women of *Herland*, Rupert's women have had to fight men to achieve their utopia and have slaughtered most in the ensuing conflict.[97] They now keep only a few fertile men for breeding purposes (after subjecting them to rigorous eugenic testing). When Lucile, the visitor from our world, asks the women's leader, Mavia, why they haven't invented parthenogenesis and abolished men altogether, she is told, "We did try it and you should have seen the results. Perfect monstrosities. We did not want our race to deteriorate, so we went back to the age-old method."

However, maintaining the quality of the race is not the only reason for the decision; Mavia reveals that the women also keep a stock of sterilized men for recreational purposes (as she puts it, for whenever "the old biological urge returns"). Lucile finds this notion "downright immoral," but Mavia argues that morality is relative: "What we consider proper would probably be condemned as immoral in your sphere."[98] As Sharp argues, Rupert's fiction largely rejected the Frankenstein-inspired prohibition against interfering with reproduction; the women happily modify nature to serve their needs.[99] However, Rupert did not entirely abandon tradi-

95. Brittain 1929, 75, 77–78.
96. Squier 1994, 68.
97. Sharp 2018, 122–24.
98. Rupert 1930, 378; Sharp 2018, 124–25.
99. Sharp 2018, 124–25.

tional gender roles; Lucile rescues one of the oppressed males, but finds he has been so cowed by his upbringing that he makes a most unsatisfactory husband. She finally loses her temper with him, shouting "Remember you are a man" and "Women aren't anything to be afraid of." In fact, she adds, "You are in every way superior to a woman" (at which point Lucile records the private thought, "May my sisters in feminism forgive the lies. I had to be drastic"). Lucile's father (the scientist whose work has made this inter-dimensional adventure possible), expresses his surprise at her speech: "I thought you wanted the men to admit the women's superiority," to which Lucile responds, "It all depends on who the man is!"[100] As with the maternalistic focus of *Herland*, her decision could be interpreted as conservatism, a symptom of the author's inability to fully shake of the spell of the natural, but it seems more playful and interesting than that: Lucile asserts her right to sexually select and is happy to tell a few anti-feminist lies to get the kind of mate she wants.

Harris and Rupert are just a couple of examples of the women writers who helped mold early SF, whose importance to the genre's early years is only gradually becoming recognized.[101] For a relatively brief period, women were welcomed in the pulps, and they sometimes offered alternatives to the genre's largely uncritical celebration of scientific progress. (Women writers would of course re-emerge—after a long, testosterone-poisoned interregnum—in the sixties and seventies to redefine SF in ways that allowed the genre's full imaginative and political possibilities to be recognized.)[102] Women's distinctive contributions to early SF are a useful reminder that the genre was never as monolithic as traditional, prescriptive definitions might imply; more historically nuanced definitions illustrate that SF was a cloth woven from many threads, as the weft of earlier genres (notably gothic, utopian, and adventure fiction) was woven through the warp of new scientific theories, new technologies (particularly in printing), and an expanding, competitive publishing market. Defining SF in a historical (rather than an essentialist) way reveals how much it had in common with the writing considered in earlier chapters, much of which could be considered as forms of speculative fiction. (As Mikhail Bakhtin has argued, "the boundaries between fiction and non-fiction, between literature and non-literature and so forth are not laid up

100. Rupert 1930, 420.

101. Roberts 1993, 40–65; Yaszek and Sharp 2016, xvii–xix; Sharp 2018; Holquist 2002.

102. For examples, see Lefanu 1988; Frank, Stine, and Ackerman 1994; Donawerth 1994; 1997; Merrick 2003; Yaszek and Sharp 2016; Yaszek 2018.

in heaven.")[103] Textbooks could contain images that might remind readers of SF, as could literary fiction or journalism, and the threads that made an argument for commercializing biological inventions could also be woven into a socialist's dream of a better future. If texts were like cloths, woven from many different threads, they could also become patchworks, as authors and readers cut them up and rejoined them—selecting and combining different texts and appropriating them for their own purposes.

Some women writers used the imaginative resources provided by biology to fashion unambiguously dystopian visions of where male scientific control over reproduction would lead, as in Charlotte Haldane's *Man's World* (1927) or Katherine Burdekin's *Swastika Night* (1937). Other women used new scientific theories to imagine utopias characterized by a more questioning tone than was exhibited by many of their male contemporaries. One of their critical strategies was refusing to completely abandon the category of the "natural"; as Fran Bigman has argued, for writers like Charlotte Haldane and Brittain, the simple right to give birth could be a way of protesting against the technocratic fantasy of the artificial womb.[104] Biology has often been used to constrain women, to imprison them within the narrow cage of their supposedly natural functions, but in the early twentieth century it could also provide ways of resisting the male imagination, which often regarded women's bodies as just another of the natural resources that science would soon be able to master and exploit.

103. "Epic and Novel," quoted in Tattersdill 2016 (23).
104. Bigman 2016.

Conclusion

BRAVER, NEWER WORLDS?

In 1932, Joseph Needham, reader in biochemistry at Cambridge University, reviewed the most recent biotopia. He explained that it predicted an "autocratic dictatorship" which would use "the resources of a really advanced biological engineering" to create a completely stable society. The first step would be "sorting out the [human] eggs into groups of known inherited characteristics and then setting each group, when adult, to do the work for which it is fitted." Needham suspected that the book would be rejected because readers would assume "the biology is all wrong, it couldn't happen." However, such reactions would bring a "sardonic smile" to the faces of biologists, because in fact "the biology is perfectly right"; the book contained only "legitimate extrapolations from knowledge and power that we already have."[1]

The book was, of course, Aldous Huxley's *Brave New World* (*BNW*), which was compared by Needham to a garden of "man-eating orchids in a tropical forest," an image of threatening unnaturalness that captured something of the book's mood. Needham emphasized that Huxley's "diabolical" vision was entirely possible (it would scarcely have been worth getting a scientist to review if it were not). Writer and critic Rebecca West struck a similar tone when she reviewed it, suggesting that Huxley ought to have included a preface to explain "how much solid justification he has for his horrid visions," given Haldane's similar inventions.[2] Needham commented at length on Huxley's epigraph, from the Russian philosopher Nikolai Berdiaev (Nicholas Alexandrovich Berdyaev), who had claimed that "utopias appear to be much more realisable than we used to think" (a similar claim to the *New York Times*'s claim that Wells had persuaded people that utopia was a place where we might "land and take possession";

1. Needham 1932, 77.
2. West 1932 (1975).

see p. 165). Berdiaev had no desire to land in utopia and asked, "How can we avoid their actualisation? For they can be made actual. Life is marching towards them."[3] After three decades of discussion and prediction, some readers believed experimental evolution's forecasts were moving from possible to inescapable.

BNW was a response to the biotopian tradition (particularly its Marxist-tinged British version) but was not a straightforward rejection of it. Margaret Atwood recalls that her first reading of the book (at the age of fourteen) left her with "a vivid picture of 'zippicamiknicks,' that female undergarment with a single zipper down the front that could be shucked so very easily." When she was growing up (in the era of "elasticized panty girdles," which required "an epic struggle" to either don or remove), Huxley's book "was heady stuff indeed."[4] Atwood's comment is an invaluable reminder that the book presented what was in many respects an alluring image of the future: unlimited, consequence-free sex; free drugs with no aftereffects; personal helicopters; "feelies" that let you experience the actor's physical sensations; and, despite being a wholly undemocratic society, the world's controllers were dedicated to universal happiness (at least, to their interpretation of it).[5] Despite the book's nightmarish aspects (particularly for snobbish, upper-caste intellectuals like its author), there is no denying its appeal.

Many of *BNW*'s ideas were borrowed directly from Haldane's *Daedalus*, the first fully developed biotopia. Although Huxley obviously did not share his friend's enthusiasm for biology's imagined futures, Haldane's optimism still leaked through (the World State had abolished hunger, disease, pain, suffering, poverty, and fear—at least for those outside the "savage reservations"). Huxley's novel generally ignored the food people ate and where it came from, but the expert cloning of humans would surely have been perfected via experiments on plants and animals (as Haldane and others had predicted), so the World State's agriculture would have resembled the engineered crops envisaged by the biotopians.

However, *BNW*'s most obvious debt to biotopianism was the rejection of both original sin and naturalistic ethics. In the novel's penultimate chapter, John "the Savage" confronts Mustapha Mond, one of the world controllers, and after a lengthy debate about the pros and cons of the

---

3. Quoted in Needham 1932, 76.

4. Atwood, introduction to A. Huxley, Atwood, and Bradshaw 1932 (2007), ix.

5. Kumar (1987, 264–65) notes that many American college students in the 1950s found *Brave New World* appealing. He also comments that the World State's treatment for dissidents was the same one that Wells proposed in *A Modern Utopia* (260).

FIGURE 9.1. Modernizers of plants and people: Thomas Edison, Luther Burbank, and Henry Ford, photographed at Luther Burbank's home in Santa Rosa, California, 1915.

new world, John rejects it, telling Mond: "I want God, I want poetry, I want real danger, I want freedom, I want goodness. I want sin."[6] That, ultimately, is what he finds missing from the future: the knowledge of good and evil, without which there can be no right or wrong and thus no freedom. Unhappiness, he agrees with Mond, is the price of freedom, and he is willing to pay it, whereas Mond is convinced that the only way to ensure that "everybody's happy nowadays" is to banish original sin from his new artificial Eden and let Henry Ford's gospel of efficiency take over the work of providing moral guidance.

Huxley later recorded that *Men Like Gods* had "annoyed me to the point of planning a parody," which developed into *BNW*. As Robert S. Baker has argued, Wells's novels provided "a kind of unconscious pattern

6. Baker 1990, 25.

of the future for Huxley and his generation."[7] (This point is reinforced by comparing the list of future biological inventions offered by Wells and his co-authors in *The Science of Life* [see p. 223] with those that appear in Huxley's novel.) The apparent absence of sin is one of many similarities between Huxley's book and those it is responding to. However, *BNW*'s critique was not as stark as it initially appeared. It is a truism that every utopia is someone else's dystopia, but Huxley's book is a reminder that the reverse can also be true; Haldane, Bernal, Muller, Wells, and Aldous's brother Julian would probably have been more than willing to live in the brave new world (not least because they would doubtless have pictured themselves as world controllers). Aldous Huxley could even imagine a place for himself in his imagined future; the World State punishes dissenters by exiling them to an island, where they can join a community of like-minded dissidents and live as they choose (details that found echoes in Huxley's 1962 utopia, *Island*). By contrast, *Men Like Gods* seems to offer its dissenter, Lychnis, no way to step off the treadmill of scientific progress; in that one small detail at least, Huxley's dystopia is more utopian than Wells's utopia.

BNW includes a brief and deliberately fragmented history of the World State's emergence (after "the Nine Years' War, the great Economic Collapse. There was a choice between World Control and destruction . . ."); a nightmarish "age of confusion" like the one Wells's Utopians had survived.[8] Huxley looked at the threats the world was facing—Fascism, Stalinism, economic depression and mass unemployment—and concluded that his grandfather, Thomas Henry Huxley, had been right: humans might resist the cosmic progress (or inadvertently collaborate with it by making their lives too comfortable), but the final outcome would be the same—amoral nature would stamp out human hope, thanks to the traits people had inherited from their apelike ancestors, which would lead to overpopulation, the multiplication of the unfit, and mass degeneration.

Aldous's grandfather had briefly mentioned one possible hope: the cosmic process might be beaten if "men's inheritance . . . their dose of original sin, is rooted out by some method at present unrevealed" (see p. 163).[9] However, for him—and for nineteenth-century biology—that was a forlorn hope; the secularized, Malthusian original sins of ingrained lust and violence might be controlled, but they could never be eradicated. The human animal's "innate tendency to self-assertion"—to struggle, compete

7. Huxley to Christopher Collins, 1962, quoted in Baker 1990 (25). See also Hillegas 1967, 110–23; Sherborne 2013, 303.

8. A. Huxley, Atwood, and Bradshaw 1932 (2007), 39–43.

9. Huxley 1894 (1989), 102.

and breed—made utopia impossible. Thomas Huxley's pessimism was rooted in the Victorian conception of ancestral heredity; a fortunate few might enjoy a valuable inheritance associated with good breeding (land, a title, or a healthy constitution), but for the ill-bred majority heredity was a burden (of debt, disease, or stupidity).[10] Heredity was the dead weight of the past, endlessly revisiting the iniquities of the fathers upon the children, generation after generation.

Yet just a few years after Thomas Huxley's death, experimental evolution rapidly transformed the biological understanding of heredity: instead of weighing and measuring past burdens, biologists became increasingly focused on the speculative tools with which they expected to make the future. Initially catalyzed by de Vries's mutation theory, the new biology claimed to be able to speed up, improve, and above all *control* the evolutionary process. Remaking nature meant, of course, that people could no longer turn to it for guidance (so naturalistic ethics had to go), but experimental evolution offered in exchange the hope of finally turning our collective backs on our evolutionary heritage, leaving humans free from old constraints and limitations. Exactly how that last trace of original sin would be eradicated remained obscure, but the biotopians confidently predicted that it could and would be banished. And if predictions are performative (see p. 39), the mere act of saying that people could be perfected began to change the future.

At first sight, Wells appears to be an exception to the period's optimism, mainly because he is best known for the work he produced in the 1890s, particularly *The Time Machine* and *The Island of Doctor Moreau*. Each expressed a view Wells summarized in an article ("Human Evolution: An Artificial Process" [1896]) in which he argued that humans breed too slowly for natural selection to have made much impact on us over the short span of human history, leaving the average human largely unchanged since the Stone Age. Hence the "permanence of man's inherent nature," which included the most important features of human psychology, "love of hunting and violent exercise, and his [sic] powerful sexual desires."[11] As Piers Hale has argued, the pupil was echoing his teacher Thomas Huxley's vision of tainted humanity.[12] Humanity's ineradicable, bestial inheritance was apparent in Wells's depiction of the Morlocks, and it provided the haunting ending to *Doctor Moreau*, when the narrator, Prendick, has escaped the island and returned safely to London yet shrinks

10. Müller-Wille and Brandt 2016a, 17.
11. Wells 1896b (2009), 591–93.
12. Hale 2010, 38–39.

in fear from other humans, convinced that "the animal was surging up through them."[13] In Wells's dramatizations of Huxley's arguments, our evolutionary original sins are lust and violence, marks of the beast within.

The Utopians in *Men Like Gods* have also eaten the forbidden fruit but have gained the knowledge to master nature and eliminate all traces of sin (as is implied by their return to an Edenic nudity).[14] Unsurprisingly, the Earthling who objects most vehemently to Utopia is the priest, Father Amerton (described as "dreadfully outspoken about the sins of society and all that sort of thing"). Upon learning that there is no marriage in Utopia, Amerton describes it as "a hell of unbridled indulgence" where everyone lives in "bestial promiscuity!" To which, as we have seen, the sinless Utopians respond, "This man's mind is very unclean" (p. 179). There is no organized religion in Utopia and no belief in original sin (Wells had identified the rejection of original sin as the "leading principle of the Utopian religion" in his earlier book *A Modern Utopia*).[15] Even the Malthusian form of secularized original sin has been banished from Utopia, since population is carefully regulated (rationality has conquered humanity's "powerful sexual desires," just as it did in *Herland*), and so there is no struggle for existence (humanity's "innate tendency to self-assertion" has also apparently been "rooted out"). Eradicating ancestral heredity's burden has made the Utopians "like Gods."

The contrast between the pessimism of Wells's fin-de-siècle work and the optimism of *Men Like Gods* encapsulated the way biology had changed in the intervening decades. The year before Wells's novel appeared, Julian Huxley began a popular account of modern genetics by noting that in his grandfather's day, "pioneer workers in the field of heredity often came to regard its power as something very like that of the Greek *Atë*"—the goddess of remorseless destiny or fate (as Huxley put it, heredity "accumulated the sins of the forefathers within itself, to charge their grievous burden upon the children's shoulders").[16] However, the twentieth century would see the end of *Atë*'s reign when Mendel's work was rediscovered

13. Wells 1896b (2009), 128.

14. When the serpent tells Eve that if she and Adam eat the fruit of the forbidden tree of knowledge, "Ye shall not surely die: For God doth know that in the day ye eat thereof, then your eyes shall be opened, and ye shall be as gods, knowing good and evil" (Genesis 3:5).

15. Kumar 1987, 29.

16. Originally published as "Heredity and Evolution," *World's Work* 41, no. 241 (December 1922): 15–22; reprinted as Huxley 1926a (1933), where it was combined with a second article, "Recent Work on Heredity," *Discovery* 7 (July 1920): 199–203.

in 1900; his ideas "had lain unproductive for half a century, like seed in a box. Now at last they were to germinate," to serve "as the basis of all the enormous advance" that led the old, "gloomy point of view" to be rejected by modern workers ("to whom the idea of a hereditary force pushing men blindly along a predestined road is hateful").[17]

By contrast, Julian's brother Aldous remained convinced that *Até* still governed; the "doctrine of Original Sin is, scientifically, so much truer than the doctrine of natural reasonableness and virtue." Aldous averred that original sin was epitomized by the "anti-social tendencies inherited from our animal ancestors" and was thus a "familiar and observable fact." This was a reality he believed all Utopians ignored; men like Wells took refuge in a "compensatory dream" because "reality disgusts them."[18] *BNW* dramatized its author's conviction that, no matter what changes science made, his grandfather Thomas had been right; original sin—our inner ape—would always be with us. While Aldous remained gloomy, genetics made Julian optimistic. Although geneticists did not yet know "how to produce mutations," they soon would, thus enabling them "to build up improved races of animals and plants as easily as the chemist now builds up every sort and kind of substance in his laboratory."[19] He made similar claims a few years later, telling a BBC radio audience that, thanks to biology, "blind acquiescence in destiny is giving place to the hope that destiny may in large measure come to be controlled."[20] The physical sciences had previously offered control over nature, but biology underpinned the twentieth-century optimism that fate was about to be conquered. Biotopianism depended on the shift from the nineteenth-century model of ancestral heredity (Aldous and Thomas Henry Huxley's view) to the twentieth-century view of speculative heredity (to which Julian Huxley adhered).

However, early twentieth-century biology's reimagining of heredity was accompanied by important continuities with earlier thinking, the most obvious of which was the survival of stadial evolution as a model for biological predictions. The ladder of progress had shaped expectations of Darwin's *Origin*, encouraging readers to interpret it as yet another

17. Huxley 1926a (1933), 10, 1. By the 1920s, Huxley was already working on what would become Huxley 1942. By the time he was writing, most of the expert biological community had abandoned de Vries's mutation theory, hence his focus on Mendel.
18. A. Huxley 1927, 19, x.
19. Huxley 1926a, 28.
20. Huxley 1926b, 1.

progressive and optimistic evolutionary epic. Half a century later, the biotopians' optimism was evidence of the evolutionary epic's abiding influence, despite the changes wrought by experimental evolution. A belief in inevitable scientific progress also shaped the Baconian scientific utopia, which was an important influence on the biotopian style. Bacon's original goal had been persuasive—by dramatizing the idea that recent science had finally made utopia possible, he hoped to inspire a powerful prince to implement his ideals. However, not all utopians shared such goals. As Judith Shklar has argued, Thomas More's original *Utopia* was not a call to action but an object of contemplation (the Platonic form of an ideal society, not a practical blueprint for one). While Wells apparently hoped his utopias would inspire action, it is less clear that Haldane entertained similar hopes for *Daedalus*. His forecasts seem intended as thought-provoking (objects of contemplation): to persuade readers of the importance of biology rather than to encourage them to create actual purple oceans. The same could be said of Bernal's *The World . . .* , while Muller seems closer to Wells; his attempt to gain Stalin's support for his plans suggest that he shared Bacon's hopes of a powerful patron who would institute a new world (with no regard for the opinions of the existing world's inhabitants). But, of course, writers can never control the ways their texts are read; the new biotopian horizon of expectations—embodied in science-fictional thinking—enabled readers to treat objects of contemplation (like *Daedalus*) as if they were calls to action, as was reflected in the repeated claims that these biological speculations were not merely possible but might even be inevitable.

Biotopianism became most popular in America, whose supposed discovery inspired many of the most influential utopias (More's narrator, Raphael Hythloday, had sailed with Amerigo Vespucci, and in *New Atlantis* the island of Bensalem is located near America).[21] The early Puritan settlements were the first of many attempts to establish working utopian communities in the New World (many of which explicitly hoped to be free from the various sins of the old one).[22] And *BNW* was strongly influenced by Huxley's first visit to the USA, after which he had argued that the "future of America is the future of the world" and so for "good or for evil, it seems that the world must be Americanized"; eventually everyone would worship Ford, dance to the wail of "sexophones," and then copulate

21. Roemer 1976; Manuel and Manuel 1979; Kumar 1987, 23; Fender 1992, 27–28.
22. Roemer 1976.

without hesitation—or any sense of sin (the seductive allure of Huxley's book was every bit as American as its nightmarish qualities).[23]

The biotopian mood emerged just as the USA became aware of itself as an emerging world power, restlessly inventive and in a hurry to shape the new century. As the people of the Philippines and Puerto Rico were discovering, the closing of the frontier at the end of the nineteenth century would lead to the USA looking for new ways to impose its values on the rest of the world (and, Jackson Lears has argued, organicist metaphors for society were routinely combined with evolutionary ideas to make expansion seem like an inevitable force of nature). Meanwhile, new inventions like the dynamo and X-rays created a vision of seemingly limitless energy, which helped fuel a reaction against the "pinched and parochial" nineteenth century, creating a new sense of abundance in which everything from money to psychic energy was to be spent, not hoarded.[24] The result was the mass consumerism that appalled Aldous Huxley.

The distinctive form biotopianism took in America was shaped by the USA's new self-confidence, but it also contributed to it. While America's entertainers and inventors were dangling affordable abundance before their countrymen, Luther Burbank promised to bring the same qualities to America's gardens and kitchens. Burbank's name has appeared so often in this book because he became synonymous with this American form of biotopianism—a bustling, pragmatic approach to creating the future (and bringing it to market) as rapidly as possible. As we have seen, his fame ensured that the contradictory ways in which he and his work were described came almost to define science for many Americans. His homely qualities helped make the new science less intimidating, while his more mystical side helped some to reconcile science with a wide variety of religious beliefs.

However, it was Burbank's image as an altruistic huckster that had the biggest influence on American biotopianism. Despite his advocacy of capitalism (and the numerous failures and frauds his name was linked to), even some socialists came to believe that his kind of science could serve humanity. Biotopianism's lasting legacy is apparent when modern biotechnology companies claim they will make a profit for their shareholders by feeding starving people (despite the fact that famine is more often caused by a lack of money than by an absolute shortage of food). These

23. "The Outlook for American Culture: Some Reflections in a Machine Age," *Harper's Magazine* 155 (August 1927). Quoted in Kumar 1987 (246).
24. Lears 2009, 199–201, 213, 224–27.

implausible claims are apparently more readily accepted in the USA than elsewhere, which perhaps explains why Americans consume biotechnology's products—notably genetically modified foods—more willingly than most Europeans do. The Burbankian myth of blooming deserts growing cheap food still lingers.

Nevertheless, American ideas about biology's impact were self-contradictory, shaped by both the application of mass production to agriculture and by the bucolic national myth that democracy had to be built on Jeffersonian small farmers with honest dirt under their fingernails. As chapter 2 showed, the contradictions of what Leo Marx termed America's "middle landscape," a form of human-made nature—provided another key context for the emergence of biotopianism. The imaginative vision of reforesting the prairies made large-scale bioengineering seem plausible. Liberty Hyde Bailey promoted the need for horticultural transformation built on recent science. Meanwhile, his attachment to the Jeffersonian myth left him convinced that "nature is the norm," and he adopted a form of naturalistic ethics that separated him—and many other environmental thinkers—from the biotopians.

By contrast, British writers were often the products of a small, crowded, heavily cultivated island in which Malthus's gloomy prognostications seemed all too persuasive. It was Britain's Francis Bacon who had first outlined the power of empirical science, which later seemed to offer an escape from the Malthusian trap. Nevertheless, Thomas Huxley and his fellow Victorians doubted whether utopia was achievable; some offered a traditional religious explanation, others a modern evolutionary one, but most agreed that humans were too competitive and lustful to reenter Eden. By contrast, Wells, Haldane, and other early twentieth-century writers used the latest biology to imagine a radical extension of the work that centuries of British farming had achieved; nature could be remade in ways whose unashamed radicalism reflected their loss of faith in any guidance that might be drawn from either God or uncultivated wilderness.

Thomas Huxley had been typical of his countrymen when he envisaged civilization as colonizing a patch of land, putting a wall around it, exterminating the natives, and calling it a garden. Bacon, who could be considered the ultimate founder of biotopianism, was a cultivated Briton living in a period when untamed nature was still a threat to human beings. His enthusiasm for gardening was apparent when he extolled "the culture and manurance of minds," merging the still-common literal sense of *culture* (tending and cultivating nature) into a more metaphorical extension (which implicitly condemned a lack of cultivation, in either a land or its people). When he celebrated gardening, he was urging his readers to create a typical, formal Elizabethan garden—a walled, geometrical, and orderly refuge from the dirt and danger of wild nature.[25] By contrast, American environmentalism traces its origins to a desire to protect the wilderness and symbolically regain its supposedly natural virtues; Britain's wild nature was too remote to serve the same purpose.[26]

Making Sense of Science

The argument that science became crucial to the way many twentieth-century people imagined the future hardly needs making, but the vital importance of biology during these early decades is less widely appreciated. The diverse range of people who read and wrote about every aspect of heredity and evolution suggests that public biology extended far beyond the well-known example of eugenics. The period saw biology become a form of participatory culture, in which fans of all kinds engaged in bricolage as they adapted the latest discoveries to their own needs and interests (a process utterly unlike the claim of a professionalized elite creating and validating knowledge, which was then communicated to a largely passive public). Analyses of the cultures of more recent media fans offer useful tools for thinking about their early twentieth-century counterparts, but we need to remember that—unlike a TV series or a movie franchise—the object of the early twentieth-century science fans' enthusiasm—experimental evolution—was incomplete and thus demanded more intensive interpretative work from everyone who took an interest in it. The absence of a stable community of acknowledged experts allowed the fans to play a part in defining such things as the full implications of de Vries's mutation theory. Their interpretations competed and overlapped with those of formally accredited scientists, and the changing nature of both science and the media at this time meant the scientists could neither control nor ignore the imaginative uses to which their work was being put.

In 1905, Professor John Coulter of the University of Chicago gave his fellow scientists some advice about communicating with those he called the "intelligent public" (defined as "the public that thinks and brings things to pass"). He argued that most general-interest publications relied on finding a "middle man who stands between science and the magazine public," who "may simply *interpret* for the public, putting the language of science into the language of literature." The results, he believed, were so

25. Hoyles 1991.

26. For the contrasts between British and American attitudes to wildness and wilderness, see Coates 1998, 82–109. Also Cronon 2003.

unsatisfactory that scientists must "become their own interpreters," an expectation that he realized would come as a "distinct shock" to most of them. Those who took up the challenge should focus on the impact of their work; it was not enough to "describe in perfectly simple English" their experiments with "evening primroses or with pigeons" (readers "would simply wonder at the things that amuse some men"). Instead the scientist-interpreter needed to demonstrate the relevance of their experiments "upon the origin of species and upon heredity," so that they assumed "a dignity and an importance that even the public will be quick to appreciate." A key reason for undertaking this work was that it would help to "*secure endowment for research*," funding being one of the biggest problems American science faced.[27]

As we have seen, *interpreter* was used regularly to describe those who tried to guide audiences in their attempts at making sense of science. Bacon had defined scientific researchers as "interpreters of nature," while Wharton's Professor Linyard railed against "false interpreters" and the Carnegie Institution's attempts to find an "interpreter of Mr. Burbank's work." *Interpreter* is a deliberately broad term to describe writers (of fiction and nonfiction, from professional journalists to novelists, including scientists themselves), but also their readers (an equally varied group). And readers and writers overlapped, as the letters pages of the SF pulps showed particularly vividly. As the media market and the scientific community were expanding and diversifying, generalist scientific journals (such as *Nature, Science, Scientific American,* and the *Popular Science Monthly*) became vital reading for both the general public and for scientists who wanted to know what was going on outside their own disciplines.

The same processes fueled everything from the newspaper "Scientific Jottings" column (described in Wharton's "Descent of Man") to "Science News of the Month" (in Gernsback's *Science Wonder Stories*). Increasing science education promoted the demand for textbooks, but also the perception that everyone should—and could—take an interest in science. The fact that biology became the most popular American school science subject (see p. 195) both reflected and created the demand for accessible information on the topic. Focusing on interpreters—creators and audiences—illustrates how biology came to function as public culture; it was more than simply popular or widely discussed, more than just science being done "in public." Biology became a decisive factor that changed how the future was imagined. We can see how it altered the horizon of

---

27. Coulter 1905, 306–10, original emphasis. Coulter's daughter, Merle Crowe Coulter (1894–1958), was a botanist who published several papers on *Oenothera*.

expectations by analyzing three examples: eugenics, Mendelism, and the mutation theory.

✳

The best-known example of how biology become part of early twentieth-century Anglo-American public culture is, of course, eugenics, but the scholarly attention it has received tends to obscure the importance of other kinds of biology.[28] One of this book's core arguments is that the dominant view of heredity began to change around 1900 (from what I have called an ancestral to a speculative model) and part of the evidence for that change was that dissident voices began to emerge within the eugenics community. Despite the fact that mainstream eugenics continued to dominate public conversations about biology's impact, the new biology's possibilities caused a subtle yet unmistakable shift in eugenic thinking during these decades.

As its historians have shown, eugenics originated with the fear that human civilization was becoming too artificial (from its sanitary provisions to its ethical codes), which stopped nature's key law—natural selection—from operating effectively. Instead of allowing the weak and unfit to perish, modern societies increasingly protected them with well-intentioned but misguided welfare provisions. These dark fantasies lay at the heart of the fear of degeneration.[29] According to the eugenicists, the solution was to reintroduce natural selection (albeit in a supposedly humane and civilized form); as its founder, Francis Galton, put it, "What Nature does blindly, slowly, and ruthlessly, man may do providently, quickly, and kindly."[30] From the outset, eugenics promised a utopian alternative to degeneration. As Galton said: "If a twentieth part of the cost and pains were spent in measures for the improvement of the human race that is spent on the improvement of the breed of horses and cattle, what a galaxy of genius might we not create!"[31] However, his vague prescriptions for achieving this goal were primarily negative: to avoid propagating "idiots by mating *crétins*" and generally preventing the allegedly inferior from outbreeding their betters (thus ensuring that the British contributed to

28. The literature on eugenics is extensive (and still growing), but for an overview see Stern 2016; Bashford and Levine 2010; Currell and Cogdell 2006; Black 2004; Gillham 2001; Searle 1998; Soloway 1995; Mazumdar 1992; Kevles 1986.
29. Lankester 1879; Kevles 1986, 70–84; Pick 1989; Soloway 1995; Woiak 1998.
30. Galton 1904, 5.
31. Galton 1865, 165.

the process by which the "feeble nations of the world are necessarily giving way before the nobler varieties of mankind").

Eugenicists never reached a consensus as to which human qualities were best, but even if they had been able to, there was even less agreement about *how* to increase the number of those who possessed the supposedly desirable qualities. By contrast, they reached all-too-eager consensus about who the "unfit" were and what to do about them (built on nineteenth-century assumptions about heredity). As Galton explained:

> The phrase "nature and nurture" is a convenient jingle of words, for it separates under two distinct heads the innumerable elements of which personality is composed. Nature is all that a man brings with himself into the world; nurture is every influence from without that affects him after his birth.[32]

This well-known quote encapsulated ancestral heredity's core assumption; every child was indelibly marked by the past, blessed—or cursed—by whatever it inherited (as Galton commented, he had "no patience with the hypothesis . . . that babies are born pretty much alike, and that the sole agencies in creating differences between boy and boy, and man and man, are steady application and moral effort").[33] One result of the ancestral model was that mainstream eugenics focused on eliminating undesirable burdens (and those who bore them).

Daniel Kevles has argued that by the 1930s eugenics had acquired a new direction, which he termed "reform eugenics," that de-emphasized coercion and negative measures in favor of education and propaganda intended to promote voluntary, positive measures to improve humanity's genetic stock. He offered various explanations for the trend: developments within genetics suggested that eliminating the supposedly inferior through compulsory sterilization would take tens or even hundreds of generations; but there were also political shifts, partly inspired by revulsion at eugenic practices in the USA and Germany.[34]

Before evaluating Kevles's argument, it is worth noting that well before the period he discusses, there were already countercurrents within eugenics, the result of speculative heredity beginning to displace the ancestral version. For example, in 1905 the pioneering chemist, sociologist,

---

32. Galton 1874, 12.

33. Galton 1869 (1892), 14.

34. Kevles 1986, 164–73. However, Pauline Mazumdar (2002) has disputed whether "reform" eugenics even existed.

and biology teacher Ellen Swallow Richards coined the term *euthenics* to describe what she called "the science of controllable environment." She argued that people could be improved by improving their circumstances, and offered this argument as an alternative to the strategies proposed by the eugenicists.[35] (And although her term never became as widely used as *eugenics*, it nevertheless occurred in more than 10 percent of the textbooks considered in chapter 6 and was particularly prominent in those affiliated with the civic-biology movement.[36]) A similar shift in emphasis was evident two years later, when the term *positive eugenics* was coined by British doctor Caleb Williams Elijah Saleeby, who emphasized methods for making better people (rather than reducing the numbers of the supposedly inferior).

Both euthenics and positive eugenics were sometimes presented as complementing mainstream eugenics, but Saleeby—like Richards—offered his alternative approach in overt opposition to some of the core tenets of mainstream eugenics. Saleeby had become an ardent publicist for eugenics after hearing Galton in 1904 (and he referred to Galton as "my master" from then on).[37] Nevertheless, he distinguished positive from negative eugenics in order to criticize what he called the "ludicrously inept" strategy of sterilizing the supposedly unfit. Instead he endorsed Alfred Russel Wallace's proposal to promote women's equality (to allow female choice to flourish; see chapter 7).[38]

Saleeby grew increasingly aggressive in his attacks on the view that "young people are the trustees of a certain type of germ plasm *which nothing can alter*," which he called a "perversion" of eugenics, merely "the echo of dead formulae from the nineteenth century" (*germ plasm* was the earlier, widely used term for what would eventually become known as genes). By contrast, thanks to "new experimental work," twentieth-century biology had proved that "some influences may and do act upon the germ plasm . . . with most important consequences for eugenics." In addition, Saleeby argued repeatedly that eugenics would degenerate into a "meaningless farce" if it focused exclusively on heredity and ignored environmental effects. He tried to persuade his fellow eugenicists to concern themselves with improved state welfare provision, health care, nutrition, and education for "the parents of the future," because (in a flat contradiction of his "master" Galton, who was dead by the time he wrote these

35. Richards 1905, 12; 1910; Richardson 2002.
36. E.g., Hunter 1914; Atwood 1922; Linville 1923.
37. Searle 2019.
38. Saleeby 1907, 31; Galton 1908, 645; Ward 1913, 738; Searle 1976, 73.

words), Saleeby was convinced that "the nurture of the future parent may affect the nature of the offspring."[39]

Saleeby was a doctor who specialized in obstetrics, so his emphasis on the health and welfare of mothers and children is unsurprising, not least because it fitted well with contemporary campaigns to improve diet and health care (he was also an avid temperance advocate). Like the advocates of euthenics, he believed the nurture of parents and children was as important to the goals of eugenics as the selection of breeding stock.[40] However, Saleeby also offered his own, idiosyncratic interpretation of new biological theories (which included neo-Lamarckianism, but also Bergson, Mendel, and a touch of de Vries), which he combined to give his eugenic writings a distinctive tone. He averred that "I am an optimist because I am an evolutionist"; looking at humanity's achievements, he put his faith "not in any supposed inevitable law which makes for progress, but in action" (a view that echoed those of socialists like Wilshire).[41] The promotion of action, not acquiescence, was typically biotopian, and he shared the experimental evolutionists' belief that people and their germ plasm were directly manipulable (like Gilman, he adopted the Lamarckian view that if the health of would-be parents were improved, their offspring would inherit those improvements).

Saleeby's eclectic biological bricolage also incorporated the arrival-of-the-fittest problem. He argued that "natural selection *selects*; it does not originate or create," so selection of human beings—however ruthless—could never create new, improved people. The origin of new variations had once seemed intractable, but it was now known that "new species can and do arise," as new experimental work showed (he cited both Mendel and the "many facts which the mutation theory explains"). The practical implications of these new biological approaches were revealed by a "study in the eugenics of wheat," which was improved by what "my master, the founder of eugenics, Sir Francis Galton . . . called nature and nurture"—even the best strains of wheat needed adequate fertilizer.

Cambridge University's Roland Biffen had used Mendelian techniques to improve wheat, but Saleeby suggested that Mendelism might not have the "all-embracing character" that many researchers assumed. He noted that Luther Burbank (despite employing methods that were "markedly dissimilar to those of Biffen") had nevertheless achieved wonders.[42]

39. Saleeby 1914, 553, 551–52.
40. Saleeby 1909, viii–ix; Bashford and Levine 2010, 69, 217.
41. Saleeby 1906, 26–27, 117, 119, 284.
42. For Biffen, see Radick 2023, 192–93, 263–64.

Acknowledging that there was a lot more to be learned about heredity and its applications, Saleeby nevertheless concluded with the prediction that, with appropriate state funding, Mendelism and Burbank's work would be extended and combined to create "hosts of new food plants, yet in the womb of Time, for the benefit of mankind at large."[43] The ways in which his arguments (couched in vaguely science-fictional language) reflected recent biological ideas effectively positioned positive eugenics as just one of twentieth-century biology's tools for shaping the future.

Despite being rather hazy as to the details, Saleeby had grasped that experimental evolution had solved the arrival-of-the-fittest problem, thanks to some combination of Mendelism and mutation. Burbank might also play a role, and Saleeby shared his neo-Lamarckian confidence in the inheritance of acquired characters. The resulting mixture of what now appear as incongruous biological ideas led him to the same conclusion as many of his more orthodox contemporaries: both the tempo and mode of evolution had been changed by experimental evolution ("Nature does sometimes make leaps," which demolished "the old dogma" of slow, gradual change).[44] Confidence that evolution meant progress was, of course, widespread, but it gave Saleeby's eugenic arguments a biotopian tone: he condemned what he called "the motto of the impotent"—that we "have to take the world as we find it"—it was, he averred, our refusal to do so that defined us a human. By rejecting what nature gave us, he claimed that "man" could become "a creator of his world," whereas refusing to accept this challenge would mean acquiescing to the "amazing wastefulness of Nature" (echoing the views of both Burbank and Wells's Utopians).

At the heart of Saleeby's challenge to mainstream eugenics was his prediction that "human intelligence" would soon mean that what "man can do for animals and plants," he would soon be able to "do for himself" (he endorsed E. Ray Lankester's description of man as "Nature's Insurgent Son").[45] The biotopian mood was particularly apparent when he claimed that human nature was not fixed ("human nature is not the same in all ages; that it was once simian nature, once vermian nature, once lower still. The establishment of organic evolution is the establishment of the truth that progress is possible, since progress has occurred").[46]

The claim that human nature could be improved contrasted sharply with the views of most conventional eugenicists, who argued that it was

43. Saleeby 1919, 324, 327.
44. Saleeby 1906, 26–27, 117, 118–20, 284.
45. Saleeby 1909, 41.
46. Saleeby 1907, 30.

FIGURE 9.2. An image of biotopia? Luther Burbank looks at the orderly, profitable garden that he and his fans dreamed of creating.

fixed and more potent than the effects of nurture. Galton's disciples tended to pick out the most uncompromisingly hereditarian of his pronouncements (for example, it is "in the most unqualified manner that I object to pretensions of natural equality"). Karl Pearson (first occupant of the Galton-funded chair of eugenics at University College London) ignored the complexities of Galton's thinking when he argued that if the government were to improve the living conditions of the working class (for example, "limit the hours of labour to eight-a-day" and provide "leisure to watch two football matches a week"), they would simply "find that the unemployable, the degenerates and the physical and mental weaklings increase rather than decrease."[47] By contrast, Saleeby accepted that the struggle for existence was a "terrible fact of Nature," but it was "only a means to an end. It is our destiny to command the end whilst *humanising* the means."[48] Many eugenicists might have accepted that as a summary of their goals, but Saleeby was unusual in placing greater emphasis on both positive measures and the role of nurture.

Saleeby's arguments were varied (and not always coherent), and they led him into conflict with the leaders of eugenics, particularly Pearson (who was especially irritated by Saleeby's references to Galton as "my

47. Galton 1869 (1892), 14; Pearson 1909, 21. Radick (2023, 52–57) argues that Galton was never the hardline hereditarian he is usually assumed to be.

48. Saleeby 1909, 1, 14, 41.

master").[49] As Geoffrey Searle has shown, what Saleeby termed positive eugenics had always been part of Galton's thinking but had proved much harder to implement than negative measures.[50] Nevertheless, Saleeby's dissent went much further than reminding eugenicists of the other side of Galton's legacy. He called on eugenicists to focus on the future, on the power of the new experimental biology to shape it, and to abandon their obsession with eliminating those who carried the burdens of the past. That emphasis aligned him with some of those Kevles calls reform eugenicists, such as Muller, who argued that eugenics needed to be rescued from reactionaries and transformed into a positive sense of "conscious social direction of human biological evolution," informed by feminist values (see chapter 5). Some of Saleeby's views also resemble those of Haldane and Bernal, who rejected eugenics as being far too slow. And he almost matched their radicalism when he proposed the beehive as a model for human society (a "bee-society can be completely or at all substantially imitated only by remodelling human nature on the lines of the individual bee. This is very far from impossible").[51]

However, long before Haldane, Bernal, and Muller were publishing, less well-known figures were arguing that the latest biological theories were fatal to eugenics. For example, in 1913 the *London Quarterly Review* published a detailed article by the Rev. Dr. J. Parton Milum titled "Fallacies of Eugenics," which he defined as the "application to human life of the current form of the evolution theory." Like Saleeby, he used the arrival-of-the-fittest argument to attack eugenics (noting that the "weak link in the evolution theory has been the attribution of creative power to selection"). Breeding experiments showed only modest improvements in a species, so eugenic policies based on selection would fail. Eugenics was based on an obsolete version of evolution, as had been proved by de Vries's mutation theory ("of which it may be said the more it is known the more it prevails").[52] Milum concluded that "if the Mutation Theory is true, then the possibility of producing a superman by selection is excluded"; if only "the knowledge of mutations come earlier, Eugenics would never have been born" (112). He condemned eugenics as a tool for silencing "the dreamers" and "idealists"; once people accepted that "the struggle for existence of man with man, so far from being our 'natural' state is a condition which falls below the truly human standard," they would unite

49. Searle 1976, 2–3; 2019.
50. Searle 1976, 74–88.
51. Saleeby 1910, 63.
52. Milum 1913, 108–11.

and use science to produce enough of the "necessities of life" to satisfy everyone's needs (115, 114).

Milum's article was picked up by several American journals, including as part of an ingenious piece of bricolage titled "Violent Science in State Legislatures: The Reaction of the Experts against Eugenics." The title derived from the British geneticist William Bateson's condemnation of US state eugenic legislation as being "too violent" and based on outmoded science. The writer noted that Bateson's views had not initially received much expert approval, but the "tide has since turned completely" in favor of those who argued "that eugenics has been carried too far." The evidence was an assemblage of selections from various other articles, which included Milum's claims (particularly those concerning mutation). After quoting Milum's detailed account of de Vries's work, the writer noted that the mutation theory had received "convincing support from Mendel's discovery," and the article concluded with a quote from Henry Smith Williams (leading promoter of the Luther Burbank Society; see p. 107), which explained the basic Mendelian ratio, a "simple formula" that explains "many observed facts of heredity that were formerly mysterious." To support the claim that Bateson's and Milum's views represented a new mood, the journal quoted another recent piece (from the British medical journal the *Lancet*), which argued that many supposedly degenerate traits were not in fact inherited.

Nevertheless, the magazine writer was not entirely anti-eugenics, and they tried to convey the sense that eugenics had been misrepresented. They quoted Pearson ("its greatest living authority") on the popular nonsense increasingly associated with eugenics (which had "become a subject for buffoonery on the stage and in the cheap press. We are treated to 'eugenic' marriages and to 'eugenic' babies, and to 'eugenic' plays").[53] The article was a curious patchwork but perhaps no more eclectic than those put together by Gilman, Saleeby, and others. It would have left readers with the (fairly) clear impression that new biological ideas had created a new optimism about solving social problems. Eugenics of some kind would continue, but in new and perhaps more progressive directions.

Mainstream, hereditarian eugenics remained the dominant form, but the subtle influence of experimental evolution—and the dissent it created—has been marginalized by existing histories of eugenics. To return to Kevles, he concluded his chapter on the popularity of eugenics with the following quote from American zoologist and geneticist Herbert Spencer Jennings:

53. "Violent Science in State Legislatures" 1914.

Gone are the days when the biologist . . . used to be pictured in the public prints as an absurd creature, his pockets bulging with snakes and newts. . . . The world . . . is to be operated on scientific principles. The conduct of life and society are to be based, as they should be, on sound biological maxims! . . . Biology has become popular![54]

Kevles used this quote to illustrate the popularity of eugenics, but Jennings had unambiguously stated that it was *biology* that had become popular. He had prefaced the words Kevles quoted by noting that an "eagerness to apply biological science to human affairs is a marked feature of the times," which had led society to turn to biologists for "advice" and "leadership"— hence the change in the public's perception of them. Kevles's ellipses removed these words:

> The uplifter hastens to secure the endorsement of the biologist for his particular remedy for human ills. The man in the street recognizes that if his practices are not biologically sound, they are not sound at all; the biological expert must set the seal of his approval upon them. Profound changes in practice are urged upon the world as pronouncements of biological science.[55]

"Uplifters" (social reformers or "do-gooders") had been around for decades—what was new was that they now sought the imprimatur of biology.[56] (Haldane's prediction that the "poor little scrubby underpaid" biologist would soon be recognized as the "most romantic figure on earth" seemed to be coming true; see p. 172.)

Of course, eugenicists were among the biological experts that society was consulting, but Jennings was publicly critical of eugenics. Far from seeing it as the cause of biology's new popularity (as Kevles seems to imply), he explicitly attacked it as the source of significant biological "fallacies" then in circulation.[57] Some were the work of ill-informed interpreters ("the work of middlemen, near-biologists—the popular writers that have undertaken to 'sell' biology to the world"). However, others, he argued, were "traceable to the biologists themselves" (203–4). While the

54. Kevles 1986, 69; original ellipses.

55. Jennings 1930, 203.

56. The *OED* gives the earliest citation from the *Nation*, November 21, 1923: "There is nothing the matter with the United States except . . . the parlor socialists, up-lifters, and do-goods."

57. Kingsland 1987, 811; Barkan 1991.

public's interest put biologists in a potentially "Utopian situation" (almost as newly acknowledged legislators of the world), he raised "doubts and worries" about "some of the maxims that are circulated in the name of biology"—and eugenics provided most of his examples of biology being misapplied to society.

Among the eugenic shibboleths Jennings dismissed was the belief that the characteristics of organisms were determined either by heredity or environment, imagined as mutually exclusive causes. He argued that anyone who saw heredity as the "force or entity" that ensured that like begets like would become "a propagandist for eugenics as the only remedy for human ills." This notion might seem "too crude for belief," Jennings acknowledged, yet it was widely assumed (208–9, 213). He also condemned those who averred that if something were heritable "it is not alterable by the environment," a mistake that expressed itself "popularly in the notion of the inevitableness, the fatality, of what is hereditary" (that is, that heredity was *Até*, remorseless fate). To disprove this view, he offered examples of clearly Mendelian characters that were "readily modifiable or improvable by environmental changes" (214–15). Many reviewers of Jennings's book appreciated that it was anti-eugenic. One noted that it would disappoint any "extreme eugenist" who had read "many popular works on the subject" instead of following "the much more painful method of experimentation." Those who paid attention to the rigorous new world of the laboratory and experimental garden would have understood that the "'new heaven and the new earth' of the eugenist are not in the immediate future."[58] And the *New York Times* was equally scornful of those "popularizers who read a little of the new biology and then loudly proclaimed that humanitarianism was all spinach. Heredity, it was announced, was everything, environment nothing." Such views had been debunked by Jennings, who called instead for more research and massive social improvements. His approach, the *Times* suggested, "may be humanity's way out," so perhaps "compassion is not, after all, an obsolete virtue."[59]

Jennings was a prominent spokesperson for the dissident minority who interpreted recent biology as undermining eugenics, specifically its negative prescriptions. He argued that experimental evolution—particularly Mendelian genetics—provided both reasons to doubt the eugenicists' claims and more appealing alternatives. Meanwhile Saleeby drew on various aspects of recent biology in an effort to discredit the nineteenth-century view of heredity, and Milum rejected eugenics completely because

58. Kunkel 1931.
59. "Biology and the Humane Spirit" 1930.

its underlying assumptions had been discredited by the mutation theory. Naturally, mainstream eugenics did not crumble before these scattered criticisms: Milum's view was attacked in the *Atlantic Monthly* by the energetic eugenics propagandist Samuel J. Holmes, professor of zoology at the University of California, who accused Milum of misunderstanding mutation theory (quoting de Vries's claim that his idea was "in harmony" with Darwin's—because selection retained its role). However, even Holmes was not immune to the persuasive power of experimental evolution. He claimed that even if Milum were right about selection running out of variation and thus being unable to improve a species beyond a certain limit, that limitation need not matter: "When the human species has been raised to the level of its best specimens Nature will probably be kind enough to supply us with further mutations in the direction of progress."[60] Even a few of the conventional eugenicists were being swayed by the biotopian possibilities of de Vries's theory.

Obviously, the new biology only influenced a minority; many more clung to the nineteenth-century view of heredity and were thus ready to follow the prescriptions of what the *New York Times* called the "harsh and pessimistic philosophies which have been built out of a cocksure half-knowledge of modern biology."[61] Their reviewer hoped Jennings's book, with its "fine and humane spirit," would change this state of affairs, but of course most remained committed to both the older views and the harsh policies derived from them. Very few eugenicists dissented from Galton's argument that "Darwin's law of natural selection acts with unimpassioned, merciless severity. The weakly die in the battle for life; the stronger and more capable individuals are alone permitted to survive," but the species—including *Homo sapiens*—improved a result. Galton believed the human "sense of original sin" was evidence that as a result of selection we were rising from our lowly origins ("our forefathers were utter savages from the beginning; and, that, after myriads of years of barbarism, our race has but very recently grown to be civilized and religious").[62]

Nobody needs reminding of the horrors that followed from Galton's theories, but it is worth remembering that when he formulated his original schemes, the prospect of changing human nature was not even a pipe dream. Once experimental evolutionists offered an alternative, some people began to imagine that positive eugenics and euthenic measures were becoming more practical and might perhaps complement the other

60. Holmes 1915, 227.
61. Duffus 1930.
62. Galton 1865, 323, 327.

possibilities that biology was offering (including traditional negative eugenics). Evidence for such a shift is limited, but recognizing the ways new biology created a more optimistic mood may help historians of eugenics address what often seems like a paradox: given that the evils of eugenics now seem self-evident, why did it remain popular for so long? Diane Paul has argued that left-wing eugenicists (among whom she lists Muller, Haldane, and Bernal) shared both their fellow geneticists' confidence in their new science's ability to solve the world's problems and their political comrades' confidence in the value of state intervention. The result was a form of eugenics (which Kevles called reform eugenics) that mostly shunned the class and race prejudices of mainstream eugenics and instead stressed voluntary choices and argued for equalizing social conditions (partly so that biologists could determine which human characteristics were really genetic). Nevertheless, most Left eugenicists still shared the wider hereditarian consensus that human personality and ability were (at least partly) determined by people's genes. Like other eugenicists, they also disagreed as to who were the fittest and thus over which kinds of people society needed more of, which made it hard to agree on positive measures. Yet despite these disagreements, they largely agreed about the definition of *unfit*, so left-wing eugenicists often endorsed the consensus around negative eugenics. For example, the leftist biologist Lancelot Hogben (*Genetic Principles in Medicine and Social Science* [1931]) described eugenics as a way of setting national minimum standards for parenthood—as if it were no different from setting factory safety standards.[63]

However, the undeniable similarities between Left/reform eugenicists and their mainstream counterparts may distract from a more interesting point: the key figures Paul identified (Muller, Haldane, and Bernal) were also the most radical biotopians. Their most celebrated ideas (from ectogenesis to modifiable human bodies) went far beyond the minimalist consensus around negative eugenics that Paul identified (and that all three partly rejected). These elite geneticists looked for radical, positive solutions to the problem of making better people; an interest they shared with SF writers, utopian feminists, and Burbank fans of all kinds. Tracing the rise and fall of eugenics to shifts (scientific or political) within the genetic community does not address the fact that eugenics was popular with many who had little or no understanding of genetics (as is evident in the way writers like Gilman and Saleeby mixed their eugenic ideas with a dash of Lamarckianism and a touch of mutation). Wells's *Men Like Gods* was just one of several biotopias in which eugenics played a role, yet for Wells,

63. Paul 1984, 574.

eugenics was only one—comparatively unimportant—option among several tools that had been used to remake nature.

As I argued in the introduction, analyzing the diffuse but widespread influence of the biotopian mode on early twentieth-century biological thought makes it clear that the word *eugenics* was being subjected to the same kind of debates that affected *Darwinism* and *evolution*—and for many of the same reasons. To return to Koselleck's argument, the space of experience (embodied in ancestral heredity) was shrinking as the weight of the future increased (thanks to the expectations aroused by speculative heredity). It could be argued that the radical biotopians' vision of future humans was simply eugenics under a new name, but that argument seems to obscure more than it reveals. The word *eugenics* undoubtedly persisted, but it is a mistake to assume that there was therefore a stable definition or concept of eugenics. As Thomas Dixon argues in the case of altruism, changing meanings do far more than reflect or mirror broader historical changes: "words are tools that allow people to do things such as creating identities for themselves, arguing with each other, and articulating new visions of the natural and social worlds." Linguistic change can be a motor that drives other kinds of change, as the changing meanings of *Darwinism*, *evolution*, *mutation*, and *eugenics* clearly illustrate.[64]

In a well-known critique of Raymond Williams' *Keywords*, Quentin Skinner argued that the shifts Williams described were not really about the meanings of words but about their reference or application, which were matters of social convention.[65] It is useful to be reminded that *meaning* covers far more than dictionary definition; however, for historians, the socially sanctioned "correct" usage of a term (and the ways communities may seek to change that usage) are decisive when we consider the history of shifting meanings. Consider the reclamation of the word *queer* in recent decades; those who used the word in the "wrong" way were not denying the existence of the behaviors or identities it referred to, but by using it to express approbation rather than disapproval they eventually succeeded in changing its reference, and thus its real (that is, social and historical) meaning. The lexicographers eventually caught up, but they were not the group that drove this important change. The same can be said about the new meanings that attached to some key biological words analyzed here: they became tools with which to describe possible worlds—and thus made biology's imaginary futures more likely to become real.

64. Dixon 2008a, 33.
65. Skinner 1979, 210–11.

Without denying the horrifying impact of eugenics on so many lives, it was not the only way in which biology reshaped the public sphere in the early twentieth century. Eugenics was just one aspect of a much wider conversation about biology's future impact, a point that may help explain the broad nature of its long-lasting attraction. For example, the fact that some of biology's audiences had absorbed the biotopian perspective helps us understand the apparent paradox of eugenics' appeal to many whom (with the benefit of historical hindsight) it was unlikely to benefit. For example, as various historians have noted, eugenics held particular appeal for women despite the fact that they were more likely to be forcibly sterilized than men were. The wider currents of biological optimism help explain why eugenic arguments also appealed to some progressive women (including Gilman), who joined eugenic societies in particularly large numbers. One motive was to gain access to information about sex and birth control (otherwise hard to obtain), but eugenics also made women's traditional work of giving birth and raising children seem even more socially valuable.[66] Perhaps for some women, eugenics was their route to DIY evolution (as experimental gardening was for the boys), allowing them to be active participants in creating an improved future humanity. Gilman's seemingly contradictory celebration of motherhood as women's most natural role, while simultaneously ridiculing most claims about women's supposed biological limitations, makes more sense if the positive eugenics of *Herland* is seen as just one among a number of biotopian strategies.

The possibility that some saw eugenics as a way of actively assisting evolution is reinforced by the way eugenics manuals were regularly advertised in the SF pulp magazines, particularly during the 1930s (figure 9.3). The quantity of such ads (as well as numerous references to eugenics in stories) shows that SF fans were assumed to be interested in the topic (albeit perhaps for a variety of not-strictly-scientific reasons).[67] Readers could send off for books like *Safe Counsel* ("tells you the things you want to know straight from the shoulder"), which covered all aspects of sexual life (from basic anatomy to "mistakes to avoid"), as well as a chapter on "the science of eugenics."[68] Numerous other pulps carried adverts for *Modern Eugenics*, whose title was dwarfed by the headline "The Greatest Sin,"

66. Hasian 1996, 72–73; Woiak 1998; Kline 2001; Rembis 2006; Ziegler 2008; Paul 2010; Weindling 2012; Oveyssi 2015; Redvaldsen 2017.

67. Investigating the role(s) of eugenics in early SF would make an excellent PhD project, particularly because so many of the pulps have now been digitized and are thus easily searchable.

68. *Safe Counsel* (published by the Franklin Association, Chicago) was advertised repeatedly in the early pulps, e.g., *Weird Tales* (September 1927): 431.

FIGURE 9.3. Eugenics was just one more form of science fandom that attracted popular interest during this period. This advertisement for a popular eugenics manual, which appeared in *Amazing Stories*, hints at the factors that were expected to attract readers.

which was "total IGNORANCE of the most important subject in the life of every man and woman—SEX." The ad urged readers to "face the facts of sex fearlessly and frankly, sincerely and scientifically." It told women "what liberties to allow a lover" and "why husbands tire of wives," while men would learn the "secrets of greater delight" (as well as explaining more obviously eugenic matters, such as "how to have perfect children"). Part of the allure of such books was evident from the stern warning "This book will not be sold to minors" and the guarantee that it would be delivered in a "plain wrapper."[69]

The Popular Book Company of New York offered pulp readers a ten-volume encyclopedia, the *Sexual Education Series*, which covered everything from the facts of life to "eroticism and society" and "the abnormal sex life of siblings" (as well as "Sterilization and Segregation"). Its author was SF author and doctor David Keller (who wrote "Stenographer's Hands"; see p. 271). His decision not to use a pseudonym for any of his writings suggests that he was as happy to be associated with SF as with sex education.[70] Herman H. Rubin's *Eugenics and Sex Harmony* (1933) was another encyclopedic volume that offered to help readers take control of their sexual and reproductive choices; it, too, was regularly advertised in the pulps.[71] As Michael Rembis has argued, these titles are evidence of the eugenics movement's shift from coercion to persuasion during the 1930s.[72] However, the way they were advertised also suggests how eugenics could be interpreted as part of the participatory culture of DIY science; these manuals offered the same opportunity as those that offered courses in electrical engineering or telegraph operation—they offered readers the skills and knowledge they would need to play their part in shaping the future.

Dissent from classical, mainstream eugenics was echoed in *Brave New World* (whose title, like Galton's terms *nature* and *nurture*, was borrowed from Shakespeare's *Tempest*). Aldous Huxley's novel is now read in the context of genetic engineering as a dire warning of the consequences of eugenics. It is routinely invoked by the media whenever there is a new

69. Adverts for *Modern Eugenics* (Grenpark Company, NY) can be found in *Air Wonder Stories* (December 1929): 566, *Science Wonder Stories* (April 1930): 1048, and *Miracle Science and Fantasy Stories* (June/July 1931): 231.

70. The *Sexual Education Series* was also advertised regularly, e.g., *Amazing Stories* (March 1929): 1143 and *Science Wonder Quarterly* (Winter 1930): 288.

71. For example, adverts for *Eugenics and Sex Harmony* (Pioneer Publications, NY) appeared in *Amazing Stories* (December 1942): 242 and *Planet Stories* (Winter 1947): 125.

72. Rembis 2006. See also Currell 2006.

biotechnological breakthrough, on the assumption that because the book is about cloning, it must be about genetic manipulation.[73] However (like their counterparts a century earlier), today's journalists are happy to talk about books they may not have read (or haven't read recently or carefully). *BNW*'s society was built primarily on *nurture*, rather than on altering people's natures (something Aldous Huxley, like his grandfather, assumed was impossible). The novel's first chapter describes the process of creating new humans, as Henry Foster (a senior technician at the Central London Hatchery) shows new students around, explaining how ova are harvested from suitable donors, then grown in artificial wombs. If they are lower-caste embryos (from eggs provided by lower-caste donors), they are cloned to produce Bokanovsky groups (large sets of identical twins).

However genetic engineering implies the manipulation or alteration of an organism's genetic material—something that simply does not happen in *BNW*. Even though deliberately induced mutations had been under discussion for thirty years by the time Huxley wrote, they are not even mentioned in the novel. (There is only one brief discussion of unsuccessful attempts to reverse a naturally occurring mutation.) Genes are neither inserted nor deleted, and genes from other species are not used to augment the new human's capabilities. Foster tells a group of students that the hatchery's work has taken its staff "out of the realm of mere slavish imitation of nature into the much more interesting world of human invention," but the embryos' genes are left untouched. The production line allows the scientists to "predestine and condition," so that embryos are decanted (born) already adapted to the environments they will face using physical conditioning; embryos destined to be workers in hot environments pass through "hot tunnels alternated with cool tunnels. Coolness was wedded to discomfort in the form of hard X-rays. By the time they were decanted the embryos had a horror of cold" (which, incidentally, is the only mention of X-rays in the book—even though Muller's use of them to induce mutations had become well known several years earlier). When the hatchery has to produce Freemartins (sterile females), normal female embryos are simply dosed with male sex hormone—their genes are not altered. Most importantly of all, the real conditioning is psychological, beginning with aversion therapy in the "Neo-Pavlovian Conditioning Rooms," which is reinforced with hypnopaedia (sleep-teaching). As Foster says, "We condition them to thrive on heat. . . . Our colleagues upstairs will teach them to love it." Or, as the Director of the Central London Hatchery puts it, conditioning makes "people like their inescapable social destiny."

73. E.g., Kumar 1987, 275–77.

The processes described in *BNW* were, among other things, a reminder that early twentieth-century biology consisted of much more than genetics and evolution; psychology (especially its new behaviorist form—which dominated Huxley's book) also played a crucial role in imagining the future. And doing justice to the imaginative possibilities that hormones brought to early twentieth-century culture would require another book as long as this one. Nevertheless, *BNW* has come to be read as the key myth, the *Frankenstein*, of the genetic century—a book about genetic manipulation, eugenics, and biologically determined predestination—while most of its other themes (and its rich ambiguity) are ignored. It is better understood as an anti-biotopian work, aimed at refuting the views of those like Wells, Haldane, and its author's brother, Julian. The possibility that their biological utopias were possible, even unavoidable, motivated Aldous Huxley's attack—evidence of just how influential the biotopian mode had become.

✳

If eugenics might be expected to be one of the dominant themes in a book about the cultural reception of biology in the early twentieth century, historians of science would expect Mendelism to be the other. The rediscovery of Mendel's work in 1900 is often regarded as marking the start of the "Century of the Gene," which began with a bottleful of fruit flies and ended with the Human Genome Project.[74] This is not, of course, an inaccurate description of the period, but nor is it a complete one. The genetic century also began in the early twentieth century's daily newspapers and popular magazines, but a detailed study of them reveals that Mendelism never impressed the broader public as much as it did the scientific community.

As with hormones, a full account of the reception of Mendelism would require another sizeable book, but the main factors that limited public excitement included the fact that Mendelism was usually presented as a limited (rather than universal) theory of heredity, one that might not be applicable to all species. For example, the radical Manchester doctor Solomon Herbert's *First Principles of Heredity* (1910)—based on a course of lectures to "working men and others"—was typical of many early textbooks in stressing that Mendelism appeared to apply "only to certain characteristics of the organism, which form allelomorphic pairs, of which one unit is dominant, the other recessive" (Mendelism's unfamiliar and

74. Keller 2000 (2002).

off-putting jargon also limited its appeal).[75] Skepticism about Mendelism's universality might be unsurprising, given the book's date, but in 1924 Vernon Kellogg—despite being a pioneer of Mendelian research in the USA—was still arguing that "there is undoubtedly much heredity that is not Mendelian in character."[76]

In addition, Mendelism was usually described as applying to hybridization—and hence concerned with blending existing species rather than creating new ones (when Liberty Hyde Bailey introduced Mendel, he described his theory as an explanation of "uniformity and constancy of action in hybridization").[77] Most writers emphasized that Mendel had worked on hybrids and often described Mendelism as a theory of hybridization rather than of heredity. More than two decades after Bailey, the British writer and broadcaster Mary Adams concluded her account of Mendelism by noting that the "breeding pen and the pollen brush offer acid tests for the hybrid. Blood will tell, and the mongrel will stand convicted."[78] Identifying what she called "impure blood" was, of course, topical and seemed important, but it could not compare to the excitement of a supposedly new theory of evolution.

Gradually, Mendelism got more attention in the press, becoming a standard feature of almost all biology textbooks during the 1920s. However, as more was written it became clear why Mendel's theory lacked popular appeal. Mendelism was a little intimidating for most general readers—mutation theory was much easier to grasp. As Wells and his fellow authors had put it in *The Science of Life*, Mendel's work initially failed to make much impact, because "peas and arithmetic" are "not the sorts of things that cause excitement and clamour." Most textbooks insisted Mendelism was mathematical, and they usually plunged straight into the details of ratios and statistics, equations and Punnett Squares, and included a host of new terms (*recessive/dominant, heterozygous/homozygous*, and so forth). *The Science of Life* acknowledged that Mendel's ideas had stimulated "an enormous amount of experimental breeding of animals and plants," but three decades after their rediscovery, the authors believed it had still not finally resolved the Lamarckian controversy, nor proven whether or not variations had any direction.[79]

75. Herbert 1910a, 140. Radick has argued that Mendelism had to be presented (and is still taught) in a radically oversimplified form in order to be comprehensible even to science students (2023, 340–46).

76. Kellogg 1924, 131.

77. Bailey 1906, 157.

78. Adams 1929, 27.

79. Wells, Huxley, and Wells 1929–1930, 279–80.

However, the main obstacle to getting the public excited about Mendelism was that they could not initially see how it solved the arrival-of-the-fittest problem. In 1911 the *Washington Post* celebrated Mendel's work as having caused a "radical change" in scientific views, yet the article concluded that "Mendel's theories throw no light upon the origin of species" but mainly proved Darwin was wrong (species did not arise "from the accumulation of minute and almost imperceptible differences").[80] And in the same year, Alfred Russel Wallace, the grand old man of British evolution, argued that while the "Mendelian phenomena" might illuminate "the theory of heredity," they "have absolutely nothing whatever to do with the origin or modification of species."[81] At best, Mendelism might have solved the swamping problem (that was how J. Parton Milum interpreted it; new mutations were not "diluted out of existence"), but it did not appear to explain where such mutations came from.

Given the ways Mendelism was interpreted for general audiences, it would be astonishing if it *had* caught the public's imagination (so perhaps the expectation that it ought to have dominated public discussions of biology is a side effect of treating the history of science as primarily the history of scientists). And given the public's limited interest in Mendelism, it seems implausible that its supposed rise could explain shifts in the popular eugenics movement. Many early twentieth-century interpreters of biology did not see Mendelian and Lamarckian inheritance as incompatible, and some eugenics organizations remained deliberately agnostic. The problem that anyone who wanted to write a popular account of Mendelism faced was nicely summed up by the British naturalist Frederick Headley in his book *Life and Evolution*: the mutation theory was "like a bomb flung at orthodox Darwinism," whereas Mendelism concerned "the non-blending of the alternative characters (the allelomorphs)."[82] It is not hard to see why one message excited readers more than the other.

✳

As early twentieth-century biology offered key imaginative resources that shaped public conversations about the future, the mutation theory's seemingly fantastic possibilities became a common language shared by both elite scientists and their various publics. Both theosophist propaganda and dry, technical reports to the directors of the Carnegie Institution of

80. "Current Theory of Heredity" 1911.
81. Wallace 1908, 136–37.
82. Headley 1906, 244–45, 247.

Washington expressed their utopian hopes in terms of what the evening primrose seemed to offer. And mutation's influence extended from textbooks to newspapers, and from pulp science to literary fiction. And yet, as noted previously, histories of early twentieth-century biology barely mention the mutation theory—it warrants little more than a footnote (and general histories of the period seldom mention any aspect of the life sciences, apart from eugenics). If we were able to ask the early twentieth-century public what biology was and whether it mattered to them, eugenics would surely have dominated their answers, but many would also have mentioned the mutation theory (far more than would have suggested Mendelism), because the mutation theory launched "experimental evolution" and the inspiring (or alarming) possibility of engineering life itself.

Much more could be said about the mutation theory's significance, but this book's focus has been on its central role in turning biology into public culture. Experimental evolution was taking shape in public, as scientists tried both to define it and communicate its significance to various audiences—many of whom were enthusiastically offering their own views.[83] The promise of controlling mutation, and thus evolution itself, generated enormous interest, but nobody knew exactly how that promise might be fulfilled. The theory's apparently unfinished nature encouraged its audiences to complete it, to produce their own interpretations of how the mutation story might turn out. As a result, the mutation theory could be considered as a candidate science (rather as phrenology had been in the early nineteenth century); it was presenting its credentials and asking for admission into the club of respected, elite sciences, and its uncertain status encouraged both experts and laypeople to play a role in deciding the theory's fate. As a result, the boundary between the experts and the general public (which had supposedly been strengthened by the increasing professionalization of science) remained porous.

As earlier chapters have shown, the problem of explaining the origin of evolutionary novelties—and thus of species themselves—captured the attention of a broad audience, hence the widespread interest in the mutation theory's claim to have solved the problem. John Burroughs, America's most popular nature writer, regularly used the phrase "arrival of the fittest" to encapsulate the central problem that conventional Darwinism

83. My thinking on these issues is much indebted to Bruno Latour and Michel Callon's actor-network theory. See Callon 1985; Latour and Woolgar 1979; Latour 1999; Star and Griesemer 1989.

seemed unable to solve.[84] He was still arguing that the "mutation theory of De Vries is a much more convincing theory of the origin of species than is Darwin's Natural Selection" two decades after its first announcement.[85] Like so many of biology's early twentieth-century interpreters, Burroughs could not see how variation—the small, everyday differences between organisms—could offer a "handle for selection to take hold of." He became convinced that something "more radical must lead the way to new species," with large de Vriesian mutations being the most plausible candidate.[86] Expert biologists demurred; zoologist Charles Nutting, director of the University of Iowa's Museum of Natural History, used the pages of the *Scientific Monthly* to respectfully suggest that Burroughs had misunderstood key aspects of evolutionary theory.[87] However, even though Nutting denied that mutation theory disproved natural selection, he nevertheless agreed that de Vries was substantially correct. Despite the stark differences between their perspectives and interests, the theory—and the key biological words associated with it—provided the two men with a common framework within which to debate.

Burroughs's enthusiasm for the latest science did not preclude a spiritual view of nature, which he imbued with a vague sense of purpose.[88] At times he could sound like one of the biotopians (as, for example, when he wrote that "Nature is non-human, non-moral, non-religious, non-scientific"), yet he remained committed to a form of naturalized ethics (arguing that it was from nature "that we get our ideas," including those of religion and morality). Hence he concluded that "living bodies have, or have had, a purpose," and so, while he acknowledged that "Nature is blind," he nevertheless believed that "she knows what she wants and she gets it."[89] Burroughs offered a mixture of Lamarckian ideas and Bergson's creative evolution to explain the apparent purposefulness of nature—and, as we have seen, many other Americans were committed to similarly eclectic mixtures of ideas, even though the scientific community would have seen them as incompatible (or incomprehensible).

The potential difficulty of reconciling spiritual and scientific beliefs was most famously dramatized at the Scopes trial, which played an unexpected

---

84. Burroughs 1915, 197; Burroughs 1916, 197–227, 265. All republished in Burroughs 1921. For Burroughs's public role and impact, see Lupfer 2010.

85. Burroughs 1920, 246. Jeff Walker (2014, 52, 42–44) briefly notes the influence of de Vries on Burroughs.

86. Burroughs 1920, 246.

87. Nutting 1921, 127–29.

88. Burroughs 1915, 197–98.

89. Burroughs 1920, 242.

role in keeping the mutation theory in the public eye. As Adam Shapiro has demonstrated, Scopes was much broader than science vs. religion, but it addressed a series of clashes that have been central to the making of the United States (North vs. South, urban vs. rural, state vs. federal, and Black vs. White). It encompassed both local issues (mandatory textbooks being used to impose Yankee values on Southern children) and broader ones (the appropriate roles of church and state)—but all were played out in terms of biology. The trial was directly sparked by the newly self-confident biologists arguing that they now had the expertise to guide their fellow Americans into the future, and the textbook at the heart of the trial argued that the biologist's newfound confidence was partly based on the way that "Hugo de Vries, the Dutchman[,] . . . recently showed that in some cases plants arise as new species by sudden and great variations known as *mutations.*"[90] A couple of years later, Hunter produced an updated *New Civic Biology*, which placed even greater emphasis on de Vries's achievement, noting that an important part of "the work of the plant breeder is to discover, isolate, and breed useful mutations."[91] New plants (and perhaps even animals or people) were the practical fruits of mutation theory. As another 1920s textbook argued, the new knowledge would allow "the farmers of to-morrow how to grow two blades of grass where one formerly grew," which would make "to-morrow a healthier, happier, and more complete day."[92] By promising to change both the tempo and mode of evolutionary change, de Vries's theory transformed evolution into a tool for shaping better futures.

Although the word *mutation* was initially defined by a scientist, it rapidly acquired an expanded significance through dialogue, gaining and shedding meanings as it was put to work by different communities of readers in Britain and the USA.[93] The same was true for the term *Darwinism*, which also became the subject of intense debate (suggesting that, far from being eclipsed, Darwinism was more important than ever). The slightly different ways in which scientists, socialists, and SF fans used *mutation* helped to expand its meaning and reference. The same is true for many other important scientific terms (notably *evolution*); their significance is

90. Hunter 1914, 253. There are parallels with the anti-Darwinian backlash that followed the 1959 Darwin centennial celebrations in the USA; see Smocovitis 1999.

91. Hunter 1926, 385.

92. Wheat and Fitzpatrick 1929, 9.

93. Bakhtin 2010; Holquist 2002. See also Fish 1976 (2010), 1982–83; Suvin 1979 (2016); Chartier 1989; Kramer 1989; Baker 1990; Alfaro 1996; Darnton 1996.

created and sustained through dialogues between members of scientific communities and nonscientific ones.

Of course, neither *scientific* nor *nonscientific* is a stable term—each defines the other to some extent. Twentieth-century evolution's paraliterature allows us to analyze how the cultural and historical significance of key terms emerged. As Delany stressed, paraliterature serves to create a social definition of literature by exclusion (see p. 30), and the scientific paraliterature has a parallel function of constructing the boundaries that are supposed to separate science from pseudoscience. However, that work becomes even more vexed than usual when a novel theory is not yet able to defend its territory with science's standard tools (such as recognized peer-reviewed journals or university departments dedicated to its study).

Against a background of shifting borders and definitions, various forms of authority are accrued by specific groups, whose accepted interpretations gain credibility, but authority and credibility often prove short-lived. Fifteen years after the mutation theory had first been announced, the British scientific journal *Nature* observed that "since the publication of de Vries's classic work the Oenotheras have attracted more scientific attention than almost any other plant or animal."[94] For a decade and half, it was perfectly respectable for scientists who studied heredity to express the belief that new species emerged virtually overnight, as a result of mutation periods, and that understanding their causes would allow new kinds of organisms to be created. However, just a few years after *Nature's* review appeared, Thomas Hunt Morgan and his colleagues at Columbia published *The Physical Basis of Heredity* (1919), which convinced most of the genetics community that Mendel was right, de Vries was wrong, and that *Drosophila*, not *Oenothera*, was the best organism to study.

For scientists (and many of their historians), the story of the mutation theory was over by 1919, but as earlier chapters showed, the story of its cultural influence and long-term historical significance had barely begun. And even during *Oenothera's* ascendancy, "the scientific community" was not homogeneous (as we have seen, even expert geneticists were using *mutation* in different senses during this period). Scientific communities cannot fully control the meanings of scientific terms: they never have and probably never will. And Delany's point about paraliterature is a reminder that if a sufficiently large group read particular writings as examples of a particular genre, they become examples of that genre (SF is whatever is read as SF). And if a community of readers assigns a particular meaning to a word (such as *mutant*), then that *is* what it means, at least for that group,

94. "The Mutation Factor in Evolution" 1915, 668.

who may choose to ignore the scientific community's definition (or to appropriate it selectively). No matter how one defines *scientific community*, nonscientists always outnumber scientists, so historians of science have to accept that the majority's understanding of a specific term simply *is* what that word meant—for particular groups at particular times.[95]

However, acknowledging that the meanings of scientific ideas are defined or disseminated in this way does not entail any kind of destructive, anti-scientific relativism. Even if a majority of people were to reject the validity of a key scientific concept because they refused to adopt the scientific community's definitions, that would not imply that "anything goes" or that, for example, the reality of climate change (or the value of pi) ought to be settled by popular vote. Opposition to scientific ideas (or of political programs based on those ideas) seldom involves disputing the definition of scientific terms; anti-vaccination campaigners, for example, do not generally deny the existence of germs, disease, or even of vaccines—the dispute is about the relative value of the claimed benefits versus the costs of alleged side effects. Any while some people still deny that humans cause climate change, they generally do not deny that carbon dioxide exists.

Additionally, and perhaps more importantly, the specific social and cultural meanings that a scientific idea acquires will affect the ways it is used in public discourse of every kind. Those who wish to understand—and perhaps change—public attitudes toward science need to know how the public uses scientific ideas and how the processes of appropriation and interpretation shape their beliefs and attitudes. (Not least because the freedom of speech on which democracy relies is predicated on a parallel obligation to listen—even to those we might disagree with.) The popularity of eugenic ideas among people whose values we would classify as progressive (for example, socialists and feminists) often strikes historians as perplexing, given our current identification of eugenic thinking with reactionary ideas. However, recognizing the overlap and similarities between biotopianism and some forms of eugenics allows us to more fully understand its historical appeal (and may perhaps help ensure that it never becomes popular again). Perhaps understanding more about how science has been received and utilized by different audiences in the past might even help create a more realistic model of how these processes continue today. In its crudest form, the top-down, diffusionist model embodies the assumption that the scientific community knows the truth and everyone outside needs to simply shut up and listen. The perception that science is the work of an arrogant elite often prompts resistance—whether from

95. See Endersby 2009.

anti-evolutionists, climate change deniers, or anti-vaxxers. However, if we understand such groups as interpretive communities bound together by a common set of readings of particular scientific claims, it becomes possible to critically interrogate the culture that creates and sustains them.

As science is published, read, talked about, and interpreted, it begins to function as public culture, providing everyone with metaphors and analogies, practical examples, and ideas with which to make sense of the world around them. The social, cultural, political—and thus historical—significance of science is largely determined by the publicly forged meanings that emerge from interpretative communities. Those meanings shape policy and funding decisions, and they help to determine what kinds of science does (and doesn't) get done. And of course, people don't need to actually read, much less accept, new scientific ideas to become part of these processes (refusing to even engage with a particular scientific idea or practice also helps define its meaning). As a result, popular ideas can be hugely influential despite (or even because) they differ markedly from those approved by recognized authorities. Histories of science generally focus on scientists; individuals, ideas, and institutions are their usual focus, but that approach has obscured experimental evolution's wider cultural, social, and political importance. Understanding its historical impact requires paying less attention to scientific literature and more to its paraliterature, from SF to textbooks. It is arguably only when science escapes from the lab, allowing wider communities to start interpreting and appropriating it, that it becomes part of history, part of the wider narratives by which we make sense of our collective pasts and through which we imagine possible futures.

Epilogue

UNNATURAL?

Was nature an obstacle to human survival and progress, or a guide to the good life? Such long-standing questions reverberated more urgently through the twentieth century's early decades as new biology promised to redefine what (if anything) was natural. It provided the dominant question for many volumes in the To-Day and To-Morrow series, which had begun with Haldane's enthusiastic embrace of biological invention (and a resultant dismissal of naturalistic ethics). Among the responses was Anthony Ludovici's *Lysistrata* (1924), which argued that feminism was a morbid symptom of the "body-despising values" of modern culture and that "the claim of sexual equality" was "a manifest and transparent absurdity." The only thing that could save the world was a reassertion of natural biological roles; to "put woman back in her place," as he bluntly put it, and restrict her to giving birth.

Haldane had used our species' willingness to drink the milk of other animals to show that any perversion would come to seem natural in time. By contrast, Ludovici argued that the sight of "a human baby at the dug of a cow, a goat, or an ass, as you sometimes see them placed in semi-civilized countries" was "an offence to the eyes, a humiliation of our racial pride," because "instinctively we feel and intellectually we know that Nature makes the wisest provision for her needs."[1] He was neither the first nor the last to invoke Nature's supposed wisdom in support of his reactionary views. Diane Paul has argued that *Daedalus* was a "scathing critique" of the argument against tampering with nature that would come to be called the "wisdom of repugnance."[2] The term was coined by American doctor Leon Kass as part of an argument against human cloning, which he called a "pollution and perversion" to which the only "fitting response can

1. Ludovici 1924, 33, 106–7, 70. See Squier 1994, 68; McLaren 2012, 33–34.
2. Paul 2004, 125–26.

only be horror and revulsion."[3] In Kass's widely cited view, most people have a natural revulsion to certain things—from incest to the smell of excrement—and he believed this instinct was one we should trust. What is colloquially known as the "yuck factor" continues to be offered as an argument against some forms of interference with nature.

Kass's argument was prefigured by some of biotopia's critics, notably the Christian writer C. S. Lewis, whose *Cosmic Trilogy* (1938–1945) was a conscious riposte to Wells, to *Daedalus* (its main character, the scientist Weston, was modeled on Haldane), and to biotopian science more generally (its concluding volume, *That Hideous Strength*, features a new species directly inspired by Bernal—bodiless brains kept alive artificially and thus potentially immortal). In the first book, *Out of the Silent Planet*, Weston tells the ruler of Malacandra (Mars) that humans must remake their minds and bodies so that they can go on "claiming planet after planet, system after system, till our posterity—whatever strange form and yet unguessed mentality they have assumed—dwell in the universe wherever the universe is habitable."[4] The Malacandrans find the proposal to rebuild humans into these perverse forms repugnant, seeing each species' body as the harmonious expression of the way it had been fitted by God to its place in the cosmos. In response to the *Cosmic Trilogy*, an unrepentant Haldane criticized Lewis, "not for his attack on my profession, but for his attack on my species," and accused him of wanting people to become servile saints. Haldane asserted that human freedom and creativity would ensure that "each generation makes newer and greater possibilities of good and evil"— and only humans could determine which was which.[5]

Daedalus, like other biotopian works, firmly rejected the idea that Nature knows best and enthusiastically embraced all biology's possibilities. As a result, it now appears like a manifesto for genetic engineering, whose proponents generally reject the wisdom-of-repugnance argument (as the psychiatrist and bioethicist Willard Gaylin argued: "I not only think that we will tamper with Mother Nature, I think Mother wants us to").[6] This kind of claim was foreshadowed in many earlier works. For example Wells's "Man of the Year Million" presented an image of the future human that seems deliberately repellent, and Krishan Kumar has argued that Wells was effectively asking, "What's so great about human bodies

3. Kass 1997.

4. Lewis 1938, 155–56. See Parrinder 2011, 255; Alt 2015, 36; Bowler 2017b, 25.

5. Haldane 1946.

6. "What's So Special about Being Human?," in *The Manipulation of Life*, ed. Robert Esbjornson (1983), 53. Quoted in Paul 2004 (139).

that we should wish to keep them?" The same thought was apparent in Bernal's admission that his proposals would be met with "distaste and hatred," feelings he acknowledged having experienced himself.[7] Biotopia entailed overcoming the "wisdom of repugnance," ignoring the instinctive "yuck" and simply asking, "Why not?" Ludovici's misogynist tract is a sharp reminder of where the supposed wisdom of repugnance can lead; he condemned as unnatural the idea of women working, being educated, or choosing when and whether to have children.

Anyone tempted to put their faith in nature should first consider how often the word *unnatural* has been used to condemn other peoples' choices—and the inhumane consequences that invariably follow. Like most Christian apologists of his day, C. S. Lewis unhesitatingly condemned "perversions of the sex instinct" that were "numerous, hard to cure, and frightful" and proposed the alternatives of monogamous, heterosexual marriage for those committed to what he regarded as natural (that is, reproductive sex)—and chastity for everyone else. However, Haldane noted that while Lewis condemned sexual perversions absolutely, he was ambiguous about other behaviors that had once been considered sins, such as usury. Haldane suspected that this ambiguity arose only because lending money at interest was fundamental to capitalism and had thus (like drinking cow's milk) come to seem both sacred and natural. Always ready to shock, Haldane speculated that if "sodomy were an important part of our social system, as it was of some past systems, Mr. Lewis would presumably wonder whether sodomy was absolutely wrong."[8] The aggressive policing of people's sexual preferences by the self-appointed guardians of the supposedly natural is a sharp reminder of why the wisdom of repugnance is not to be trusted.

Part of biotopia's appeal to radical and progressive writers was that it undermined the claim that "you can't change human nature," the familiar cry of those whose power and privilege is under threat. As Frederic Jameson has argued, one of the most fundamental questions about a potentially perfect future is "whether the impulse to Utopia" was "already grounded in human nature" or whether it was impossible without "a mutation in human nature and the emergence of whole new beings."[9] At their most radical, the biotopians were convinced that humans could indeed mutate and remake the whole of nature, including themselves, to

7. Kumar 1987, 179; Bernal 1929 (1970), 56–57.

8. Lewis 1943, 26; Haldane 1946. Lewis did not in fact mention sodomy (hardly surprising, given the date when these radio talks were given).

9. Jameson 2005, 168; M. Saunders 2019, 56.

produce a world that reflected human standards—ethical and aesthetic. It was a vision that appealed to many, but it is important to remember why others resisted it. As earlier chapters showed, some feminist writers resisted the biotopian project, for example by retaining the idea of motherhood as a natural function. Such views might seem reactionary but can also be understood as a way of resisting those men who wanted to control women's bodies and their power to reproduce. Others resisted biotopia on what would now be called environmental grounds (for example, Bailey's biocentric philosophy urged recognition that humans needed to work in harmony with nonhuman nature). Perhaps the most poignant protest came from the character Lychnis in Wells's *Men Like Gods*, who expressed her discontent with utopia by simply noting "there is no rest"; the desire to innovate and keep innovating made the future feel like an endless, Sisyphean struggle.

Many of those who resisted the kinds of utopias that biology offered were objecting to the idea of handing control over life (especially its most intimate aspects) to unelected and unaccountable scientists. (One did not need to be a reactionary to have doubts about the wisdom of letting a Wellsian techno-scientific elite take over the running of the world.) Many of the British authors discussed in this book were unabashed elitists, who made little (if any) effort to conceal their contempt for the majority of their countrymen.[10] Haldane, Bernal, Muller, and H. G. Wells all espoused some version of socialism, which ranged from bureaucratic to Stalinist, and they all saw themselves as a biological vanguard, confident (to the point of arrogance) of their ability to lead humanity into the future. Meanwhile, American biotopianism was personified in Luther Burbank, by his supposedly domestic virtues and the illusion that market mechanisms allowed free choice. His vision of the future—fields of profitable, spineless cacti—was less challenging than those offered by the British authors (or by Muller, who was unsympathetic to mainstream American values), who were blithely oblivious to the question of whether their fellow citizens could ever be persuaded to accept purple oceans or babies in bottles. It is more than a little chilling to learn that Muller hoped Stalin would support his plans (see p. 312), or to read J. B. S. Haldane arguing that a workers' state might be a better master for science, because "a robust and selfish labour party" could sweep away the "sentimentalism" that so often hinders biological research.[11] Nevertheless, the illusion of choice in Burbank's

---

10. Carey 1992. See also Bradshaw 1995, xii–xxi; Kemp 1996, 215–20.
11. Haldane 1924, 6–7.

vision should not distract from the fact that his future was potentially as perverse and unnatural as those of the more elitist biotopians.

Diane Paul has argued that Haldane, Bernal, Muller, and the other eugenicists were "all genetic imperialists"; regardless of the differences between their individual politics, they saw theirs as the master science that could solve every problem.[12] Her comment is a timely reminder that one of biotopianism's roots was the dream of Bacon, who saw science as primarily a means to power—a goal that inspired many of his successors. And Thomas Huxley's vision of the Baconian garden, a colony from which indigenous life-forms had been "extirpated," was central to the vision that most biotopians shared: inconvenient species would be exterminated without hesitation, just as other planets would be colonized with no thought for whoever/whatever currently inhabited them. As has been highlighted in earlier chapters, biotopianism was often racist, imperialist, misogynist, and ecocidal (and indifferent—at best—to the rights of people with disabilities). Nevertheless, historians need not share the past's values to understand why they once prevailed, and understanding their appeal may help us avoid repeating past mistakes.

Biotopianism was a both a product of and a contributor to early twentieth-century optimism, especially as biology gained new respect in the postwar period. Wells and his fellow authors did not coin the term *Science of Life*, but they helped make it famous; their conclusion that "man" would "survive and triumph" by taking "control not only of his own destinies but of the whole of life" probably marked the high-water mark of biotopianism.[13] Just two years later *Brave New World* appeared, and it has proven to be biotopianism's lasting legacy; insofar as the early twentieth-century idea of a biologically engineered future is remembered at all, it is as a dystopian idea. The century's early optimism faded in the face of the Depression, fascism, and Stalinism—and biotopianism faded with it, helped by critics like Aldous Huxley, whose book ensured that later readers would imagine genetics as creating an unbreakable prison of inequality. These are, of course, real threats, and people are right to be concerned about them; however, resistance to twenty-first-century genetic engineering (and related technologies) often takes the same form that resistance to biotopianism took a century earlier—*trust the yuck factor* and *assume nature knows best*. I would argue that those of us who worry about the impact of biotechnology—and about climate change and other environmental disasters—need to resist the temptation to put our trust

12. Paul 1984, 588.
13. Wells, Huxley, and Wells 1929–1930, 973.

in nature. Such misplaced trust empowers those who want to persecute whomever they consider unnatural, but it also misses the obvious point that there is nothing "natural" about any aspect of human life in the last ten thousand years.

There are interesting parallels between early twentieth-century bio-topian dreams and current work in such fields as synthetic biology. As numerous scholars have shown, recent efforts to create new life-forms has emerged out of similar interactions between new biological theories, commercial imperatives, and the need to solve urgent social and environmental problems.[14] There are striking parallels with the early twentieth-century cultures considered here—synthetic biology even has its own fan-based participatory culture in the form of "biopunks" and "biohackers."[15]

The dream of biotopia still matters. Controlling nature is one of the most ancient human dreams, but the dark side of biotopia is a timely reminder of the effects of power without democratically exercised responsibility. These early twentieth-century fantasies serve as warnings against leaving the power of science in the hands of unelected elites, or of allowing the short-sighted desires of the market to shape the world. Biology continues to provide crucial tools with which to transform the way the world is gardened, but biotopianism offers an important reminder that science offers little or no guidance as to *how* its power should be used. Science enables us to spray the earth with chemicals to maximize food and profits for the wealthiest members of our species, but it can also offer guidance on how to garden organically and sustainably, recognizing the needs and rights of other species—and our dependence on them. If we desire a different kind of garden, we have to adopt very different ethical, political, and economic standards than those that prevail today.

Finally, and perhaps most importantly, biotopianism is a reminder that every biological invention is a perversion, a violation of the natural order. Those violations have included curing once-fatal diseases, rotting milk to enjoy alongside fermented grape juice, and learning all kinds of enjoyable ways to have sex without reproducing (if we choose not to). Given how often the purely cultural category of "Nature" has been used to galvanize various lynch mobs, a more unnatural, artificial, and perverse future seems like the only kind worth fighting for.

14. Campos 2012; Hilgartner 2015; Roosth 2017.
15. Wohlsen 2012.

Acknowledgments

This project has been going on for so long that a complete list of everyone who has helped would require a second volume (so, if I have omitted anyone, I can only ask them to forgive me).

I am especially grateful to the following people for reading sections and offering advice, critical support, and encouragement throughout: Thomas Broman, Luis Campos, Ronald Ladouceur, Katherine Pandora, Diane Paul, Charlotte Sleigh, Betty Smocovitis, Pam Thurschwell, and Paul White.

Thanks also to all my current and former colleagues at Sussex, especially Iain McDaniel, Lynne Murphy, Joanne Paul, Lucy Robinson, Darrow Schecter, Eric Schneider, Chris Warne, and Clive Webb. And to all my students, especially the Genetic Centurions.

Many thanks to the libraries and archives I consulted, and to all their staff. I am particularly indebted to the American Philosophical Society; Luther Burbank Home and Gardens; the University of California, Berkeley; the Cambridge University Library; Carnegie Science (especially John Strom); the Lilly Library at Indiana University; the John Innes Centre (especially Sarah Wilmot); the Royal Botanic Gardens, Kew; the New York Botanical Garden; the Tamiment Library and Robert F. Wagner Labor Archives at New York University; the University of California, San Diego (especially Harold Colson); the Sutro Library in San Francisco (especially Steenalisa Tilcock); and the University of Sussex Library.

Thanks also to Felice Belman, Krister Bykvist, Constance Clark, Angela Creager, Helen Curry, Lisa Ganobcsik-Williams, Piers Hale, Steve Harvey, Andrew Hinderaker, Nick Jardine, Sharon Kingsland, Rob Kohler, Mark Largent, Bruce V. Lewenstein, Wendy McElroy, Erika Milam, George R. Morgan, Anna Neill, Don Opitz, Patrick Parrinder, Marsha Richmond, Max Saunders, Perry Sayles, Beth Sutton-Ramspeck, Jon Turney, and Erik Zevenhuizen. And to the team at the University of Chicago Press, particularly Karen Merikangas Darling, Fabiola Enríquez Flores, and freelance copyeditor Jessica Wilson.

Last (and *very* far from least), my family, particularly Pam Thurschwell, Max Endersby, and Katya Endersby (who wasn't even born when I started thinking about this project)—they have all been endlessly patient and supportive throughout.

Appendix

Over 150 textbooks of various kinds (see chapter 6) were consulted, some in more than one edition (see table A.2), but those that were straight reprints, minor revisions of earlier editions, or too technical/specialized are omitted from the table. "Place" of publication is for the edition consulted (UK/US signifies the British edition of a text also published in the US, while US/UK signifies the opposite).

Table A.1 illustrates the crisis in Darwinism very clearly; almost half the textbooks argued that natural selection was insufficient to explain the "arrival of the fittest" (and eleven of them—identified with an asterisk—used the phrase *arrival of the fittest* to express skepticism). As table A.2 shows, this skepticism was most pronounced in the century's first decade and faded slowly in the second. Enthusiasm for de Vries's original mutation theory was widespread, rivaling or surpassing interest in Mendel and Mendelism throughout this period, although the balance of interest was clearly shifting toward genetics by the end of the period (figures A.1–A.3).

TABLE A.1 **Overview of textbook presentations of biological theories**

Numbers in the first four columns are the number of textbooks in each category.

TOPIC	POSITIVE	BRIEF/AMBIGUOUS	NEGATIVE	OMITTED	NOTES
Natural selection	62		66		*Natural selection* refers to the arrival-of-the-fittest problem: a positive mention means natural selection was considered sufficient to explain the origin of species; a negative means significant doubts were expressed.
De Vries/mutation theory	53	62	7	17	Brief/ambiguous references mainly described de Vries's work as subsidiary to Mendel's or used *mutation* in the later, Mendelian sense.
Mendel/genetics	77	37	0	22	Brief/ambiguous references mainly described Mendel's work as subsidiary to de Vries's original mutation theory.
Biotopia	42	22	—	—	*Biotopia* is an imprecise category but indicates confidence in a better future through controlling evolution, especially via experimental/laboratory techniques. Too nebulous to identify opposition.
Burbank	44	5	—	—	

TABLE A.2 **Textbooks consulted**

Abbreviated column heads signify the following: NS = natural selection; Mut. = mutation (de Vriesian sense); Men. = Mendel (Mendelism, genetics); Bio. = biotopia; Bur. = Burbank. The numbers in these columns indicate whether the specific topics were covered (1) positively, (2) negatively, (3) briefly or ambiguously, or (4) not at all; (5) indicates "not applicable." An asterisk signifies that the phrase "arrival of the fittest" was used.

DATE	AUTHOR	PLACE	TITLE	NS	MUT.	MEN.	BIO.	BUR.	NOTES
1900	H. W. Conn	US/UK	The Method of Evolution	3	5	5	4	4	
1900	F. W. Headley	UK	Problems of Evolution	2	5	5	4	4	
1901	P. Geddes and J. A. Thomson	UK	The Evolution of Sex	5	5		4	4	
1902	L. H. Bailey	US/UK	Plant-Breeding	1	5	3	4	3	
1902	W. Bateson	UK/US	Mendel's Principles of Heredity	4	3	1	1	4	De Vries's full publication not yet available, but keenly awaited.
1903	T. H. Morgan	US/UK	Evolution and Adaptation	2	1	1	4	4	
1903	C. L. Redfield	US	Control of Heredity	1	4	4	1	4	
1903	C. Snyder	US/UK	New Conceptions in Science	4	4	4	3	4	
1904	E. D. Cope	US/UK	The Primary Factors of Organic Evolution	2	4	4	4	4	Orthogenesis (directional variation) explains arrival.
1904	E. Dennert	US	At the Deathbed of Darwinism: A Series of Papers	2	1	4	4	4	*Mutation* mentioned by O'Hara in intro, not by Dennert.
1904	F. E. Lloyd and M. A. Bigelow	US/UK	The Teaching of Biology in the Secondary School	3	3	4	3	4	
1904	M. M. Metcalf	US/UK	An Outline of the Theory of Organic Evolution: With a Description of Some of the Phenomena Which It Explains	2	1	1	4	4	

(continued)

TABLE A.2 (*continued*)

DATE	AUTHOR	PLACE	TITLE	NS	MUT.	MEN.	BIO.	BUR.	NOTES
1904	A. Weismann	UK	The Evolution Theory	2	2	3	4	4	
1905	H. De Vries	US/UK	Species and Varieties: Their Origin by Mutation; Lectures Delivered at the University of California	2*	1	3	1	1	
1905	É. Metchnikoff	US/UK	The Nature of Man: Studies in Optimistic Philosophy	2	1	4	3	4	
1905	W. J. V. L. Osterhout	UK	Experiments with Plants	2	1	4	1	1	Reviews: Plenty. Translated into Dutch and Russian.
1905	R. C. Punnett	UK	Mendelism	2	3	1	1	4	
1906	L. H. Bailey	US/UK	Plant-Breeding: Being Six Lectures upon the Amelioration of Domestic Plants	2	1	1	1	1	
1906	L. H. Bailey	US/UK	The Survival of the Unlike: A Collection of Evolution Essays Suggested by the Study of Domestic Plants	2	4	4	4	4	
1906	W. Boelsche [Bölsche]	US	The Triumph of Life	1	4	4	4	4	
1906	C. Guenther	UK	Darwinism and the Problem of Life: A Study of Familiar Animal Life	1	2	4	4	4	
1906	F. W. Headley	UK	Life and Evolution	1	1	1	4	4	
1906	R. H. Lock	UK	Recent Progress in the Study of Variation, Heredity, and Evolution	2	1	1	1	2	
1906	M. M. Metcalf	US/UK	An Outline of the Theory of Organic Evolution: With a Description of Some of the Phenomena Which It Explains	1	1	3	4	4	

1906	T. H. Montgomery	US	The Analysis of Racial Descent in Animals	2	1	3	4	4
1906	G. A. Reid	UK	The Principles of Heredity: with Some Applications	1	1	1	4	4
1906	C.W. Saleeby	UK/US	Evolution: The Master-Key	2	1	1	4	4
1907	C. C. Curtis	US	Nature and Development of Plants	2	1	3	4	4
1907	H. De Vries	US	Plant-Breeding	2	1	4	1	1
1907	G. W. Hunter	US	Elements of Biology: A Practical Text-Book Correlating Botany, Zoology, and Human Physiology	1	4	4	4	4
1907	D. S. Jordan, E. G. Conklin, F. M. McFarland, and J. P. Smith	US	Foot-Notes to Evolution: A Series of Popular Addresses on the Evolution of Life	1	3	4	4	4
1907	D. S. Jordan and V. L. Kellogg	US	Evolution and Animal Life: An Elementary Discussion of Facts, Processes, Laws and Theories Relating to the Life and Evolution of Animals	2	1	3	1	1
1907	V. L. Kellogg	US	Darwinism To-Day: A Discussion of Present-Day Scientific Criticism of the Darwinian Selection Theories, Together with a Brief Account of the Principal Other Proposed Auxiliary and Alternative Theories of Species-Forming	2*	1	3	1	1
1907	E. R. Lankester	UK	The Kingdom of Man	4	3	3	1	4

(continued)

TABLE A.2 (*continued*)

DATE	AUTHOR	PLACE	TITLE	NS	MUT.	MEN.	BIO.	BUR.	NOTES
1907	T. H. Morgan	UK	Experimental Zoology	2	1	1	1	5	
1907	R. C. Punnett	UK	Mendelism (2nd ed.)	1	1	1	3	4	
1908	W. Bateson	UK	The Methods and Scope of Genetics	1	3	1	4	4	
1908	O. F. Cook	US	Methods and Causes of Evolution	2	3	3	4	4	Promotes Cook's own kinetic theory.
1908	É. Metchnikoff	US	The Prolongation of Life: Optimistic Studies	1	3	4	4	1	
1908	E. B. Poulton	UK	Essays on Evolution, 1889–1907	1	2	3	4	4	
1908	J. A. Thomson	US/UK	Heredity	2*	1	3	1	1	
1909	L. H. Bailey and W. M. Coleman	US/UK	First Course in Biology	1	3	4	4	4	
1909	W. Bateson	UK	Mendel's Principles of Heredity	2	3	1	4	4	
1909/ 10	H. De Vries	US	The Mutation Theory: Experiments and Observations on the Origin of Species in the Vegetable Kingdom. Volume 1: The Origin of Species by Mutation	2	1	3	1	1	
1909	D. Dewar and F. Finn	UK	The Making of Species	2	3	3	1	3	
1909	Various	US	Fifty Years of Darwinism	1	1	1	1	4	Edited volume, so diverse views included.
1909	E. B. Poulton	UK	Charles Darwin and the Origin of Species: Addresses, etc., in America and England in the Year of the Two Anniversaries	1	2	3	4	4	
1909	R. C. Punnett	US	Mendelism (American ed.)	1	1	1	1	4	Punnett himself relegated de Vries's theory to a minor role, but Wilshire's preface offered a prominent endorsement.

Year	Author	Country	Title						Notes
1909	C. W. E. Saleeby	UK/US	Parenthood and Race Culture: An Outline of Eugenics	1	4	1	1	4	
1909	C. E. Stackpole	US	Biology	2	1	1	4	4	
1909	W. H. Thomson	US	What Is Physical Life: Its Origin and Nature	2	3	4	4	4	Mostly about spontaneous generation, not evolution.
1910	J. Beard	UK	Philosophical Biology	2*	1	3	3	4	
1910	D. B. Hart	UK	Phases of Evolution and Heredity	2	3	3	4	4	
1910	L. Doncaster	UK	Heredity: In the Light of Recent Research	1	4	1	4	4	
1910	H. Drinkwater	UK	A Lecture on Mendelism	1	3	1	3	4	
1910	S. Herbert	UK	The First Principles of Heredity	1	1	3	3	4	
1910	J. W. Judd	UK	Coming of Evolution	1	4	4	4	4	
1910	J. McFarland	US	Biology, General and Medical	2	1	1	4	4	
1910	G. A. Reid	UK	The Laws of Heredity	1*	2	3	4	4	Phrase *arrival of the fittest* used, but only to dismiss it as a non-problem.
1910	J. A. Thomson	UK	Darwinism and Human Life	2*	3	1	1	1	Used the phrase *arrival of the fittest*, but argued the objection was not grave.
1910	C. E. Walker	UK	Hereditary Characters and Their Modes of Transmission	3	2	3	4	4	
1911	M. A. Bigelow and A. N. Bigelow	US	Applied Biology: An Elementary Textbook and Laboratory Guide	1	4	4	4	4	
1911	H. E. Crampton	US	The Doctrine of Evolution: Its Basis and Its Scope	1	3	1	4	4	
1911	P. Geddes and J. A. Thomson	US/UK	Evolution	2	3	1	1	1	

(continued)

TABLE A.2 (*continued*)

DATE	AUTHOR	PLACE	TITLE	NS	MUT.	MEN.	BIO.	BUR.	NOTES
1911	M. M. Metcalf	US/UK	An Outline of the Theory of Organic Evolution: With a Description of Some of the Phenomena Which It Explains	2	3	1	4	4	
1911	R. C. Punnett	US	Mendelism: Third Edition, Entirely Rewritten and Much Enlarged	2	3	1	4	4	
1912	W. E. Castle, J. M. Coulter, C. B. Davenport, E. M. East, and W. L. Tower	US	Heredity and Eugenics	1	3	1	1	1	Edited volume, so differing views (esp. of mutation). Coulter most positive; Castle skeptical; Tower argued that mutation theory was vital to real experimental work, but key details of de Vries's theory were wrong.
1912	H. W. Conn	US	Biology: An Introductory Study, for Use in Colleges	1	3	3	4	4	
1912	A. D. D. Darbishire	UK/US	Breeding and the Mendelian Discovery	4	1	1	3	4	De Vries's work summarized in detail, but *mutation* defined in purely Mendelian sense.
1912	A. Dendy	UK	Outlines of Evolutionary Biology	1	3	3	4	1	Primarily an enthusiastic account of Weismann's theories.
1912	R. E. Lloyd	UK/US	Growth of Groups in the Animal Kingdom	2	1	3	4	5	
1912	J. A. S. Watson	UK	Heredity	4	3	1	4	4	
1912	H. S. Williams and E. H. Williams	US/UK	Modern Development of the Chemical and Biological Sciences	1	4	4	4	4	
1913	W. Bateson	US/UK	Problems of Genetics	2	3	1	4	4	
1913	J. G. Coulter	US	Plant Life and Plant Uses	4	4	4	3	3	

Year	Author	Country	Title						Notes
1913	Y. Delage	US	The Theories of Evolution	2	3	3	4	1	
1913	W. F. Ganong	US	The Living Plant: A Description and Interpretation of Its Functions and Structure	2	1	1	1	1	
1913	S. Herbert	UK	The First Principles of Evolution	2*	3	3	4	4	
1913	S. C. Schmucker	US	The Meaning of Evolution	2	3	4	4	4	
1913	H. S. Williams	US	Miracles of Science	3	1	1	1	1	
1914	J. F. Abbott	US	Elementary Principles of General Biology	2	3	1	4	4	
1914	J. M. Coulter	US	Fundamentals of Plant-Breeding	2	3	1	1	1	
1914	G. W. Hunter	US	A Civic Biology: Presented in Problems	1	1	1	4	1	
1914	F. G. Jewett	US	The Next Generation	1	3	1	3	1	
1914	G. H. Parker	US	Biology and Social Problems	2	1	1	1	4	
1914	H. E. Walter	US	Genetics: An Introduction to the Study of Heredity	2	1	3	3	1	
1915	L. H. Bailey and A.W. Gilbert	US	Plant Breeding	1	1	1	1	1	
1915	L. M. Bristol	US/UK	Social Adaptation: A Study in the Development of the Doctrine of Adaptation as a Theory of Social Progress	2*	1	1	4	4	
1915	E. G. Conklin	US/UK	Heredity and Environment in the Development of Men	2	1	1	1	1	
1915	W. A. Locy	US	Biology and its Makers (3rd ed.)	1	1	1	1	4	
1915	P. C. Mitchell	UK	Evolution and the War	1	4	1	5	4	Largely concerned with refuting the idea that war is natural selection in action and will improve humanity.
1916	J. M. Coulter	US	Evolution, Heredity and Eugenics	2	1	3	4	1	

(continued)

TABLE A.2 (*continued*)

DATE	AUTHOR	PLACE	TITLE	NS	MUT.	MEN.	BIO.	BUR.	NOTES
1916	C. S. Gager	US	Fundamentals of Botany	2	1	1	1	4	
1916	T. H. Morgan	US/UK	A Critique of the Theory of Evolution	2	3	1	4	4	
1916	J. Wilson	UK	A Manual of Mendelism	4	4	1	4	4	
1917	G. N. Calkins	US	Biology	1	3	1	1	4	
1917	A. D. Darbishire and H. Darbishire	US	An Introduction to Biology: And Other Papers	2	1	1	5	4	Celebration of Bergson and Butler; dismissed other biologists as materialistic.
1917	R. S. Lull	US	Organic Evolution, a Text Book	1	3	3	4	1	
1917	W. B. Scott	US	The Theory of Evolution: with Special Reference to the Evidence upon Which It Is Founded	2	3	3	5	4	
1918	E.B. Babcock and R. E. Clausen	US/UK	Genetics in Relation to Agriculture	1	3	1	4	1	
1918	C. F. Hodge and J. Dawson	US	Civic Biology: Textbook of Problems, Local and National, that Can Be Solved Only by Civic Coöperation	3	3	1	1	4	
1918	J. M. Macfarlane	US	The Causes and Course of Organic Evolution	1	3	3	1	4	
1918	H. F. Osborn	UK	The Origin and Evolution of Life: On the Theory of Action, Reaction and Interaction of Energy	2	3	3	3	4	
1919	B. C. Gruenberg	US	Elementary Biology: An Introduction to the Science of Life	2	3	1	1	1	
1919	T. H. Morgan	US	The Physical Basis of Heredity	1	3	1	3	4	
1920	W. E. Castle	US/UK	Genetics and Eugenics (2nd ed.)	1	3	1	3	4	

Year	Author	Country	Title						Comments
1920	C. S. Gager	US	Heredity and Evolution in Plants	2	1	1	1	4	Updating of his *Fundamentals of Botany*. Allegiance to de Vries undiminished.
1920	W. Lochhead	Unknown	An Introduction to Heredity and Genetics: A Study of the Modern Biological Laws and Theories Relating to Animal and Plant Breeding	2*	3	1	3	1	
1920	W. Patten	US	The Grand Strategy of Evolution	2	4	3	5	4	
1920	W. M. Smallwood, I. L. Reveley, and G. A. Bailey	US	Biology for High Schools	1	4	4	4	4	Darwin and evolution barely mentioned.
1921	L. M. Bristol	US/UK	Social Adaptation	2*	1	1	4	4	Biology inadequate to provide a basis for a social/religious evolutionary philosophy.
1921	W. M. Coleman	Canada	Beginners' Zoology	1	3	4	4	4	
1921	R. R. Gates	UK	Mutations and Evolution	1	4	4	4	1	Too specialized and technical.
1921	M. J. Gauvin	US	Illustrated Story of Evolution	1	4	4	5	1	Recommended by Queen Silver. Strongly secularist.
1921	A. L. Hagedoorn and A. C. Hagedoorn-Vorstheuvel la Brand	Netherlands	The Relative Value of the Processes Causing Evolution	1	3	1	4	1	
1921	T. J. Moon	US	Biology for Beginners	1	2	1	4	1	
1921	H. H. Newman	US	Readings in Evolution, Genetics and Eugenics	2*	1	1	3	4	

(continued)

TABLE A.2 (*continued*)

DATE	AUTHOR	PLACE	TITLE	NS	MUT.	MEN.	BIO.	BUR.	NOTES
1922	W. H. Atwood	US	Civic and Economic Biology	1	3	1	3	1	
1922	L. L. Burlingame	US	General Biology	1	3	1	4	1	
1922	E. G. Conklin	US/UK	Heredity and Environment in the Development of Men	2	1	1	1	1	As in first edition, but second clarified that de Vries's mutants were not mutants in the modern, post-*Drosophila* sense.
1922	E. G. Conklin	US	The Direction of Human Evolution	2	1	3	3	1	
1922	J. A. Thomson (ed.)	US	The Outline of Science	2	1	1	4	1	
1922	J. C. Willis	UK	Age and Area: A Study in Geographical Distribution and Origin of Species	2	1	3	4	4	
1922	L. L. Woodruff	US	Foundations of Biology	1	3	1	4	4	
1923	W. H. Atwood	US	Problems, Projects, and Experiments in Biology						Companion to *Civic and Economic Biology*. Practical suggestions, not relevant for this research.
1923	A. Dendy	UK	Outlines of Evolutionary Biology	1	3	1	4	1	Revised and enlarged 3rd edition.
1923	R. R. Gates	UK	Heredity and Eugenics	1	4	1	4	4	
1923	H. R. Linville	US	The Biology of Man and Other Organisms	1	1	3	1	1	
1923	G. H. Trafton	US	Biology of Home and Community	1	1	3	4	1	
1924	V. L. Kellogg	US	Evolution	2	3	1	3	1	
1925	B. C. Gruenberg	US	Biology and Human Life	1	3	1	3	1	
1925	More	US	The Dogma of Evolution	2	3	4	4	4	Example of mutation theory used to attack evolution on religious grounds.

Year	Author	Country	Title						Notes
1925	H. H. Newman	US	Evolution, Genetics and Eugenics	2*	1				More nuanced and skeptical account of mutation; included recent criticisms.
1925	A. E. Shipley	UK	Life: A Book for Elementary Students						No evolution or heredity, not relevant.
1926	S. J. Holmes	US	An Introduction to General Biology	1	3	1	1	3	
1926	J. S. Huxley	UK/US	The Stream of Life.	1	3	1	1	1	
1926	J. G. Kerr	UK	Evolution	1	3	1	4	4	
1926	A. C. Kinsey	US	An Introduction to Biology	1	3	1	4	4	
1926	H. H. Newman	US	The Gist of Evolution	3	3	1	3	1	
1928	F. Mason (ed.)	US	Creation by Evolution	1	3	3	4	4	
1929	M. Adams	UK	Six Talks on Heredity	1	3	1	4	4	
1929	B. C. Gruenberg	US	Story of Evolution	2	3	1	1	1	
1929	F. M. Wheat and E. T. Fitzpatrick	US	Advanced Biology	1	1	1	1	1	
1929–1930	H. G. Wells, J. Huxley, and G. P. Wells	UK	The Science of Life: A Summary of Contemporary Knowledge about Life and Its Possibilities	1	1		1	1	
1930	O. H. Latter	UK	Readable School Biology	1	1	1	3		
1931	J. A. Thomson and P. Geddes	UK	Life: Outlines of General Biology	1*	3	1	1	4	Quoted de Vries: "Natural selection may explain the survival of the fittest, but it does not explain their arrival," but disagreed with his claim.
1932	J. B. S. Haldane	UK	The Causes of Evolution	1	3	1	3	4	

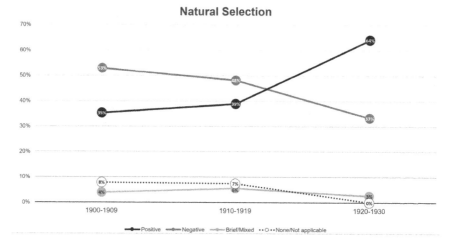

FIGURE A.1. Darwin's theory of natural selection was never eclipsed as completely as existing accounts suggest, but it received increasingly positive mentions after the *Origin*'s fiftieth-anniversary celebrations (1909), as seen in a sampling of textbooks from the time.

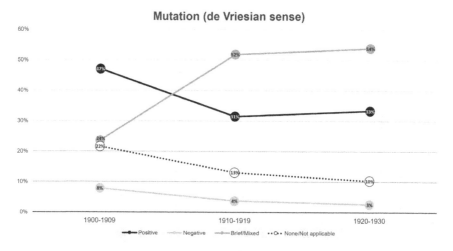

FIGURE A.2. Enthusiasm declined for the de Vriesian sense of *mutation* from its initial peak; however, it continued to be widely covered throughout the first three decades of the twentieth century, as seen in a sampling of textbooks from the time.

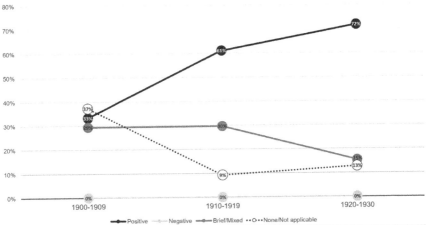

FIGURE A.3. *Mendelism* and synonymous terms became increasingly important over the decades considered in this book, as seen in a sampling of textbooks from the time.

References

Abir-Am, Pnina G. 1982. "The Discourse of Physical Power and Biological Knowledge in the 1930s: A Reappraisal of the Rockefeller Foundations 'Policy' in Molecular Biology." *Social Studies of Science* 12: 341–82.

Adams, Mark B. 1970. "Towards a Synthesis: Population Concepts in Russian Evolutionary Thought." *Journal of the History of Biology* 3, no. 1: 107–29.

———. 2000. "Last Judgment: The Visionary Biology of J. B. S. Haldane." *Journal of the History of Biology* 33: 457–91.

Adams, Mary. 1929. *Six Talks on Heredity: A Handbook to the Cradle*. Cambridge: W. Heffer and Sons.

Aîné, J.-H. Rosny. 2012. *Three Science Fiction Novellas: From Prehistory to the End of Mankind*. Translated by Danièle Chatelain and George Slusser. Middletown, CT: Wesleyan University Press.

Aldridge, Susan. 1996. *The Thread of Life: The Story of Genes and Genetic Engineering*. Cambridge: Cambridge University Press.

Alfaro, María Jesús Martínez. 1996. "Intertextuality: Origins and Development of the Concept." *Atlantis* 18, nos. 1–2: 268–85.

Allen, Garland E. 1969. "Hugo de Vries and the Reception of the 'Mutation Theory.'" *Journal of the History of Biology* 2, no. 1: 55–87.

———. 1975. "The Introduction of *Drosophila* into the Study of Heredity and Evolution, 1900–1910." *Isis* 66, no. 3: 322–33.

———. 1978. *Thomas Hunt Morgan: The Man and His Science*. Princeton, NJ: Princeton University Press.

———. 2013. "'Culling the Herd': Eugenics and the Conservation Movement in the United States, 1900–1940." *Journal of the History of Biology* 46, no. 1: 31–72.

Alt, Christina. 2015. "Extinction, Extermination, and the Ecological Optimism of H. G. Wells." In *Green Planets: Ecology and Science Fiction*, edited by Gerry Canavan and Kim Stanley Robinson, 25–39. Middletown, CT: Wesleyan University Press.

Altman, Rick. 1999. *Film/Genre*. London: BFI.

"America's Wrong Notion of Luther Burbank's Achievement." 1907. Review of de Vries, *Plant-Breeding*. *Current Literature* 43, no. 1: 97–98.

Andrews, E. Benjamin. 1894. "Science a Natural Ally of Religion." *New World: A Quarterly Review of Religion, Ethics and Theology* 3, no. 12 (December): 658–70.

Anguelo, Jaime. 1909. "Evolution and Revolution." *Ogden Standard*, July 10: 4.

Anker, Peder. 2002. *Imperial Ecology: Environmental Order in the British Empire, 1895–1945*. Cambridge, MA: Harvard University Press.

A.R. 1923. "Mr. Wells in Utopia." *Bookman* 64, no. 379: 32–34.

Armitage, Kevin C. 2009. "'The Science-Spirit in a Democracy': Liberty Hyde Bailey, Nature Study, and the Democratic Impulse of Progressive Conservation." In *Natural Protest: Essays on the History of American Environmentalism*, edited by Michael Egan and Ejeff Crane, 89–116. New York: Routledge.

Ashley, Mike, and Robert A. W. Lowndes. 2004. *The Gernsback Years: A Study of the Evolution of Modern Science Fiction from 1911 to 1936*. Holicong, PA: Wildside Press.

Attebery, Brian. 2003. "The Magazine Era: 1926–1960." In *The Cambridge Companion to Science Fiction*, edited by Edward James and Farah Mendlesohn, 32–47. Cambridge: Cambridge University Press.

Atwood, William Henry. 1922. *Civic and Economic Biology*. Philadelphia, PA: P. Blakiston's Son.

Avrich, Paul. 1980. "Kropotkin in America." *International Review of Social History* 25, no. 1: 1–34.

Ayala, Francisco J., Walter M. Fitch, and Francisco Jose Ayala. 1995. *Tempo and Mode in Evolution: Genetics and Paleontology 50 Years after Simpson*. Washington, DC: National Academies Press.

Babcock, Ernest Brown, and Roy Elwood Clausen. 1918. *Genetics in Relation to Agriculture*. New York: McGraw-Hill.

Baccolini, Raffaella, and Tom Moylan, eds. 2003. *Dark Horizons: Science Fiction and the Dystopian Imagination*. New York: Routledge.

Bache, René. 1905. "The Riddle of Heredity: Some Scientific Guesses at the Answer." *Saturday Evening Post* 177, no. 51: 13–14.

"Backward America." 1913. *Arizona Republican*, January 31: 4.

Bacon, Francis. 1620 (2000). *The New Organon*. Edited by Lisa Jardine and Michael Silverthorne. Cambridge Texts in the History of Philosophy. Cambridge: Cambridge University Press.

———. 1627 (1989). *New Atlantis and the Great Instauration*. Edited by Jerry Weinberger. Wheeling, IL: Harlan Davidson.

Bacon-Smith, Camille. 1992. *Enterprising Women: Television Fandom and the Creation of Popular Myth*. Philadelphia: University of Pennsylvania Press.

Bagger, Eugene S. 1924. "Haldane Looks into the Future: What Marvels Science Will Achieve for Human Life—Impending Urbanization of the World." *New York Times*, April 6: BR1, 24, 27.

Bailey, Liberty Hyde. 1901a. "A Maker of New Fruits and Flowers." *World's Work* 2, no. 5: 1209–14.

———. 1901b. "The Revolution in Farming." *World's Work* 2, no. 3: 945–48.

———. 1902. *Plant-Breeding: Being Five Lectures upon the Amelioration of Domestic Plants*. 2nd ed. New York: Macmillan.

———. 1903a. "Mendel's Law: A Newly Discovered Contribution to the Discussion of Heredity." *Independent* 55, no. 2825 (January 22): 179–82.

———. 1903b. "Some Recent Ideas on the Evolution of Plants." *Science* 17, no. 429: 441–54.

———. 1905. *The Outlook to Nature*. London: Macmillan.

———. 1906. *Plant-Breeding: Being Six Lectures upon the Amelioration of Domestic Plants*. 4th ed. New York: Macmillan.

———. 1910. "Making New Plants: The Creation of Improved Varieties." *Collier's: The National Weekly* 46, no. 1: 27, 36.

———. 1916. *The Holy Earth*. New York: Charles Scribner and Sons.

———. 2019. "The Garden Fence." In *The Liberty Hyde Bailey Gardener's Companion: Essential Writings*, edited by John A. Stempien and John Linstrom, 227–62. Ithaca, NY: Cornell University Press.

Bailey, Liberty Hyde, and Arthur W. Gilbert. 1915. *Plant-Breeding: Being Six Lectures upon the Amelioration of Domestic Plants*. 5th ed. New York: Macmillan.

Bailey, Liberty Hyde, John A. Stempien, and John Linstrom. 2019. *The Liberty Hyde Bailey Gardener's Companion: Essential Writings*. Cornell Scholarship Online. Ithaca, NY: Cornell University Press.

Baker, Robert S. 1990. *Brave New World: History, Science, and Dystopia*. Boston, MA: Twayne.

Bakhtin, Mikhail M. 2010. "From *Discourse in the Novel*." In *The Norton Anthology of Theory and Criticism*, edited by Vincent B. Leitch, William E. Cain, Laurie A. Finke, and Barbara E. Johnson, 1076–106. New York: W. W. Norton.

Baldwin, James Mark, ed. 1902. *Development and Evolution: Including Psychophysical Evolution, Evolution by Orthoplasy, and the Theory of Genetic Modes*. New York: Macmillan.

Bancel, Nicolas, Thomas David, and Dominic Thomas. 2014. *The Invention of Race: Scientific and Popular Representations*. Routledge Studies in Cultural History 28. London: Routledge.

Banks, Harlan P. 1994. "Liberty Hyde Bailey." In *Biographical Memoirs: National Academy of Sciences*. Washington, DC: National Academies Press.

Barabba, Vincent P., ed. 1975. *Historical Statistics of the United States: Colonial Times to 1970*. 2 vols. Washington, DC: US Bureau of the Census.

Barkan, E. 1991. "Reevaluating Progressive Eugenics: Herbert Spencer Jennings and the 1924 Immigration Legislation." *Journal of the History of Biology* 24, no. 1: 91–112.

Barnett, Richard. 2006. "Education or Degeneration: E. Ray Lankester, H. G. Wells and the Outline of History." *Studies in History and Philosophy of Science, Part C: Studies in History and Philosophy of Biological and Biomedical Sciences* 37: 203–29.

Barron, Leonard. 1907. "How to Do What Burbank Does." *New York Herald*, August 4: 15 (magazine section).

Bashford, Alison. 2022. *An Intimate History of Evolution: The Story of the Huxley Family*. London: Allen Lane.

Bashford, Alison, and Philippa Levine, eds. 2010. *The Oxford Handbook of the History of Eugenics*. Oxford: Oxford University Press.

Bateson, William. 1902. *Mendel's Principles of Heredity: A Defence*. Cambridge: Cambridge University Press.

———. 1909. *Mendel's Principles of Heredity*. 2nd ed. Cambridge: Cambridge University Press.

Bavel, Cornelius van. 2000. *Hugo de Vries: Travels of a Dutch Botanist in America 1904–1912*. Pamphlets on Biography of APS Members 630. Center Point, TX: Pecan Valley Press.

Beer, George Louis 1907. "Democracy, Nationalism and Imperialism." *Putnam's* 2, no. 6: 741–47.

Beer, Gillian. 1999a. "'The Death of the Sun': Victorian Solar Physics and Solar Theory." In *Open Fields: Science in Cultural Encounter*. Oxford: Oxford University Press.

———. 1999b. *Open Fields: Science in Cultural Encounter*. 1st paperback ed. Oxford: Oxford University Press.

Bell, Alexander Graham. 1912. "Sheep-Breeding Experiments on Beinn Bhreagh." *Science* 36, no. 925: 378–84.

Bellamy, Edward. 1888 (1917). *Looking Backward: 2000–1887*. Edited by Heywood Broun. Boston, MA: Houghton Mifflin.

Belloc, Hilaire. 1926. *A Companion to Mr. Wells's Outline of History*. London: Sheed and Ward.

Bender, Bert. 1996. *The Descent of Love: Darwin and the Theory of Sexual Selection in American Fiction, 1871–1926*. Philadelphia: University of Pennsylvania Press.

Benjamin, Ruha. 2016. "Racial Fictions, Biological Facts: Expanding the Sociological Imagination through Speculative Methods." *Catalyst: Feminism, Theory, Technoscience* 2, no. 2: 1–28.

Beresford, John Davys. 1911. *The Hampdenshire Wonder*. London: Sidgwick and Jackson.

———. 1914. *H. G. Wells*. Writers of the Day. London: Nisbet.

———. 1917. *The Wonder*. 1st American ed. New York: George H. Doran.

Bernal, John Desmond. 1929 (1970). *The World, the Flesh and the Devil: An Inquiry into the Future of the Three Enemies of the Rational Soul*. 2nd ed. London: Jonathan Cape.

Berneri, Marie Louise. 1950 (1982). *Journey through Utopia*. London: Freedom Press.

Berry, Riley M. Fletcher. 1908. "The Recent Work of Luther Burbank." *Scientific American* 98, no. 15: 260–62.

Bigman, Fran. 2016. "Pregnancy as Protest in Interwar British Women's Writing: An Antecedent Alternative to Aldous Huxley's *Brave New World*." *Medical Humanities* 42, no. 4: 265–70.

Binner, Oscar E. 1911. *Luther Burbank: How His Discoveries Are to Be Put into Practical Use. How the Results of His Forty Years of Experiments Are to Be Placed within the Easy Reach of the Farmers of the World*. Chicago: Oscar E. Binner, Luther Burbank's Publishers.

"Biology and the Humane Spirit." 1930. *New York Times*, April 13: 56.

"Biology a Science of Experiment." 1907. Review of Morgan, *Experimental Zoology*. *Independent* 68, no. 3060 (July 25): 218.

"Biology for All." 1929. Review of Wells, *The Science of Life*. *Nature* 123, no. 3099: 442–43. https://doi.org/10.1038/123442a0.

Black, Edwin. 2004. *The War against the Weak: Eugenics and America's Campaign to Create a Master Race*. 1st paperback ed. New York: Thunder's Mouth Press.

Blavatsky, H. P. 1877 (2012). *Isis Unveiled: A Master-Key to the Mysteries of Ancient and Modern Science and Theology*, vol. 1, *Science*. Cambridge Library Collection—Spiritualism and Esoteric Knowledge. Cambridge: Cambridge University Press.

Bleiler, Everett Franklin. 1948. *The Checklist of Fantastic Literature: A Bibliography of Fantasy, Weird and Science Fiction Books Published in the English Language*. Edited by Melvin Korshak. Chicago: Shasta.

Blinks, Lawrence Rogers. 1974. "Winthrop J. V. Osterhout, 1871–1964." In
Biographical Memoirs, edited by National Academy of Sciences, 212–49.
Washington, DC: National Academies Press.

Bliss, Arthur J. 1906. "Hybridisation and Plant Breeding." *Monthly Review* 24,
no. 70: 84–102.

Bowler, Peter J. 1978. "Hugo De Vries and Thomas Hunt Morgan: The Mutation
Theory and the Spirit of Darwinism." *Annals of Science* 35, no. 1: 55–73.

———. 1983 (1992). *The Eclipse of Darwinism: Anti-Darwinian Evolution Theories
in the Decades around 1900*. 1st paperback ed. Baltimore, MD: Johns Hopkins
University Press.

———. 2006. "Experts and Publishers: Writing Popular Science in Early Twentieth-
Century Britain, Writing Popular History of Science Now." *British Journal for the
History of Science* 39, no. 2: 159–87.

———. 2009. *Science for All: The Popularization of Science in Early Twentieth-
Century Britain*. Chicago: University of Chicago Press.

———. 2017a. "Alternatives to Darwinism in the Early Twentieth Century." In *The
Darwinian Tradition in Context: Research Programs in Evolutionary Biology*,
edited by Richard G. Delisle, 195–217. Cham, Switzerland: Springer.

———. 2017b. *A History of the Future: Prophets of Progress from H. G. Wells to Isaac
Asimov*. Edited by Richard G. Delisle. Cambridge: Cambridge University Press.

Bowler, Peter J., and Iwan Rhys Morus. 2005. *Making Modern Science: A Historical
Survey*. Chicago: University of Chicago Press.

Bradshaw, David, ed. 1995. *The Hidden Huxley*. 1st paperback ed. London: Faber and
Faber.

Brannigan, Augustine. 1979. "The Reification of Mendel." *Social Studies of Science* 9,
no. 4: 423–54.

Brewster, E. T. 1908. "Nature against Nurture." *Atlantic Monthly* 102, no. 1: 120–23.

Brittain, Vera. 1929. *Halcyon; or, The Future of Monogamy*. To-Day and To-Morrow.
London: Kegan Paul, Trench, Trubner.

Broman, Thomas. 1998. "The Habermasian Public Sphere and 'Science in the
Enlightenment.'" *History of Science* 36, no. 2: 123–49.

———. 2012. "Metaphysics for an Enlightened Public: The Controversy over
Monads in Germany, 1746–1748." *Isis* 103, no. 1: 1–23.

Brooke, John Hedley. 1991. *Science and Religion: Some Historical Perspectives*.
Cambridge: Cambridge University Press.

Brooke, John Hedley, and Geoffrey N. Cantor. 1998. *Reconstructing Nature: The
Engagement of Science and Religion*. Edinburgh: T. and T. Clark.

Brown, Andrew. 2005. *J. D. Bernal: The Sage of Science*. Oxford: Oxford University
Press.

Brown, Enos. 1905. "Luther Burbank and Plant Breeding." *Scientific American* 93,
no. 2: 220–21.

Brown, Nik, Brian Rappert, Andrew Webster, and Barbara Adam, eds. 2000.
Contested Futures: A Sociology of Prospective Techno-Science. Florence, UK:
Taylor and Francis Group.

Bryan, William Jennings. 1922. "God and Evolution." *New York Times*, February 26: 2.

Bulmer, Michael. 2004. "Did Jenkin's Swamping Argument Invalidate Darwin's
Theory of Natural Selection?" *British Journal for the History of Science* 37, no. 3:
281–97.

Burbank, Luther. 1895. "How to Produce New Trees, Fruits and Flowers." In *Proceedings of the 24th Session of the American Pomological Society, Sacramento, CA, 16–18 January 1895*, edited by Horticultural Society of New York, 59–66. Topeka, KS: American Pomological Society.

———. 1901. "How to Produce New Flowers and Fruits." *Pacific Rural Press*, July 6: 5–6.

———. 1902–1903. Administrative Files. Carnegie Institution of Washington. Folder 4/29, part 7 of 8.

———. 1904. "Fundamentals of Plant-Breeding." In *Proceedings: International Conference on Plant Breeding and Hybridization, 1902*, edited by Horticultural Society of New York. New York: Horticultural Society of New York.

———. 1904–1905. Administrative Files. Carnegie Institution of Washington. Folder 4/28, part 6 of 8.

———. 1906. "The Training of the Human Plant." *Century* 72, no. 1: 127–37.

———. 1907. *The Training of the Human Plant.* New York: Century.

———. 1908. Administrative Files. Carnegie Institution of Washington. Folder 4/25, part 3 of 8.

———. 1913. *How Nature Makes Plants to Our Order.* Santa Rosa, CA: Luther Burbank Society.

———. 1921. *How Plants Are Trained to Work for Man.* 8 vols. New York: P. F. Collier and Son.

———. 1922. *Half-Hour Experiments with Plants.* New York: P. F. Collier and Son.

———. 1923. "What Plants Have Taught Me about Men." *Hearst's International* 43, no. 1: 66–69, 114.

———. 1926. "Prodigal Mother Nature." *Scientific American* 134, no. 6: 365–66.

Burbank, Luther, and Wilbur Hall. 1927. *The Harvest of the Years: An Autobiography, Edited, with a Biographical Sketch, by W. Hall.* Boston, MA: Houghton Mifflin.

Burbank, Luther, John Whitson, Robert John, Henry Smith Williams, and Luther Burbank Society. 1914. *Luther Burbank: His Methods and Discoveries and Their Practical Application.* 12 vols. New York: Luther Burbank Press.

"Burbank, the Horticultural Wizard." 1905. *Current Literature* 39, no. 2: 212–13.

Burkhardt, Richard W. 1995. *The Spirit of System: Lamarck and Evolutionary Biology.* 2nd ed. Cambridge, MA: Harvard University Press.

Burroughs, John. 1915. "The Arrival of the Fit." *North American Review* 201, no. 711: 197–201.

———. 1916. *Under the Apple-Trees.* Boston, MA: Houghton Mifflin.

———. 1920. "A Critical Glance into Darwin." *Atlantic Monthly* 126, no. 2: 237–47.

———. 1921. *Burroughs's Complete Works.* Boston, MA: Houghton Mifflin.

Burt, Donald C. 1981. "The Well-Manicured Landscape: Nature in Utopia." In *America as Utopia*, edited by Kenneth M. Roemer, 175–85. New York: B. Franklin.

Cairns, William B. 1907–1921 (2000). "Later Magazines." In *The Cambridge History of English and American Literature in 18 Volumes (1907–21)*, edited by Adolphus William Ward, William Peterfield Trent, Alfred Rayney Waller, John Erskine, Stuart Pratt Sherman, and Carl Van Doren. New York: G. P. Putnam's Sons.

California State Board of Trade. 1905. *Complimentary Banquet in Honor of Luther Burbank: Given by the California State Board of Trade, Palace Hotel, San Francisco, 14 September 1905.* San Francisco: California State Board of Trade.

Callon, Michel. 1985. "Some Elements of a Sociology of Translation: Domestication of the Scallops and the Fishermen of St. Brieuc Bay." In *Power, Action and Belief*, edited by John Law, 196–233. London: Routledge and Kegan Paul.

Campos, Luis. 2010. "Mutant Sexuality: The Private Life of a Plant." In *Making Mutations: Objects, Practices, Contexts*, edited by Luis Campos and Alexander von Schwerin, 49–70. Berlin: Max-Planck-Institut für Wissenschaftsgeschichte.

———. 2012. "The BioBrick™ Road." *BioSocieties* 7, no. 2: 115–39.

———. 2015. *Radium and the Secret of Life*. Chicago: University of Chicago Press.

———. 2017. "Dialectics Denied: Muller, Lysenkoism, and the Fate of Chromosomal Mutation." In *The Lysenko Controversy as a Global Phenomenon*, vol. 2, *Genetics and Agriculture in the Soviet Union and Beyond*, edited by William de Jong-Lambert and Nikolai Krementsov, 161–84. Cham, Switzerland: Springer International.

Campos, Luis, and Alexander von Schwerin. 2016. "Transatlantic Mutants: Evolution, Epistemics, and the Engineering of Variation, 1903–1930." In *Heredity Explored: Between Public Domain and Experimental Science, 1850–1930*, edited by Staffan Müller-Wille and Christina Brandt, 143–66. Cambridge, MA: MIT Press.

Canaday, Margot. 2009. *The Straight State: Sexuality and Citizenship in Twentieth-Century America*. Princeton, NJ: Princeton University Press.

Canales, Jimena. 2010. *A Tenth of a Second: A History*. Chicago: University of Chicago Press.

———. 2015. *The Physicist and the Philosopher: Einstein, Bergson, and the Debate That Changed Our Understanding of Time*. Princeton, NJ: Princeton University Press.

Cantor, Paul A., and Peter Hufnagel. 2006. "The Empire of the Future: Imperialism and Modernism in H. G. Wells." *Studies in the Novel* 38, no. 1: 36–56.

Carey, John. 1992. *The Intellectuals and the Masses: Pride and Prejudice among the Literary Intelligentsia, 1880–1939*. Harmondsworth, UK: Penguin.

Carlson, Elof Axel. 1974. "The 'Drosophila' Group: The Transition from Mendelian Unit to Individual Gene." *Journal of the History of Biology* 7: 31–48.

———. 1981. *Genes, Radiation, and Society: The Life and Work of H. J. Muller*. Ithaca, NY: Cornell University Press.

———. 1995. "The Parallel Lives of H. J. Muller and J. B. S. Haldane—Geneticists, Eugenists, and Futurists." In *Haldane's Daedalus Revisited*, edited by Krishna R. Dronamraju and J. B. S. Haldane, 90–101. Oxford: Oxford University Press.

———. 2011. *Mutation: The History of an Idea from Darwin to Genomics*. Cold Spring Harbor, NY: Cold Spring Harbor Laboratory Press.

Carnegie Institution of Washington. 1902. *Year Book No. 1 1902*. Washington, DC: Carnegie Institution of Washington.

———. 1905. *Year Book No. 4 1905*. Washington, DC: Carnegie Institution of Washington.

———. 1907. *Year Book No. 6 1907*. Washington, DC: Carnegie Institution of Washington.

Carpenter, Frank G. 1906. "Telephone Bell Has New Breed of Sheep and a New Flying Machine." *Omaha Daily Bee*, August 26: 24.

Carrington, André M. 2016. *Speculative Blackness: The Future of Race in Science Fiction*. Minneapolis: University of Minnesota Press.

Carter, Paul Allen. 1977. *The Creation of Tomorrow: Fifty Years of Magazine Science Fiction*. New York: Columbia University Press.

Cashbaugh, Sean. 2016. "A Paradoxical, Discrepant, and Mutant Marxism: Imagining a Radical Science Fiction in the American Popular Front." *Journal for the Study of Radicalism* 10, no. 1: 63–106.

Castle, William E. 1917. "The Rôle of Selection in Evolution." *Journal of the Washington Academy of Sciences* 7, no. 12: 369–87.

Cattell, James McKeen. 1901. "Two Remarks Concerning the Monthly." *Popular Science Monthly* 59: 510–11.

[————]. 1904. Review of Bailey, *Plant-Breeding*, 3rd ed. *Botanical Gazette* 37, no. 6: 471–72.

Ceplair, Larry, ed. 1991. *Charlotte Perkins Gilman: A Nonfiction Reader*. New York: Columbia University Press.

Certeau, Michel de. 1984 (1988). *The Practice of Everyday Life*. Translated by Steven Rendall. Berkeley: University of California Press.

Chalaby, Jean K. 1998. *The Invention of Journalism*. London: Macmillan.

Chartier, Roger. 1989. "Texts, Printing, Readings." In *The New Cultural History*, edited by Lynn Avery Hunt, 154–75. Berkeley: University of California Press.

Cheng, John. 2012. *Astounding Wonder: Imagining Science and Science Fiction in Interwar America*. Philadelphia: University of Pennsylvania Press.

Chew, Carissa. 2019. "The Ant as Metaphor: Orientalism, Imperialism and Myrmecology." *Archives of Natural History* 46, no: 2: 347–61.

Child, C. M. 1907. Review of Morgan, *Experimental Zoology*. *Science* 26, no. 676: 824–29.

Christensen, Andrew G. 2017. "Charlotte Perkins Gilman's *Herland* and the Tradition of the Scientific Utopia." *Utopian Studies* 28, no. 2: 286–304.

Clark, Constance Areson. 2001. "Evolution for John Doe: Pictures, the Public, and the Scopes Trial Debate." *Journal of American History* 87, no. 4: 1275–303.

————. 2008. *God—or Gorilla: Images of Evolution in the Jazz Age*. Baltimore, MD: Johns Hopkins University Press.

Clark, George Archibald. 1905. "Luther Burbank: The High Priest of Horticulture Who Has Worked Marvels in Transforming and Improving Plant Life and Products." *Success* 8, no. 134: 455–58.

Clayton, Jay. 2016. "The Modern Synthesis: Genetics and Dystopia in the Huxley Circle." *Modernism/Modernity* 23, no. 4: 875–96.

Cleland, Ralph E. 1972. *Oenothera: Cytogenetics and Evolution*. London: Academic Press.

Coates, Peter. 1998. *Nature: Western Attitudes since Ancient Times*. Cambridge: Polity Press.

Cockerell, T. D. A. 1904. "Criticisms of Darwin." Review of Morgan, *Evolution and Adaptation* and Semi-Darwinian *Doubts about Darwinism*. *Dial: A Semi-Monthly Journal of Literary Criticism, Discussion, and Information*, March 16: 196–97.

Cohen, Joseph E. 1910. *Socialism for Students*. Chicago: Charles H. Kerr.

Cohn, Gene. 1925. "Battlecry of Peace Times Should Be 'Toward the New Utopia,' Says Scientist." *Battle Creek Enquirer and Evening News*, July 3: 8.

Cohn, Raymond L. 2017. "Immigration to the United States." *EH.Net Encyclopedia*, edited by Robert Whaples. Economic History Association. https://eh.net/encyclopedia/immigration-to-the-united-states/.

Coleman, Walter M. 1921. *Beginners' Zoology*. Toronto: Macmillan.

Collini, Stefan, Richard Whatmore, and Brian Young, eds. 2000. *History, Religion, and Culture: British Intellectual History 1750–1950*. Cambridge: Cambridge University Press.

Colp, Ralph. 1974. "The Contacts between Karl Marx and Charles Darwin." *Journal of the History of Ideas* 35, no. 2: 329–38.

———. 1982. "The Myth of the Darwin-Marx Letter." *History of Political Economy* 14, no. 4: 461–82.

"Concerning the Mode of Evolution." 1904. Review of Morgan, *Evolution and Adaptation*. *Independent* 57, no. 2912 (September 22): 677–78.

Conklin, Edwin Grant. 1908. Review of Morgan, *Experimental Zoology*. *Science* 27, no. 682: 139–40.

———. 1922. *Heredity and Environment in the Development of Men*. 5th ed. Northwestern University N. W. Harris Lectures for 1914. Princeton, NJ: Princeton University Press.

Cook, Orator Fuller. 1901. "A Kinetic Theory of Evolution." *Science* 13, no. 338: 969–78.

———. 1903. "Stages of Vital Motion." *Popular Science Monthly* 63: 14–24.

———. 1904a. "Evolution Not the Origin of Species." *Popular Science Monthly* 64, no. 25: 445–56.

———. 1904b. "Natural Selection in Kinetic Evolution." *Science* 19, no. 483: 549–50.

———. 1907. "Aspects of Kinetic Evolution." *Proceedings of the Washington Academy of Sciences* 8: 197–403.

———. 1908. *Methods and Causes of Evolution*. Edited by Beverly T. Galloway. USDA Bureau of Plant Industry Bulletin 136. Washington, DC: US Department of Agriculture.

"Cook, Orator Fuller." 1953. In *The National Cyclopædia of American Biography: Being the History of the United States as Illustrated in the Lives of the Founders, Builders, and Defenders of the Republic, and of the Men and Women Who Are Doing the Work and Moulding the Thought of the Present Time*, edited by James Terry White and George Derby, 369–70. New York: James T. White.

Cooter, Roger. 1984. *The Cultural Meaning of Popular Science: Phrenology and the Organization of Consent in Nineteenth-Century Britain*. Cambridge: Cambridge University Press.

Cope, Edward Drinker. 1896 (1904). *The Primary Factors of Organic Evolution*. Chicago: Open Court Publishing.

"Cosmic Rays and Evolution." 1928. *New York Times*, April 26: 26.

Cotkin, George. 1981. "'They All Talk Like Goddam Bourgeois': Scientism and the Socialist Discourse of Arthur M. Lewis." *ETC: A Review of General Semantics* 38, no. 3: 272–84.

———. 1984. "The Socialist Popularization of Science in America, 1901 to the First World War." *History of Education Quarterly* 24, no. 2: 201–14.

Coulter, John Merle. 1905. "Public Interest in Research." *Popular Science Monthly* 67: 306–12.

———. 1907. Review of de Vries, *Plant-Breeding*. *Botanical Gazette* 44, no. 2: 147–51.

———. 1914. *Fundamentals of Plant-Breeding*. New York: D. Appleton.

Cowles, Henry Chandler. 1902. "The Mutation Theory." Review of de Vries, *Die Mutationstheorie. Versuche und Beobachtungen über die Entstehung von Arten im Pflanzenreich. Botanical Gazette* 33, no. 3: 236–39.

Craig, Patricia. 2005. *The Department of Plant Biology.* Centennial History of the Carnegie Institution of Washington 4. Cambridge: Cambridge University Press.

Crane, Jonathan Mayo. 1909. "Frank Criticisms of Thompson's Work." Review of Thompson, *New Reading of Evolution. Washington Herald,* August 29: 5.

Crèvecoeur, J. Hector St. John. 1782 (1904). *Letters from an American Farmer.* New York: Fox, Duffield.

Crew, F. A. E. (Francis Albert Eley). 1967. "Reginald Crundall Punnett (1875–1967)." *Biographical Memoirs of Fellows of the Royal Society* 13: 309–26.

Crew, F. A. E., C. D. Darlington, J. B. S. Haldane, C. Harland, L. T. Hogben, J. S. Huxley, H. J. Muller, J. Needham, G. P. Child, P. C. Koller, P. R. David, W. Landauer, G. Dahlberg, H. H. Plough, T. H. Dobzhansky, B. Price, R. A. Emerson, J. Schultz, C. Gordon, A. G. Steinberg, J. Hammond, C. H. Waddington, and C. L. Huskins. 1939. "Social Biology and Population Improvement." *Nature* 144, no. 3646: 521–22. https://doi.org/10.1038/144521a0.

Cronin, Helena. 1991. *The Ant and the Peacock: Altruism and Sexual Selection from Darwin to Today.* Cambridge: Cambridge University Press.

Cronon, William. 2003. *Changes in the Land: Indians, Colonists, and the Ecology of New England.* 1st rev. ed. New York: Hill and Wang.

Crow, James F. 2001. "Plant Breeding Giants: Burbank, the Artist; Vavilov, the Scientist." *Genetics* 158, no. 4: 1391–95.

Csicsery-Ronay, Istvan. 1991. "The SF of Theory: Baudrillard and Haraway." *Science Fiction Studies* 18, no. 3: 387–404.

Cuddy, Lois A., and Claire M. Roche, eds. 2003. *Evolution and Eugenics in American Literature and Culture, 1880–1940.* Lewisburg, PA: Bucknell University Press.

Currell, Susan. 2006. "Eugenic Decline and Recovery in Self-Improvement Literature of the Thirties." In *Popular Eugenics: National Efficiency and American Mass Culture in the 1930s,* edited by Susan Currell and Christina Cogdell, 44–69. Athens: Ohio University Press.

Currell, Susan, and Christina Cogdell, eds. 2006. *Popular Eugenics: National Efficiency and American Mass Culture in the 1930s.* Athens: Ohio University Press.

"Current Theory of Heredity: Radical Change in the Views of Scientist Caused by Mendel's Experiments." 1911. *Washington Post,* January 15: MS4.

Curry, Helen Anne. 2016. *Evolution Made to Order: Plant Breeding and Technological Innovation in Twentieth-Century America.* Chicago: University of Chicago Press.

Darnton, Robert. 1996. *The Forbidden Best-Sellers of Pre-Revolutionary France.* London: Fontana Press.

Darwin, Charles. 1845. *Journal of Researches into the Natural History and Geology of the Countries Visited during the Voyage of H.M.S. Beagle round the World, under the Command of Capt. Fitz Roy, R.N.* 2nd ed. London: John Murray.

———. 1859. *On the Origin of Species by Means of Natural Selection; or, The Preservation of Favoured Races in the Struggle for Life.* London: John Murray.

———. 1871. *The Descent of Man, and Selection in Relation to Sex.* 2 vols. London: John Murray.

Daston, Lorraine. 2001. "Scientific Objectivity with and without Words." In *Little Tools of Knowledge: Historical Essays on Academic and Bureaucratic Practices,* edited by Peter Becker and William Clark, 259–84. Ann Arbor: University of Michigan Press.

———. 2014. "The Naturalistic Fallacy Is Modern." *Isis* 105, no. 3: 579–87.

Daston, Lorraine, and Peter Galison. 1992. "The Image of Objectivity." *Representations* 40: 81–128.

Daston, Lorraine, and Fernando Vidal, eds. 2004. *The Moral Authority of Nature.* Chicago: University of Chicago Press.

Davenport, Charles Benedict, W. R. T. Jones, John S. Billings, and Hugo De Vries. 1905. "Addresses at Opening of the Station for Experimental Evolution, June 11, 1904." In *Year Book No. 4 1905,* 33–34. Washington, DC: Carnegie Institution of Washington.

Davis, J. Colin. 1984. "Science and Utopia: The History of a Dilemma." In *Nineteen Eighty-Four: Science between Utopia and Dystopia,* edited by Everett Mendelsohn and Helga Nowotny, 21–48. Dordrecht: Reidel.

Davis, William Harper. 1904. "The International Congress of Arts and Sciences." *Popular Science Monthly* 66: 5–32.

Dean, Bashford. 1904. Review of Morgan, *Evolution and Adaptation. Science* 19, no. 475: 221–25.

———. 1908. Review of Kellogg, *Darwinism To-Day. Science* 27, no. 689: 421–23.

"Declined the Grant—Luther Burbank Not Denied Public Assistance." 1903. Press cutting. Administrative Files. Carnegie Institution of Washington. Folder 4/29, part 7 of 8.

Deegan, Mary Jo. 1997. "Introduction: Gilman's Sociological Journey from *Herland* to *Ourland.*" In *With Her in Ourland: Sequel to Herland,* edited by Mary Jo Deegan and Charlotte Perkins Gilman, 1–57. Westport, CT: Praeger.

Delany, Samuel R. 1994. "K. Leslie Steiner Interview." In *Silent Interviews: On Language, Race, Sex, Science Fiction, and Some Comics,* 269–86. Hanover, NH: Wesleyan University Press.

———. 1999. *Shorter Views: Queer Thoughts and the Politics of the Paraliterary.* Hanover, NH: Wesleyan University Press.

Dennert, Eberhardt. 1904. *At the Deathbed of Darwinism: A Series of Papers.* Translated by E. V. O'Harra and John H. Peschges. Burlington, IA: German Literary Board.

De Rooy, Piet. 1998. "The Natural Selection of Evolutionary Theory: Darwinism in the Netherlands 1850–1900." *Acta botanica neerlandica* 47, no. 4: 419–25.

Deutscher, Penelope. 2004. "The Descent of Man and the Evolution of Woman." *Hypatia* 19, no. 2: 35–55.

De Vries, Hugo. 1901–1903. *Die Mutationstheorie: Versuche und Beobachtungen über die Entstehung von Arten im Pflanzenreich / von Hugo de Vries.* Leipzig: Verlag Von Veit.

———. 1902a. "My Primrose Experiments." *Independent* 54, no. 2808 (September 25): 2285–87.

———. 1902b. "The Origin of Species by Mutation." *Science* 15, no. 384: 721–29.

———. 1903. "On the Origin of Species." *Popular Science Monthly* 62, no. 6: 481–96.

———. 1904. "The Evidence of Evolution." *Science* 20, no. 508: 395–401.

———. 1905a. *Naar Californië: Reisherinneringen*. Haarlem [Netherlands]: H. D. Tjeenk Willink and Zoon.

———. 1905b. "A New Conception Concerning the Origin of Species." *Harper's Monthly* 110, no. 656: 209–13.

———. 1905c. *Species and Varieties: Their Origin by Mutation; Lectures Delivered at the University of California*. Edited by Daniel Trembly MacDougal. Chicago: Open Court Publishing.

———. 1905d. "A Visit to Luther Burbank." *Popular Science Monthly* 67, no. 4: 329–47.

———. 1906a. "Burbank's Production of Horticultural Novelties." *Open Court* 20, no. 606: 641–53.

———. 1906b. "Personal Impressions of Luther Burbank." *Independent* 60, no. 2998 (May 17): 1134–40.

———. 1907a. "Luther Burbank's Ideas on Scientific Horticulture." *Century* 73, no. 5: 674–81.

———. 1907b. *Plant-Breeding: Comments on the Experiments of Nilsson and Burbank*. Chicago: Open Court Publishing.

———. 1909. *The Mutation Theory: Experiments and Observations on the Origin of Species in the Vegetable Kingdom*. Translated by J. B. Farmer and A. D. Darbishire. 2 vols. Chicago: Open Court Publishing Company.

"Distinguished Botanist Plans for Experiments." 1904. *San Francisco Call*, February 15: 6.

Dixon, Thomas. 2008a. *The Invention of Altruism: Making Moral Meanings in Victorian Britain*. British Academy Postdoctoral Fellowship Monographs. Oxford: Oxford University Press.

———. 2008b. *Science and Religion: A Very Short Introduction*. Oxford: Oxford University Press.

Donawerth, Jane. 1994. "Science Fiction by Women in the Early Pulps, 1926–1930." In *Utopian and Science Fiction by Women: Worlds of Difference*, edited by Jane L. Donawerth and Carol A. Kolmerten, 137–52. Syracuse, NY: Syracuse University Press.

———. 1997. *Frankenstein's Daughters: Women Writing Science Fiction*. Syracuse, NY: Syracuse University Press.

———. 2009. "Feminisms." In *The Routledge Companion to Science Fiction*, edited by Mark Bould, Andrew M. Butler, Adam Roberts, and Sherryl Vint, 214–24. Abingdon, UK: Routledge.

Drayton, Richard H. 2000. *Nature's Government: Science, Imperial Britain and the "Improvement" of the World*. New Haven, CT: Yale University Press.

Dreyer, Peter. 1985. *A Gardener Touched with Genius: The Life of Luther Burbank*. Rev. ed. Berkeley: University of California Press.

Drinkwater, Harry. 1910. *A Lecture on Mendelism*. London: J. M. Dent and Sons.

Dronamraju, Krishna. 2016. *Popularizing Science: The Life and Work of JBS Haldane*. Oxford: Oxford University Press.

Dronamraju, Krishna R., and J. B. S. Haldane, eds. 1995. *Haldane's Daedalus Revisited*. Oxford: Oxford University Press.

Dubrow, Heather. 1982. *Genre*. London: Methuen.

Duffus, R. L. 1930. "Modern Biological Science and the Future of the Race: Professor H. S. Jennings Elucidates the New Theories and Shows Their Applications." *New York Times*, April 13: 66.

Durant, Alan. 2006. "Raymond Williams's Keywords: Investigating Meanings 'Offered, Felt for, Tested, Confirmed, Asserted, Qualified, Changed.'" *Critical Quarterly* 48, no. 4: 1–26.

Durant, John R. 1979. "Scientific Naturalism and Social Reform in the Thought of Alfred Russel Wallace." *British Journal for the History of Science* 12, no. 1: 31–58.

"Dutch Botanist Here to Study." 1906. *Salt Lake Herald*, August 8: 10.

Dyson, Freeman. 1995. "Daedalus after Seventy Years." In *Haldane's Daedalus Revisited*, edited by Krishna R. Dronamraju, 55–63. Oxford: Oxford University Press.

Eder, M. D. 1908. "Good Breeding or Eugenics." *New Age: A Weekly Review of Politics, Literature, and Art* 3, no. 715: 67.

[Edge], H. T. 1902. "The Mutation Theory in Evolution." *New Century* 5, no. 46: 4.

Edson, Milan C. 1900. *Solaris Farm: A Story of the Twentieth Century*. Washington, DC: Published by the Author.

Ellingson, Ter. 2001. *The Myth of the Noble Savage*. Berkeley: University of California Press.

Ellis, William T. 1925. "Today's Tremendous Testing (International Sunday School Lesson)." *Washington Post*, January 31: 10.

Endersby, Jim. 2003. "Darwin on Generation, Pangenesis and Sexual Selection." In *The Cambridge Companion to Darwin*, edited by M. J. S. Hodge and G. Radick, 69–91. Cambridge: Cambridge University Press.

———. 2007. *A Guinea Pig's History of Biology: The Plants and Animals Who Taught Us the Facts of Life*. London: William Heinemann.

———. 2009. "Editor's Introduction." In *On the Origin of Species by Means of Natural Selection; or, The Preservation of Favoured Races in the Struggle for Life*, edited by Jim Endersby, xi–lxv. Cambridge: Cambridge University Press.

———. 2013. "Mutant Utopias: Evening Primroses and Imagined Futures in Early Twentieth-Century America." *Isis* 104, no. 3: 471–503.

———. 2016. "Deceived by Orchids: Sex, Science, Fiction and Darwin." *British Journal for the History of Science* 49, no. 2: 205–29.

———. 2018. "A Visit to Biotopia: Genre, Genetics and Gardening in the Early Twentieth Century." *British Journal for the History of Science* 51, no. 3: 423–55.

Eshun, Kodwo. 2003. "Further Considerations on Afrofuturism." *CR: The New Centennial Review* 3, no. 2: 287–302.

Esposito, Maurizio. 2011. "Utopianism in the British Evolutionary Synthesis." *Studies in History and Philosophy of Science, Part C: Studies in History and Philosophy of Biological and Biomedical Sciences* 42, no. 1: 40–49.

———. 2017. "Expectation and Futurity: The Remarkable Success of Genetic Determinism." *Studies in History and Philosophy of Science, Part C: Studies in History and Philosophy of Biological and Biomedical Sciences* 62: 1–9.

"The Experimental Method in Biology." 1907. Review of Morgan, *Experimental Zoology. Dial: A Semi-Monthly Journal of Literary Criticism, Discussion, and Information*, April 1: 228–29.

Fairchild, David Grandison. 1938. *The World Was My Garden: Travels of a Plant Explorer*. New York: Charles Scribner and Sons.

Fairchild, David Grandison. 1903. Application for grant in aid of research. Administrative Files. Carnegie Institution of Washington. Folder 4/29, part 7 of 8.

Fara, Patricia. 1996. *Sympathetic Attractions: Magnetic Practices, Beliefs, and Symbolism in Eighteenth-Century England*. Princeton, NJ: Princeton University Press.

Fender, Stephen. 1992. *Sea Changes: British Emigration and American Literature*. Cambridge: Cambridge University Press.

Ferns, Chris. 1999. *Narrating Utopia: Ideology, Gender, Form in Utopian Literature*. Edited by David Seed. Liverpool Science Fiction Texts and Studies. Liverpool: Liverpool University Press.

Ferri, Enrico. 1905 (1909). *Socialism and Positive Science: Darwin—Spencer—Marx*. Translated by Edith C. Harvey. Edited by James Ramsay MacDonald. 5th ed. Socialist Library. London: Independent Labour Party.

F.H.K. 1904. Review of Bailey, *Plant-Breeding*. *Plant World* 7, no. 7: 188.

Firestone, Shulamith. 1970 (2015). *The Dialectic of Sex: The Case for Feminist Revolution*. London: Verso.

"First Glances at New Books." 1929. Review of Wells, *The Science of Life*. *Science News-Letter*, October 19: 247–48.

Fish, Stanley E. 1976 (2010). "Interpreting the *Variorum*." In *The Norton Anthology of Theory and Criticism*, edited by Vincent B. Leitch, William E. Cain, Laurie A. Finke, and Barbara E. Johnson, 1974–92. New York: W. W. Norton.

———. 1980. *Is There a Text in This Class? The Authority of Interpretive Communities*. Cambridge, MA: Harvard University Press.

Fisher, Clyde. 1929. "Garrett P. Serviss: One Who Loved the Stars." *Popular Astronomy* 37, no. 7: 365–69.

Fletcher, Walter Morley. 1933. "Biology as a Foundation for Education." In *Biology in Education: A Handbook Based on the Proceedings of the National Conference on the Place of Biology in Education, Organised by the British Social Hygiene Council*, edited by James Gerald Crowther, 128–32. London: William Heinemann.

Flury, Henry. 1936. "Scientists." Review of Muller, *Out of the Night*. *Washington Post*, August 9: B8.

"For All Those Who Love the Outdoor World." 1905. *Dial: A Semi-Monthly Journal of Literary Criticism, Discussion, and Information*, November 16: 312.

Forman, Henry James. 1923. "H. G. Wells Skids into Utopia." Review of Wells, *Men Like Gods*. *New York Times*, May 27: BR1–BR2.

Frank, Janrae, Jean Stine, and Forrest J. Ackerman, eds. 1994. *New Eves: Extraordinary Fiction about the Extraordinary Women of Today and Tomorrow*. Stamford, CT: Longmeadow Press.

Friends and Relatives of Luther Burbank. 1906. *General Information for the Public*. Santa Rosa, CA. Administrative Files. Carnegie Institution of Washington. Folder 4/30, part 8 of 8: Burbank, Luther (Miscellaneous).

F.T.L. 1908. "Evolution and Heredity." Review of Kellogg, *Darwinism To-Day*. *American Naturalist* 42, no. 493: 58–60.

Fyfe, Aileen, and Bernard V. Lightman. 2007. *Science in the Marketplace: Nineteenth-Century Sites and Experiences*. Chicago: University of Chicago Press.

Gager, C. Stuart. 1920. *Heredity and Evolution in Plants*. Philadelphia, PA: P. Blakiston's Son.

Galloway, Beverly T. 1902. "Applied Botany, Retrospective and Prospective." *Science* 16 (new series), no. 393: 49–59.

Galton, Francis. 1865. "Hereditary Talent and Character." *Macmillan's* 12: 157–66, 318–27.

———. 1869 (1892). *Hereditary Genius*. 2nd ed. London: Macmillan.

———. 1874. *English Men of Science: Their Nature and Nurture*. London: Macmillan.

———. 1904. "Eugenics: Its Definition, Scope, and Aims." *American Journal of Sociology* 10, no. 1: 1–25.

———. 1908. "Local Associations for Promoting Eugenics." *Nature* 78, no. 2034: 645–47. https://doi.org/10.1038/078645a0.

Ganobcsik-Williams, Lisa. 1999. "The Intellectualism of Charlotte Perkins Gilman: Evolutionary Perspectives on Race, Ethnicity, and Class." In *Charlotte Perkins Gilman: Optimist Reformer*, edited by Jill Rudd and Val Gough, 16–41. Iowa City: University of Iowa Press.

Garrett, Elizabeth. 1924. "Muller Talks upon Subject of 'The Promise of Biology.'" *Austin Daily Texan* 24, no. 155: 1.

Gates, Henry Louis. 1986. *"Race," Writing and Difference*. Chicago: University of Chicago Press.

Gauvin, Marshall Jerome. 1921. *The Illustrated Story of Evolution*. New York: Peter Eckler.

Geddes, Patrick, and J. Arthur Thomson. 1901. *The Evolution of Sex*. 2nd, rev. ed. London: Walter Scott.

———. 1911. *Evolution*. New York: Henry Holt.

Gerber, Richard. 1955. *Utopian Fantasy: A Study of English Utopian Fiction since the End of the Nineteenth Century*. London: Routledge and Kegan Paul.

Ghent, William James. 1910. *Socialism and Success: Some Uninvited Messages*. New York: John Lane.

Gianquitto, Tina, and Lydia Fisher, eds. 2014. *America's Darwin: Darwinian Theory and U.S. Literary Culture*. Athens: University of Georgia Press.

Gilbert, Sandra M., and Susan Gubar. 1989. "Home Rule: The Colonies of the New Woman." In *No Man's Land: The Place of the Woman Writer in the Twentieth Century*, vol. 2, *Sexchanges*, 47–82. New Haven, CT: Yale University Press.

———. 1999. "'Fecundate! Discriminate!' Charlotte Perkins Gilman and the Theologizing of Maternity." In *Charlotte Perkins Gilman: Optimist Reformer*, edited by Jill Rudd and Val Gough, 200–216. Iowa City: University of Iowa Press.

Giles, Chauncey. 1887. *The True and False Theory of Evolution*. Philadelphia: William H. Alden. Available at http://www.swedenborgdigitallibrary.org/evolution/evotc.htm.

Gillham, Nicholas Wright. 2001. *A Life of Sir Francis Galton: From African Exploration to the Birth of Eugenics*. Oxford: Oxford University Press.

Gilman, Charlotte Perkins. 1895. *In This Our World, and Other Poems*. San Francisco, CA: James H. Barry and John H. Marble.

———. 1908. "A Suggestion on the Negro Problem." *American Journal of Sociology* 14, no. 1: 78–85.

———. 1910. "Comment and Review." *Forerunner* 1, no: 3: 28–29.

———. 1915. *Women and Economics: A Study of the Economic Relation between Men and Women as a Factor in Social Evolution*. 7th ed. Boston, MA: Small, Maynard.

———. 1916a. "As to Parthenogenesis and Humanity." *Forerunner* 7, no. 3: 83.

———. 1916b. "Assisted Evolution." *Forerunner* 7, no. 1: 5.

———. 1916c (1997). *With Her in Ourland: Sequel to Herland.* Westport, CT: Praeger.

———. 1923 (1991). "Is America Too Hospitable?" In *Charlotte Perkins Gilman: A Nonfiction Reader,* edited by Larry Ceplair, 288–95. New York: Columbia University Press. Original edition, *Forum* 70 (October 1923): 1983–89.

———. 1999. *Herland, The Yellow Wall-Paper, and Selected Writings.* Edited by Denise D. Knight. Penguin Twentieth-Century Classics. New York: Penguin.

———. 2013. *Herland and Related Writings.* Edited by Beth Sutton-Ramspeck. Peterborough, ON: Broadview Press.

Gissis, Snait B., and Eva Jablonka, eds. 2011. *Transformations of Lamarckism: From Subtle Fluids to Molecular Biology.* Cambridge, MA: MIT Press.

Give the Boy His Chance. 1913. Santa Rosa, CA: Luther Burbank Society.

Glass, Bentley. 1980. "The Strange Encounter of Luther Burbank and George Harrison Shull." *Proceedings of the American Philosophical Society* 124, no. 2: 133–53.

Gliboff, Sandor 2011. "The Golden Age of Lamarckism, 1866–1926." In *Transformations of Lamarckism: From Subtle Fluids to Molecular Biology,* edited by Snait B. Gissis and Eva Jablonka, 45–56. Cambridge, MA: MIT Press.

Glick, Thomas F. 1988. *The Comparative Reception of Darwinism.* Chicago: University of Chicago Press.

Golinski, Jan. 1999. *Science as Public Culture: Chemistry and Enlightenment in Britain, 1760–1820.* 1st paperback ed. Cambridge: Cambridge University Press.

Goodwin, Barbara, ed. 2001. *The Philosophy of Utopia.* Abingdon, UK: Routledge and Kegan Paul.

Gossett, Thomas F. 1997. *Race: The History of an Idea in America.* New York: Oxford University Press.

Gould, Stephen Jay. 1974. "The Origin and Function of 'Bizarre' Structures: Antler Size and Skull Size in the 'Irish Elk,' *Megaloceros giganteus.*" *Evolution* 28, no. 2: 191–220.

———. 1977 (1987). *Ever Since Darwin: Reflections in Natural History.* Harmondsworth, UK: Penguin.

———. 1983a. *Hen's Teeth and Horse's Toes: Further Reflections in Natural History.* Harmondsworth, UK: Penguin.

———. 1983b. "Nonmoral Nature." In *Hen's Teeth and Horse's Toes,* 32–45. Harmondsworth, UK: Penguin.

———. 1988. *Time's Arrow, Time's Cycle: Myth and Metaphor in the Discovery of Geological Time.* Harmondsworth, UK: Penguin.

———. 1990. *Wonderful Life: The Burgess Shale and the Nature of History.* London: Hutchinson Radius.

———. 1991. *Bully for Brontosaurus: Reflections in Natural History.* London: Hutchinson Radius.

———. 1996. *Dinosaur in a Haystack: Reflections in Natural History.* London: Jonathan Cape.

———. 1997. *The Mismeasure of Man.* 2nd, rev. and exp. ed. Harmondsworth, UK: Penguin.

Graham, Amanda. 1998. "*Herland*: Definitive Ecofeminist Fiction?" In *A Very Different Story: Studies on the Fiction of Charlotte Perkins Gilman,* edited by Val Gough and Jill Rudd, 115–28. Liverpool: Liverpool University Press.

Graham, Jean. 1905. "A Wizard in the Garden." *Canadian Magazine of Politics, Science, Art and Literature* 25, no. 2: 176–77.

Greene, John C. 1990. "The Interaction of Science and World View in Sir Julian Huxley's Evolutionary Biology." *Journal of the History of Biology* 23, no. 1: 39–55.

Greg, William R. 1868. "On the Failure of 'Natural Selection' in the Case of Man." *Fraser's Magazine for Town and Country* 78, no. 465: 353–62.

Gregory, Thomas Jefferson. 1911. *History of Sonoma County, California, with Biographical Sketches of the Leading Men and Women of the County, Who Have Been Identified with Its Growth and Development from the Early Days to the Present Time.* Los Angeles, CA: Historic Record Company.

Grinnell, Joseph. 1908. Review of Jordan and Kellogg, *Evolution and Animal Life.* *Condor* 10, no. 1: 52–53.

Gronlund, Laurence. 1884. *The Coöperative Commonwealth in Its Outlines: An Exposition of Modern Socialism.* Boston, MA: Lee and Shepard.

Grove, Richard H. 1995. *Green Imperialism: Colonial Expansion, Tropical Island Edens, and the Origins of Environmentalism, 1600–1860.* Cambridge: Cambridge University Press.

Gruenberg, Benjamin Charles. 1919. *Elementary Biology: An Introduction to the Science of Life.* Boston, MA: Ginn.

———. 1929. *The Story of Evolution: Facts and Theories on the Development of Life.* Garden City, NY: Garden City Publishing.

Hagan, William T. 1985. "Adjusting to the Opening of the Kiowa, Comanche, and Kiowa-Apache Reservation." In *The Plains Indians of the Twentieth Century,* edited by Peter Iverson. Norman: University of Oklahoma Press.

Hagedoorn, Arend Lourens, and Anna Cornelia Hagedoorn-Vorstheuvel la Brand. 1921. *The Relative Value of the Processes Causing Evolution.* The Hague: Martinus Nijhoff.

Haldane, J. B. S. 1924. *Daedalus; or, Science and the Future.* London: Kegan Paul, Trench, Trubner.

———. 1925. *Callinicus: A Defence of Chemical Warfare.* 2nd ed. London: Kegan Paul, Trench, Trubner.

———. 1927 (1930). *Possible Worlds and Other Essays.* London: Chatto and Windus.

———. 1928. *Science and Ethics.* London: Watts.

———. 1940. *Possible Worlds and Other Essays.* London: Chatto and Windus.

———. 1946. "Auld Hornie, F.R.S." *Modern Quarterly* (Autumn): 32.

Hale, Piers J. 2010. "Of Mice and Men: Evolution and the Socialist Utopia; William Morris, H. G. Wells, and George Bernard Shaw." *Journal of the History of Biology* 43, no. 1: 17–66.

———. 2014. *Political Descent: Malthus, Mutualism, and the Politics of Evolution in Victorian England.* Chicago: University of Chicago Press.

Hall, Stuart, Jennifer Daryl Slack, and Lawrence Grossberg. 2016. *Cultural Studies 1983: A Theoretical History.* Durham, NC: Duke University Press.

Hamilton, Edmond. 1927. "Evolution Island." *Weird Tales* 9, no. 3: 337–54.

———. 1931. "Ten Million Years Ahead." *Weird Tales* 17, no. 3: 304–19, 426–32.

———. 1977. *The Best of Edmond Hamilton.* Edited by Leigh Brackett. New York: Del Rey/Ballantine.

Hamlin, Kimberly A. 2014a. *From Eve to Evolution: Darwin, Science and Women's Rights in Gilded Age America.* Chicago: University of Chicago Press.

———. 2014b. "Sexual Selection and Marriage: 'Female Choice' in the Writings of Edward Bellamy and Charlotte Perkins Gilman." In *America's Darwin: Darwinian Theory and U.S. Literary Culture*, edited by Tina Gianquitto and Lydia Fisher, 151–80. Athens: University of Georgia Press.

Hansen, Niels Ebbesen. 1905. "A Visit to Luther Burbank." *Proceedings of the Society for Horticultural Science*, December 27: 32–34.

Haraway, Donna. 1989. *Primate Visions: Gender, Race and Nature in the World of Modern Science*. London: Routledge.

———. 1991. "A Cyborg Manifesto." In *Simians, Cyborgs and Women*, edited by Donna Haraway, 149–82. London: Routledge.

———. 1997. *Modest_Witness@Second_Millennium. FemaleMan©_Meets_ OncoMouse™: Feminism and Technoscience*. New York: Routledge.

Harding, John W. 1905. "Dr. Macdougal's Botanical Feat Threatens Evolution Theories." *New York Times*, December 24: SM1–2.

Harding, Sandra. 1993. "Rethinking Standpoint Epistemology: 'What Is Strong Objectivity?'" In *Feminist Epistemologies*, edited by Linda Alcoff and Elizabeth Potter, 437–70. New York: Routledge.

Harley, Gail M. 1991. "Emma Curtis Hopkins: 'Forgotten Founder' of New Thought." PhD diss., College of Arts and Sciences, Florida State University. https://www.proquest.com/pqdtglobal1/dissertations-theses/emma-curtis -hopkins-forgotten-founder-new-thought/docview/303949560/sem-2.

Harris, Clare Winger. 1928. "The Miracle of the Lily." *Amazing Stories* 3, no. 1: 48–55.

———. 1929. "The Evolutionary Monstrosity." *Amazing Stories Quarterly* 2, no. 1: 70–77.

———. 1930. "The Ape Cycle." *Science Wonder Quarterly* 1, no. 3: 388–405.

———. 1931. "Possible Science Fiction Plots." *Wonder Stories* 3, no. 3: 426–27.

Harris, James Arthur. 1904. "A New Theory of the Origin of Species." *Open Court* 18, no. 575: 193–202.

Harrison, Peter. 2017. "Science and Secularization." *Intellectual History Review* 27, no. 1: 47–70.

Harvey, David. 1990. *The Condition of Postmodernity: An Enquiry into the Origins of Cultural Change*. Malden, MA: Blackwell.

Harwood, William Sumner. 1905a. "Every Man His Own Burbank." *Country Calendar* 1, no. 1: 21–23.

———. 1905b. *New Creations in Plant Life: An Authoritative Account of the Life and Work of Luther Burbank*. New York: Macmillan.

———. 1905c. "A Wonder-Worker of Science: An Authoritative Account of Luther Burbank's Unique Work in Creating New Forms of Plant Life (First Paper)." *Century* 69, no. 5: 656–72.

———. 1905d. "A Wonder-Worker of Science: An Authoritative Account of Luther Burbank's Unique Work in Creating New Forms of Plant Life (Second Paper)." *Century* 69, no. 6: 821–37.

———. 1906. *The New Earth: A Recital of the Triumphs of Modern Agriculture in America*. New York: Macmillan.

———. 1907. "How Luther Burbank Creates New Flowers: The Story of One of the Most Wonderful Men in America." *Ladies' Home Journal* 24, no. 6: 11–12.

Hasian, Marouf A. 1996. *The Rhetoric of Eugenics in Anglo-American Thought.* Athens: University of Georgia Press.

Hassall, C. V., and Mark Pottle. 2004. "Marsh, Sir Edward Howard (1872–1953)." *Oxford Dictionary of National Biography Online.* https://doi.org/10.1093/ref:odnb/34892.

Hausman, Bernice L. 1998. "Sex before Gender: Charlotte Perkins Gilman and the Evolutionary Paradigm of Utopia." *Feminist Studies* 24, no. 3: 488–510.

Hayes, Sandra Chrystal. 1996. "No Woman's Zone: Edith Wharton's Revolutionary Writing." PhD diss., University of Notre Dame. https://www.proquest.com/pqdtglobal1/dissertations-theses/no-womans-zone-edith-whartons-revolutionary/docview/304363146/sem-2.

Hays, Willet Martin. 1910. "Efficiency Records in People." *American Breeders* 1, no. 3: 222–23.

Headley, Frederick Webb. 1906. *Life and Evolution.* London: Duckworth.

Helleberg, Victor E. 1908. Review of Thompson, *New Reading of Evolution: A Study Plan Correlating the Known Facts of Nature and Forming a Scientific Basis for a Synthetic Philosophy of Individual and Social Life. American Journal of Sociology* 14, no. 1: 122–23.

Helmreich, Stefan, and Sophia Roosth. 2016. "Life Forms: A Keyword Entry." In *Sounding the Limits of Life: Essays in the Anthropology of Biology and Beyond,* edited by Stefan Helmreich, Sophia Roosth, and Michele Ilana Friedner, 19–34. Princeton, NJ: Princeton University Press.

Herbert, Solomon. 1910a. *The First Principles of Heredity.* London: Adam and Charles Black.

———. 1910b. "Socialism in the Making." *Socialist Review* 5: 32–37.

Heyck, T. W. 1982. *The Transformation of Intellectual Life in Victorian England.* Edited by Richard Price. Croom Helm Studies in Society and History. London: Croom Helm.

Hilgartner, Stephen. 2015. "Capturing the Imaginary: Vanguards, Visions and the Synthetic Biology Revolution." In *Science and Democracy: Knowledge as Wealth and Power in the Biosciences and Beyond,* edited by Stephen Hilgartner, Clark Miller, and Rob Hagendijk, 33–55. Abingdon, UK: Routledge.

Hillegas, Mark Robert. 1967. *The Future as Nightmare: H. G. Wells and the Anti-Utopians.* Carbondale: Southern Illinois University Press.

Hillquit, Morris. 1965. *History of Socialism in the United States.* 5th ed. New York: Russel and Russell.

Hine, Donald W., Joseph P. Reser, Wendy J. Phillips, Ray Cooksey, Anthony D. G. Marks, Patrick Nunn, Susan E. Watt, Graham L. Bradley, and A. Ian Glendon. 2013. "Identifying Climate Change Interpretive Communities in a Large Australian Sample." *Journal of Environmental Psychology* 36: 229–39.

Hodge, Clifton Fremont, and Jean Dawson. 1918. *Civic Biology: Textbook of Problems, Local and National, That Can Be Solved Only by Civic Coöperation.* Boston, MA: Ginn.

Hofstadter, Richard. 1944. *Social Darwinism in American Thought, 1860–1915.* Philadelphia: University of Pennsylvania Press.

Hollinger, Veronica. 2014. "Genre vs. Mode." In *The Oxford Handbook of Science Fiction,* edited by Rob Latham, 139–51. Oxford: Oxford University Press.

Holmes, S. J. 1915. "Some Misconceptions of Eugenics." *Atlantic Monthly* 115, no. 2: 222–27.

Holquist, Michael. 2002. *Dialogism: Bakhtin and His World*. 2nd ed. New Accents. London: Routledge.

Howard, Norman. 1906. "Dr. Luther Burbank, the Magician of Plants: The Life-Story of an Explorer into the Infinite." *Quiver: An Illustrated Magazine for Sunday and General Reading* 220: 451–57.

Howard, Walter L. 1945. "Luther Burbank: A Victim of Hero Worship." *Chronica Botanica* 9, no. 5–6: 299–506.

Howells, William Dean. 1904. "Editor's Easy Chair." Review of Wallace, *Man's Place in the Universe*. *Harper's Monthly* 108, no. 646: 640–44.

Hoyles, Martin. 1991. *The Story of Gardening*. London: Journeyman.

Hubrecht, Ambrosius Arnold Willem. 1904. "Hugo De Vries's Theory of Mutations." *Popular Science Monthly* 65: 205–23.

Hudak, Jennifer. 2003. "The 'Social Inventor': Charlotte Perkins Gilman and the (Re) Production of Perfection." *Women's Studies* 32: 455–77.

Hudson, Thomson Jay. 1900. *The Divine Pedigree of Man; or, The Testimony of Evolution and Psychology to the Fatherhood of God*. Chicago: A. C. McClurg.

Hughan, Jessie Wallace. 1912. *American Socialism of the Present Day*. New York: John Lane.

Hughes, David Y. 1977. "The Garden in Wells' Early Science Fiction." In *H. G. Wells and Modern Science Fiction*, edited by Darko Suvin and Robert M. Philmus, 48–69. Lewisburg, PA; London: Bucknell University Press; Associated University Presses.

Hull, David L, ed. 1973. *Darwin and His Critics: The Reception of Darwin's Theory of Evolution by the Scientific Community*. Cambridge, MA: Harvard University Press.

Hunter, George William. 1907. *Elements of Biology: A Practical Text-Book Correlating Botany, Zoology, and Human Physiology*. New York: American Book Company.

———. 1914. *A Civic Biology: Presented in Problems*. New York: American Book Company.

———. 1926. *New Civic Biology: Presented in Problems*. New York: American Book Company.

Hus, Henri Theodore Antoine de Leng. 1906. Review of Osterhout, *Experiments with Plants*. *American Naturalist* 40, no. 470: 146–48.

Hutchinson, Woods. 1898. *The Gospel According to Darwin*. Chicago: Open Court Publishing.

Huxley, Aldous. 1927. *Proper Studies*. London: Chatto and Windus.

Huxley, Aldous, Margaret Atwood, and David Bradshaw. 1932 (2007). *Brave New World*. New edition, with introductions by Margaret Atwood and David Bradshaw. Vintage Classics. London: Vintage.

Huxley, Julian S. 1923. "Biology in Utopia." Review of Wells, *Men Like Gods*. *Nature* 111, no. 2792: 591–94. https://doi.org/10.1038/111591a0.

———. 1926a (1933). *Essays in Popular Science*. Reprint ed. London: Chatto and Windus.

———. 1926b. *The Stream of Life*. London: Watts.

———. 1927. "The Tissue-Culture King." *Amazing Stories* 2, no. 5: 451–59.

———. 1942. *Evolution: The Modern Synthesis*. 3rd ed. London: George Allen and Unwin.

————. 1970 (1978). *Memories I.* Harmondsworth, UK: Penguin.

Huxley, Thomas Henry, ed. 1870. *Lay Sermons, Addresses, and Reviews.* New York: D. Appleton.

————. 1894 (1989). "Evolution and Ethics." In *Evolution and Ethics: T. H. Huxley's "Evolution and Ethics" with New Essays on Its Victorian and Sociobiological Context,* edited by James G. Paradis and George C. Williams, 59–174. Princeton, NJ: Princeton University Press.

Huxley, Thomas Henry, James G. Paradis, and George C. Williams, eds. 1989. *Evolution and Ethics: T. H. Huxley's "Evolution and Ethics" with New Essays on Its Victorian and Sociobiological Context.* Princeton, NJ: Princeton University Press.

Hyde, William J. 1956. "The Socialism of H. G. Wells in the Early Twentieth Century." *Journal of the History of Ideas* 17, no. 2: 217–34.

Inkster, Ian. 2003. "Patents as Indicators of Technological Change and Innovation—an Historical Analysis of the Patent Data 1830–1914." *Transactions of the Newcomen Society* 73, no. 2: 179–208.

Jack, Jordynn. 2006. "Chronotopes: Forms of Time in Rhetorical Argument." *College English* 69, no. 1: 52–73.

Jack, Zachary Michael. 2008a. "Introducing Sower and Seer, Liberty Hyde Bailey." In *Liberty Hyde Bailey: Essential Agrarian and Environmental Writings,* edited by Zachary Michael Jack, 1–37. Ithaca, NY: Cornell University Press.

————, ed. 2008b. *Liberty Hyde Bailey: Essential Agrarian and Environmental Writings.* Ithaca, NY: Cornell University Press.

James, Henry. 1908. *The Novels and Tales. Volume 13: The Reverberator.* London: Macmillan.

James, William. 1902 (2011). *The Varieties of Religious Experience: A Study in Human Nature.* Cambridge Library Collection—Philosophy. Cambridge: Cambridge University Press.

Jameson, Fredric. 1982. "Progress versus Utopia; or, Can We Imagine the Future?" *Science Fiction Studies* 9, no. 2: 147–58.

————. 1991. *Postmodernism; or, The Cultural Logic of Late Capitalism.* Durham, NC: Duke University Press.

————. 2005. *Archaeologies of the Future: The Desire Called Utopia and Other Science Fictions.* New York: Verso.

Jasanoff, Sheila, and Sang-Hyun Kim, eds. 2015. *Dreamscapes of Modernity: Sociotechnical Imaginaries and the Fabrication of Power.* Chicago: University of Chicago Press.

[Jenkin], Fleeming. 1867 (1973). "The Origin of Species (from *North British Review*)." In *Darwin and His Critics,* edited by David Hull, 302–50. Cambridge, MA: Harvard University Press.

Jenkins, Henry. 1992 (2013). *Textual Poachers: Television Fans and Participatory Culture.* Updated 20th anniversary ed. New York: Routledge.

Jennings, Herbert Spencer. 1930. *The Biological Basis of Human Nature.* London: Faber and Faber.

"Jews and Primroses." 1902. Editorial. *Independent* 54, no. 2791 (May 29): 1317–18.

Jones, Greta. 2004. "Spencer and His Circle." In *Herbert Spencer: The Intellectual Legacy,* edited by Greta Jones and Robert Peel, 1–16. London: Galton Institute.

Jordan, David Starr. 1905. "Some Experiments of Luther Burbank." *Popular Science Monthly* 66: 201–25.

———. 1907. "The Present Status of Darwinism." Review of Kellogg, *Darwinism To-Day*. *Dial: A Semi-Monthly Journal of Literary Criticism, Discussion, and Information*, September 16, 161–63.

———. 1909. "Some Experiments of Luther Burbank." In *The Scientific Aspects of Luther Burbank's Work*, edited by David Starr Jordan and Vernon L. Kellogg, 1–81. San Francisco, CA: A. M. Robertson.

———. 1921. "Prefatory Note." In *How Plants Are Trained to Work for Man*, edited by Luther Burbank, 21–35. New York: P. F. Collier and Son.

Jordan, David Starr, and Vernon Lyman Kellogg. 1907. *Evolution and Animal Life: An Elementary Discussion of Facts, Processes, Laws and Theories Relating to the Life and Evolution of Animals*. New York: D. Appleton.

———. 1920. *Evolution and Animal Life: An Elementary Discussion of Facts, Processes, Laws and Theories Relating to the Life and Evolution of Animals*. 2nd ed. New York: D. Appleton.

Jordanova, Ludmilla J. 1984. *Lamarck*. Oxford: Oxford University Press.

Jordheim, Helge. 2012. "Against Periodization: Koselleck's Theory of Multiple Temporalities." *History and Theory* 51, no. 2: 151–71.

Kaempffert, Waldemar. 1905. "New Varieties of Butterfly." *Appleton's Booklovers* 6, no. 1: 85–86.

———. 1928. "The Superman: Eugenics Sifted; New Discoveries in Biology and Chemistry Lead Fanciful Minds to Predict Such Control of Human Evolution That Genius Can Be Produced at Will; but the Scientists Are Still Skeptical the Superman Is Still Far in the Distance." *New York Times*, May 27: 72.

———. 1936. "A Biologist's View of Man's Future." Review of Muller, *Out of the Night*. *New York Times*, March 15: BR4.

Karpenko, Lara, and Shalyn Claggett, eds. 2017. *Strange Science: Investigating the Limits of Knowledge in the Victorian Age*. Ann Arbor: University of Michigan Press.

Kass, Leon R. 1997. "The Wisdom of Repugnance." *New Republic* 216, no. 22: 17–26.

Kateb, George. 1963. *Utopia and Its Enemies*. London: Free Press of Glencoe.

Kautsky, Karl. 1902. *The Social Revolution*. Translated by A. M. Simons and May Wood Simon. Chicago: Charles H. Kerr.

Keller, David Henry. 1928. "Stenographer's Hands." *Amazing Stories Quarterly* 1, no. 4: 522–29, 569.

Keller, Evelyn Fox. 1984. "Science and Power for What?" In *Nineteen Eighty-Four: Science between Utopia and Dystopia*, edited by Everett Mendelsohn and Helga Nowotny, 261–72. Dordrecht: Reidel.

———. 2000 (2002). *The Century of the Gene*. 1st paperback ed. Cambridge, MA: Harvard University Press.

Kelley, Donald R. 2003. *Fortunes of History: Historical Inquiry from Herder to Huizinga*. New Haven, CT: Yale University Press.

Kellogg, Vernon Lyman. 1906. "Scientific Aspects of Luther Burbank's Work." *Popular Science Monthly* 69, no. 19: 363–74.

———. 1907. *Darwinism To-Day: A Discussion of Present-Day Scientific Criticism of the Darwinian Selection Theories, Together with a Brief Account of the Principal Other Proposed Auxiliary and Alternative Theories of Species-Forming*. New York: Henry Holt.

———. 1909. "Scientific Aspects of Luther Burbank's Work." In *The Scientific Aspects of Luther Burbank's Work*, edited by David Starr Jordan and Vernon L. Kellogg, 84–115. San Francisco, CA: A. M. Robertson.

———. 1924. *Evolution: The Way of Man*. New York: D. Appleton.

Kemp, Peter. 1996. *H. G. Wells and the Culminating Ape: Biological Imperatives and Imaginative Obsessions*. Basingstoke, UK: Macmillan.

Kenny, Neil. 2004. *The Uses of Curiosity in Early Modern France and Germany*. Oxford: Oxford University Press.

Keogh, Luke. 2020. *The Wardian Case: How a Simple Box Moved Plants and Changed the World*. Chicago: University of Chicago Press.

Kevles, Daniel J. 1986. *In the Name of Eugenics: Genetics and the Uses of Human Heredity*. 1st paperback ed. Harmondsworth, UK: Penguin.

———. 1995. *In the Name of Eugenics: Genetics and the Uses of Human Heredity*. 2nd ed. Cambridge, MA: Harvard University Press.

Kilgore, De Witt Douglas. 2003. *Astrofuturism: Science, Race, and Visions of Utopia in Space*. Philadelphia: University of Pennsylvania Press.

Kimmelman, Barbara A. 1983. "American Breeders Association: Genetics and Eugenics in an Agricultural Context, 1903–13." *Social Studies of Science* 13, no. 2: 163–204.

Kincaid, Paul. 2005. "On the Origins of Genre." In *Speculations on Speculation: Theories of Science Fiction*, edited by James E. Gunn and Matthew Candelaria, 41–57. Lanham, MD: Scarecrow Press.

Kingsland, Sharon. 1987. "A Man out of Place: Herbert Spencer Jennings at Johns Hopkins, 1906–1938." *American Zoologist* 27, no. 3: 807–17.

———. 1991. "The Battling Botanist: Daniel Trembly MacDougal, Mutation Theory, and the Rise of Experimental Evolutionary Biology in America, 1900–1912." *Isis* 82: 479–509.

Kinsey, Alfred C. 1938. *New Introduction to Biology (Revised)*. 3rd ed. Philadelphia: J. B. Lippincott.

Kline, Wendy. 2001. *Building a Better Race: Gender, Sexuality, and Eugenics from the Turn of the Century to the Baby Boom*. Berkeley: University of California Press.

Kneeland, M. B. 1892. "Boston Monday Lectures: The Bible and Modern Discoveries: Prelude, Evolution or Revolution in Theology." *Advance* 25, no. 1875 (February 25): 153.

Knittel, Janna. 2006. "Environmental History and Charlotte Perkins Gilman." *Foundation: International Review of Science Fiction* 35, no. 96: 49–67.

Kober, William. 1932. "Charles L. Campbell and Taine's 'Seeds of Life' ["Your Viewpoint," lead letter]." *Amazing Stories Quarterly* 5, no. 2: 288.

Koch, Lene. 2006. "Past Futures: On the Conceptual History of Eugenics—a Social Technology of the Past." *Technology Analysis and Strategic Management* 18, no. 3–4: 329–44.

Koerner, Lisbet. 1999. *Linnaeus: Nature and Nation*. Cambridge, MA: Harvard University Press.

Kohler, Robert E. 1991. *Partners in Science: Foundations and Natural Scientists, 1900–1945*. Chicago: University of Chicago Press.

———. 1994. *Lords of the Fly: Drosophila Genetics and the Experimental Life*. Chicago: University of Chicago Press.

Kohlstedt, Sally Gregory. 2010. *Teaching Children Science: Hands-On Nature Study in North America, 1890–1930*. Chicago: University of Chicago Press.

Koselleck, Reinhart. 1985. *Futures Past: On the Semantics of Historical Time*. Studies in Contemporary German Social Thought. Cambridge, MA: MIT Press.

Kramer, Lloyd S. 1989. "Literature, Criticism, and Historical Imagination: The Literary Challenge of Hayden White and Dominick LaCapra." In *The New Cultural History*, edited by Lynn Avery Hunt, 97–128. Berkeley: University of California Press.

Krementsov, Nikolai. 2014. *Revolutionary Experiments: The Quest for Immortality in Bolshevik Science and Fiction*. New York: Oxford University Press.

———. 2015. "Between Science and Fiction." *Histories of the Future*, website created in conjunction with a workshop held at Princeton University, Princeton, NJ, in February 2015. http://histscifi.com/essays/krementsov/.

Kropotkin, Petr Alexeyevich. 1901. "Recent Science." *Nineteenth Century and After, a Monthly Review* 50, no. 295: 417–38.

———. 1902 (1904). *Mutual Aid: A Factor of Evolution*. Rev ed. London: Heinemann.

———. 1923. *Modern Science and Anarchism*. 2nd ed. London: Freedom Press.

Kumar, Krishan. 1987. *Utopia and Anti-Utopia in Modern Times*. Oxford: Basil Blackwell.

———. 2003. "Aspects of the Western Utopian Tradition." *History of the Human Sciences* 16, no. 1: 63–77.

Kunkel, Beverley. 1931. Review of Jennings, *Biological Basis of Human Nature*. *Saturday Review of Literature* 8, no. 2: 24.

Kupperman, Karen Ordahl, ed. 1995. *America in European Consciousness, 1493–1750*. Chapel Hill: University of North Carolina Press.

Ladouceur, Ronald. 2008a. "'All with Stories to Sell': Carleton S. Coon, Bentley Glass, Marston Bates, and the Struggle by Life Scientists in the United States to Construct a Social Mission after World War II." MA thesis, State University of New York, Empire State College. https://www.proquest.com/pqdtglobal1/dissertations-theses/all-with-stories-sell-carleton-s-coon-bentley/docview/304824969/sem-2.

———. 2008b. "Ella Thea Smith and the Lost History of American High School Biology Textbooks." *Journal of the History of Biology* 41, no. 3: 435–71.

LaFollette, Marcel C. 1990. *Making Science Our Own: Public Images of Science, 1910–1955*. Chicago: University of Chicago Press.

La Monte, Robert Rives. 1909. "Science and Revolution." *Wilshire's Monthly* 13, no. 6: 8–9.

Lane, Ann J. 1997. *To Herland and Beyond: The Life and Work of Charlotte Perkins Gilman*. Charlottesville: University Press of Virginia.

Lankester, Edwin Ray. 1879. "Degeneration: A Chapter in Darwinism." In *The Advancement of Science: Occasional Essays and Addresses*, 3–59. London: Macmillan.

———. 1880. *Degeneration: A Chapter in Darwinism*. London: Macmillan.

———. 1907. *The Kingdom of Man*. London: Archibald Constable.

Largent, Mark A. 2000. "'These Are Times of Scientific Ideals': Vernon Lyman Kellogg and Scientific Activism, 1890–1930." PhD diss., University of Minnesota. https://www.proquest.com/pqdtglobal1/dissertations-theses/these-are-times-scientific-ideals-vernon-lyman/docview/304610309/sem-2.

———. 2009. "The So-Called Eclipse of Darwinism." *Transactions of the American Philosophical Society* 99, no. 1: 3–21.

Latour, Bruno. 1993. *We Have Never Been Modern*. Cambridge, MA: Harvard University Press.

———. 1999. "Circulating Reference: Sampling the Soil in the Amazon Forest." In *Pandora's Hope: Essays on the Reality of Science Studies*, edited by Bruno Latour, 24–79. Cambridge, MA: Harvard University Press.

Latour, Bruno, and Steve Woolgar. 1979. *Laboratory Life: The Social Construction of Scientific Facts*. Beverly Hills, CA: Sage Publications.

Latter, Oswald Hawkins. 1930. *Readable School Biology*. Edited by E. J. Holmyard. Bells' Natural Science Series. London: G. Bell and Sons.

Lears, Jackson. 2009. *Rebirth of a Nation: The Making of Modern America, 1877–1920*. London: Harper Perennial.

Ledger, Sally, and Roger Luckhurst, eds. 2000. *The Fin de Siècle: A Reader in Cultural History, c. 1880–1900*. Oxford: Oxford University Press.

Lefanu, Sarah. 1988. *In the Chinks of the World Machine: Feminism and Science Fiction*. London: Women's Press.

Leiserowitz, Anthony. 2007. "Communicating the Risks of Global Warming: American Risk Perceptions, Affective Images, and Interpretive Communities." In *Creating a Climate for Change: Communicating Climate Change and Facilitating Social Change*, edited by Lisa Dilling and Susanne C. Moser, 44–63. Cambridge: Cambridge University Press.

LeMahieu, Daniel L. 1988. *A Culture for Democracy: Mass Communication and the Cultivated Mind in Britain between the Wars*. Oxford: Clarendon Press.

———. 1992. "The Ambiguity of Popularization." In *Julian Huxley: Biologist and Statesman of Science*, edited by C. Kenneth Waters and Albert Van Helden, 252–56. Houston, TX: Rice University Press.

Lévi-Strauss, Claude. 1962 (1974). *The Savage Mind*. New ed. Nature of Human Society Series. London: Weidenfeld and Nicolson.

Levitas, Ruth. 2011. *The Concept of Utopia*. Reissue with new preface by the author. Bern, Switzerland: Peter Lang GmbH, Internationaler Verlag der Wissenschaften.

Lewis, Arthur Morrow. 1908. *Evolution: Social and Organic*. 3rd ed. Chicago: Charles H. Kerr.

———. 1911. *Vital Problems in Social Evolution*. Chicago: Charles H. Kerr.

———. 1919. "The Student's Corner [De Vries' 'Mutation']." *Butte Daily Bulletin*, December 22, 23, and 24: 2, 2–3, 2.

Lewis, C. S. 1938. *Out of the Silent Planet*. London: Bodley Head.

———. 1943. *Christian Behaviour: A Further Series of Broadcast Talks*. New York: Macmillan.

Lewis, Richard Warrington Baldwin. 1975. *Edith Wharton: A Biography*. London: Constable.

Lightman, Bernard. 2007. *Victorian Popularizers of Science: Designing Nature for New Audiences*. Chicago: University of Chicago Press.

Lindberg, David C., and Ronald L. Numbers, eds. 2003. *When Science and Christianity Meet*. Chicago: University of Chicago Press.

Linville, Henry R. 1923. *The Biology of Man and Other Organisms*. New York: Harcourt, Brace.

Lipow, Arthur. 1982. *Authoritarian Socialism in America: Edward Bellamy and the Nationalist Movement*. Berkeley: University of California Press.

Livingstone, David N., Darryl G. Hart, and Mark A. Noll, eds. 1998. *Evangelicals and Science in Historical Perspective*. Religion in America Series. Oxford: Oxford University Press.

Lloyd, F. E. 1904. Review of Bailey, *Plant-Breeding*, 3rd ed. *Torreya* 4, no. 7: 109–10.

Lloyd, Francis Ernest, and Maurice Alpheus Bigelow. 1904. *The Teaching of Biology in the Secondary School*. American Teachers Series. New York: Longmans, Green.

Lock, Robert Heath. 1906. *Recent Progress in the Study of Variation, Heredity, and Evolution*. London: John Murray.

Loeb, Jacques. 1904. "The Recent Development of Biology." *Science* 20, no. 519: 777–86.

Long, Pamela O. 2011. *Artisan/Practitioners and the Rise of the New Sciences, 1400–1600*. Corvallis: Oregon State University Press.

Luckhurst, Roger. 2005. *Science Fiction*. Cultural History of Literature. Cambridge: Polity Press.

———. 2006. "Bruno Latour's Scientifiction: Networks, Assemblages, and Tangled Objects." *Science Fiction Studies* 33, no. 1: 4–17.

———. 2010. "Science Fiction and Cultural History." *Science Fiction Studies* 37, no. 1: 3–15.

Ludovici, Anthony Mario. 1924. *Lysistrata; or, Woman's Future and Future Woman*. To-Day and To-Morrow. London: Kegan Paul, Trench, Trubner.

Lunn, Arnold. 1934. "Is Evolution True?" *English Review* (January–February): 79–90, 173–83.

Lupfer, Eric. 2010. "Becoming America's 'Prophet of Outdoordom': John Burroughs and the Profession of Nature Writing, 1856–1880." *Texas Studies in Literature and Language* 52, no. 4: 381–407.

Luther Burbank Society. n.d. [c. 1913]. Prospectus and membership application. Santa Rosa, CA: Luther Burbank Press.

Luther Burbank's Spineless Cactus. 1912. Santa Rosa, CA: Luther Burbank Company.

Lyle, Eugene P., Jr. 1902. "Plant Making in a Dutch Garden." *Everybody's* 6, no. 6: 596–602.

MacDonald, James Ramsay. 1909. *Socialism To-Day*. London: Independent Labour Party.

MacDougal, Daniel Trembly. 1902. "Professor de Vries's Experiments upon the Origin of Species." *Independent* 54, no. 2808 (September 25): 2283–84.

MacFadden, Bruce J., Luz Helena Oviedo, Grace M. Seymour, and Shari Ellis. 2012. "Fossil Horses, Orthogenesis, and Communicating Evolution in Museums." *Evolution: Education and Outreach* 5, no. 1: 29–37.

MacFie, Ronald Campbell. 1928. *Metanthropos; or, The Body of the Future*. To-Day and To-Morrow. London: Kegan Paul, Trench, Trubner.

MacLaurin, Richard C. 1910. "Some Tests of Academic Efficiency." *Popular Science Monthly* 86, no. 4: 476–94.

Malpas, P. A. 1904. "The Trend of Twentieth Century Science: Progress in the Vegetable World . . . Man Aids Nature." *New Century Path* 7, no. 41: 13.

Mandler, Peter. 2000. "'Race' and 'Nation' in Mid-Victorian Thought." In *History, Religion, and Culture: British Intellectual History 1750–1950*, edited by Stefan Collini, Richard Whatmore, and Brian Young. Cambridge: Cambridge University Press.

Manuel, Frank E., and Fritzie P. Manuel. 1979. *Utopian Thought in the Western World*. Cambridge, MA: Harvard University Press.

"The March of Events." 1901. *World's Work* 2, no. 5: 1127–44.

Marx, Leo. 1964 (2000). *The Machine in the Garden: Technology and the Pastoral Ideal in America*. Oxford: Oxford University Press.

Massey, W. F. 1907. "Burbankitis." *Country Gentleman* (Albany, NY) 72, no. 1: 621.

Matheson, Donald. 2000. "The Birth of News Discourse: Changes in News Language in British Newspapers, 1880–1930." *Media, Culture and Society* 22, no. 5: 557–73.

Matzke, Brian S. 2017. "'The Weaker (?) Sex': Women and the Space Opera in Hugo Gernsback's *Amazing Stories*." *Foundation* 46, no. 126: 6–20.

Mayr, Ernst. 1980 (1998). "Prologue: Some Thoughts on the History of the Evolutionary Synthesis." In *The Evolutionary Synthesis: Perspectives on the Unification of Biology*, edited by Ernst Mayr and William B. Provine, 1–48. Cambridge, MA: Harvard University Press.

Mayr, Ernst, and William Provine, eds. 1998. *The Evolutionary Synthesis: Perspectives on the Unification of Biology*. 2nd ed. Cambridge, MA: Harvard University Press.

Mazumdar, Pauline M. H. 1992. *Eugenics, Human Genetics and Human Failings: The Eugenics Society, Its Sources and Its Critics in Britain*. London: Routledge.

———. 2002. "'Reform' Eugenics and the Decline of Mendelism." *Trends in Genetics* 18, no. 1: 48–52.

McCaffery, Larry, and Jack Williamson. 1991. "An Interview with Jack Williamson." *Science Fiction Studies* 18, no. 2: 230–52.

McElroy, Wendy. 2000. *Queen Silver: The Godless Girl*. Amherst, NY: Prometheus Books.

McGee, Anita Newcomb. 1891. "An Experiment in Human Stirpiculture." *American Anthropologist* 4, no. 4: 319–26.

McLaren, Angus. 2012. *Reproduction by Design: Sex, Robots, Trees, and Test-Tube Babies in Interwar Britain*. Chicago: University of Chicago Press.

McLean, Steven. 2009. *The Early Fiction of H. G. Wells: Fantasies of Science*. London: Palgrave Macmillan.

McLellan, David. 1998. *Marxism after Marx: An Introduction*. 3rd ed. Basingstoke, UK: Macmillan.

Meagher, J. F. 1919. "Nadfratities." *Silent Worker* 31, no. 10: 204.

Meek, Ronald L. 1976. *Social Science and the Ignoble Savage*. Cambridge Studies in the History and Theory of Politics. London: Cambridge University Press.

Merchant, Carolyn. 1982. *The Death of Nature: Women, Ecology, and the Scientific Revolution*. London: Wildwood House.

Merriam, C. Hart. 1906. "Is Mutation a Factor in the Evolution of the Higher Vertebrates?" *Science* 23, no. 581: 241–57.

Merrick, Helen. 2003. "Gender in Science Fiction." In *The Cambridge Companion to Science Fiction*, edited by Edward James and Farah Mendlesohn, 241–52. Cambridge: Cambridge University Press.

Metcalf, Maynard Mayo. 1904. *An Outline of the Theory of Organic Evolution: With a Description of Some of the Phenomena Which It Explains*. New York: Macmillan.

———. 1911. *An Outline of the Theory of Organic Evolution: With a Description of Some of the Phenomena Which It Explains*. 3rd rev. ed. New York: Macmillan.

Michel, John B. 2017. "Mutation or Death!" In *Science Fiction Criticism: An Anthology of Essential Writings*, edited by Rob Latham, 183–87. London: Bloomsbury Academic.

Milam, Erika Lorraine. 2010. *Looking for a Few Good Males: Female Choice in Evolutionary Biology*. Baltimore, MD: Johns Hopkins University Press.

Milburn, Colin. 2010. "Modifiable Futures: Science Fiction at the Bench." *Isis* 101, no. 3: 560–69.

Millikan, R. A., and G. H. Cameron. 1928. "The Origin of the Cosmic Rays." *Physical Review* 32, no. 4: 533–57.

Milton, George F. 1925. "Testing the 'Monkey Bill': Tennessee's Anti-Evolution Law and Personal Liberty." *Independent* 114, no. 3915 (June 13): 659–61.

Milum, J. Parton. 1913. "Fallacies of Eugenics." *London Quarterly Review* 120: 107–15.

Montana Historical Society. 2013. "About *The Butte Daily Bulletin*." Chronicling America. Library of Congress. https://chroniclingamerica.loc.gov/lccn/sn83045085/.

Montgomery, Thomas Harrison. 1906. *The Analysis of Racial Descent in Animals*. New York: Henry Holt.

Moore, Randy. 2001. "The Lingering Impact of the Scopes Trial on High School Biology Textbooks." *BioScience* 51, no. 9: 790–96.

Morgan, Paul A., and Scott J. Peters. 2006. "The Foundations of Planetary Agrarianism: Thomas Berry and Liberty Hyde Bailey." *Journal of Agricultural and Environmental Ethics* 19, no. 5: 443–68.

Morgan, Thomas Hunt. 1903a. "Darwinism in the Light of Modern Criticism." *Harper's Monthly* 106, no. 633: 476–79.

———. 1903b. *Evolution and Adaptation*. New York: Macmillan.

———. 1907. *Experimental Zoology*. New York: Macmillan.

———. 1916 (1919). *A Critique of the Theory of Evolution*. 1st ed., 3rd printing. Princeton, NJ: Princeton University Press.

———. 1919. *The Physical Basis of Heredity*. Edited by J. Morgan Loeb and T. H. Osterhout, W. J. V. Monographs on Experimental Biology. Philadelphia, PA: J. B. Lippincott.

Morgan, Thomas Hunt, Alfred Henry Sturtevant, Hermann J. Muller, and Calvin Blackman Bridges. 1915. *The Mechanism of Mendelian Heredity*. London: Constable.

Morris, William. 1890 (1998). *News from Nowhere; or, An Epoch of Rest: Being Some Chapters from a Utopian Romance*. Edited by Clive Wilmer. Harmondsworth, UK: Penguin.

Mortimer, Raymond. 1928. "New Novels." Review of Wells, *Men Like Gods*. *New Statesman* 20: 695–96.

Morton, Arthur Leslie. 1952 (1978). *The English Utopia*. London: Lawrence and Wishart.

Moskowitz, Sam. 1994. "The Origins of Science Fiction Fandom." In *Science Fiction Fandom*, edited by Joseph L. Sanders, 17–36. Westport, CT: Greenwood Press.

Mott, Frank Luther. 1957. *A History of American Magazines, 1885–1905*. Cambridge, MA: Harvard University Press.

Motzkin, Gabriel. 1996. "Koselleck's Intuition of Time in History." In *The Meaning of Historical Terms and Concepts: New Studies on Begriffsgeschichte*, edited by Hartmut Lehmann and Melvin Richter, 41–45. Washington, DC: German Historical Institute.

Moylan, Tom. 1986. *Demand the Impossible: Science Fiction and the Utopian Imagination.* London: Methuen.

Muller, Hermann Joseph. 1936. *Out of the Night: A Biologist's View of the Future.* London: Victor Gollancz.

Müller-Wille, Staffan, and Christina Brandt. 2016a. "From Heredity to Genetics: Political, Medical, and Agro-Industrial Contexts." In *Heredity Explored: Between Public Domain and Experimental Science, 1850–1930,* edited by Staffan Müller-Wille and Christina Brandt, 3–25. Cambridge, MA: MIT Press.

———, eds. 2016b. *Heredity Explored: Between Public Domain and Experimental Science, 1850–1930.* Transformations: Studies in the History of Science and Technology. Cambridge, MA: MIT Press.

Müller-Wille, Staffan, and Hans-Jörg Rheinberger. 2007. *Heredity Produced: At the Crossroads of Biology, Politics, and Culture, 1500–1870.* Transformations: Studies in the History of Science and Technology. Cambridge, MA: MIT Press.

Murphy, Michelle. 2015. "How Does Technoscience Dream?" Histories of the Future, website created in conjunction with a workshop held at Princeton University, Princeton, NJ, in February 2015. http://histscifi.com/essays/murphy/.

Murray, D. L. 1937. "Return of the Martians." *Times Literary Supplement,* June 5: 427.

"The Mutation Factor in Evolution, with Particular Reference to Oenothera." 1915. *Nature* 95, no. 2390: 668–69. https://doi.org/10.1038/095668a0.

Nate, Richard. 2000. "Scientific Utopianism in Francis Bacon and H. G. Wells: From Salomon's House to the *Open Conspiracy.*" *Critical Review of International Social and Political Philosophy* 3, nos. 2–3: 172–88.

"Natural Law in the Dock." 1925. *New York Times,* May 27: 22.

Needham, Joseph. 1932. "Biology and Mr. Huxley." *Scrutiny* 1, no. 18: 76–79.

Nelkin, Dorothy, and M. Susan Lindee. 2004. *The DNA Mystique: The Gene as a Cultural Icon.* Ann Arbor: University of Michigan Press.

Nelson, Mark W. 2014. "Henry Gaylord Wilshire: At the Barricades for Socialism and *Amour.*" *Southern California Quarterly* 96, no. 1: 41–85.

Nersessian, Anahid. 2017. "Utopia's Afterlife in the Anthropocene." In *The Routledge Companion to the Environmental Humanities,* edited by Ursula K. Heise, Jon Christensen, and Michelle Niemann, 107–16. London: Routledge.

"New Light on the Origin of Species." 1901. *Youth's Companion,* August 1: 87.

Newman, Horatio Hackett. 1921. *Readings in Evolution, Genetics and Eugenics.* Chicago: University of Chicago Press.

Nicolosi, Riccardo. 2020. "The Darwinian Rhetoric of Science in Petr Kropotkin's *Mutual Aid: A Factor of Evolution* (1902)." *Berichte zur Wissenschaftsgeschichte* 43, no. 1: 141–59.

"Noted Holland Expert Tells How to Double Our Crops: Prof. Hugo de Vries of Amsterdam University, 'the Darwin of Botany,' Now Visiting This Country, Gives the Solution of America's Biggest Agricultural Problem." 1912. *New York Times,* September 29: SM14.

Numbers, Ronald L. 1998. *Darwinism Comes to America.* Cambridge, MA: Harvard University Press.

Numbers, Ronald L., and John Stenhouse, eds. 2001. *Disseminating Darwinism: The Role of Place, Race, Religion, and Gender.* Cambridge: Cambridge University Press.

Nutting, Charles Cleveland. 1921. "Is Darwin Shorn?" *Scientific Monthly* 12, no. 2: 127–36.

Nye, David E. 2003. *America as Second Creation: Technology and Narratives of New Beginnings.* Cambridge, MA: MIT Press.

Ogilvie, Caroline. 2002. "Socialist Darwinism: The Response of the Left to Darwinian Evolutionary Theory, 1880–1905." PhD diss., Faculty of Arts, Royal Holloway, University of London. https://www.proquest.com/pqdtglobal1/dissertations-theses/socialist-darwinism-response-left-darwinian/docview/1783894901/sem-2.

O'Hara, Edwin V. 1904. "Preface." In Eberhardt Dennert, *At the Deathbed of Darwinism: A Series of Papers,* translated by E. V. O'Harra and John H. Peschges, 9–25. Burlington, IA: German Literary Board.

———. 1905. "The Latest Defence of Darwinism." *Catholic World: A Monthly Magazine of General Literature and Science* 80, no. 480: 719.

Ohler, Paul. 2014. "Darwinism and the 'Stored Beauty' of Culture in Edith Wharton's Writing." In *America's Darwin: Darwinian Theory and U.S. Literary Culture,* edited by Tina Gianquitto and Lydia Fisher, 104–26. Athens: University of Georgia Press.

Ortner, Sherry B. 1972. "Is Female to Male as Nature Is to Culture?" *Feminist Studies* 1, no. 2: 5–31.

Osterhout, Winthrop John Van Leuven. 1905. *Experiments with Plants.* London: Macmillan.

Outram, Dorinda. 2019. *The Enlightenment.* 4th ed. New Approaches to European History. Cambridge: Cambridge University Press.

Oveyssi, Natalie. 2015. "Dangerous Love: 'Positive' Eugenics, Mass Media, and the Scientific Woman, 1900–1945." *Berkeley Undergraduate Journal* 28, no. 2: 1–54.

"The Pacific Coast in Brief." 1905. *Pacific,* March 9: 3–4.

Paden, Roger. 2002. "Marx's Critique of the Utopian Socialists." *Utopian Studies* 13, no. 2: 67–91.

Pandora, Katherine. 2001. "Knowledge Held in Common: Tales of Luther Burbank and Science in the American Vernacular." *Isis* 92, no. 3: 484–516.

———. 2009. "Popular Science in National and Transnational Perspective: Suggestions from the American Context." *Isis* 100, no. 2: 346–58.

Pannekoek, Anton. 1912. *Marxism and Darwinism.* Translated by Nathan Weiser. Chicago: Charles H. Kerr.

Paradis, James G. 1989. "Evolution and Ethics in Its Victorian Context." In *Evolution and Ethics: T. H. Huxley's "Evolution and Ethics" with New Essays on Its Victorian and Sociobiological Context,* edited by Thomas Henry Huxley, James G. Paradis, and George C. Williams, 3–56. Princeton, NJ: Princeton University Press.

Parrinder, Patrick. 1981. "*The Time Machine*: H. G. Wells's Journey through Death." *Wellsian* 4: 15–23.

———. 1995. *Shadows of the Future: H. G. Wells, Science Fiction and Prophecy.* Liverpool: Liverpool University Press.

———. 2011. "Satanism and Genetics: From Frankenstein to J. B. S. Haldane's *Daedalus and Beyond.*" In *Discourses and Narrations in the Biosciences,* edited by Paola Spinozzi and Brian Hurwitz, 247–58. Göttingen: V and R Unipress.

Partington, John S. 2000. "The Death of the Static: H. G. Wells and the Kinetic Utopia." *Utopian Studies* 11, no. 2: 96–111.

Paul, Diane B. 1979. "Marxism, Darwinism and the Theory of Two Sciences."
    Marxist Perspectives, no. 5: 116–43.
————. 1984. "Eugenics and the Left." *Journal of the History of Ideas* 45, no. 4:
    567–90.
————. 1995. *Controlling Human Heredity: 1865 to the Present*. New York:
    Humanities Press.
————. 2004. "Genetic Engineering and Eugenics: The Uses of History." In *Is
    Human Nature Obsolete? Genetics, Bioengineering, and the Future of the Human
    Condition*, edited by Harold W. Baillie, Timothy K. Casey, and Arthur L.
    Caplan, 123–51. Cambridge, MA: MIT Press.
————. 2010. "Wallace, Women and Eugenics." In *Natural Selection and Beyond:
    The Intellectual Legacy of Alfred Russel Wallace*, edited by Charles H. Smith and
    George Beccaloni, 263–78. Oxford: Oxford University Press.
Paul, Diane B., and Barbara A. Kimmelman. 1988. "Mendel in America: Theory
    and Practice, 1900–1919." In *The American Development of Biology*, edited by
    Ronald Rainger, Keith R. Benson, and Jane Maienschein, 281–310. Philadelphia:
    University of Pennsylvania Press.
Pauly, Philip J. 1987. *Controlling Life: Jacques Loeb and the Engineering Ideal in
    Biology*. New York: Oxford University Press.
————. 1996. "The Beauty and Menace of the Japanese Cherry Trees: Conflicting
    Visions of American Ecological Independence." *Isis* 87, no. 1: 51–73.
————. 2000. *Biologists and the Promise of American Life: From Merriweather Lewis
    to Alfred Kinsey*. Princeton, NJ: Princeton University Press.
————. 2007. *Fruits and Plains: The Horticultural Transformation of America*.
    Cambridge, MA: Harvard University Press.
Pearson, Karl. 1909. *The Groundwork of Eugenics*. Eugenics Laboratory Lecture
    Series, 2. London: University of London, Galton Laboratory for National
    Eugenics.
Penley, Constance. 1997. *NASA/Trek: Popular Science and Sex in America*. London:
    Verso.
Pesic, Peter. 2008. "Proteus Rebound: Reconsidering the 'Torture of Nature.'" *Isis*
    99, no. 2: 304–17.
Peterson, Larry. 1986. "The Intellectual World of the IWW: An American Worker's
    Library in the First Half of the 20th Century." *History Workshop Journal* 22,
    no. 1: 153–72.
Pick, Daniel. 1989. *Faces of Degeneration: A European Disorder, c. 1848–1918*. Edited
    by Quentin Skinner. Ideas in Context. Cambridge: Cambridge University Press.
Pierce, John J. 1987. *Great Themes of Science Fiction: A Study in Imagination and
    Evolution*. New York: Greenwood Press.
Piercy, Marge. 1976 (2000). *Woman on the Edge of Time*. London: Women's Press.
Pierson, Ruth Roach, and Nupur Chaudhuri. 1998. *Nation, Empire, Colony:
    Historicizing Gender and Race*. Bloomington: Indiana University Press.
Pittenger, Mark. 1993. *American Socialists and Evolutionary Thought, 1870–1920*.
    Madison: University of Wisconsin Press.
"Plant Breeding and the Origin of Species." 1907. *Dial: A Semi-Monthly Journal of
    Literary Criticism, Discussion, and Information*, July 16: 43.
"Plant Freaks to Be Shown: Wizard Burbank Will Exhibit Some Queer Ones." 1911.
    Los Angeles Times, March 16: 118.

Plekhanov, Georgi. 1907 (1976). "Fundamental Problems of Marxism." In *Selected Philosophical Works*, vol. 3: 117–83. Moscow: Progress Publishers.

———. 1929. *Fundamental Problems of Marxism*. Translated by Eden Paul and Cedar Paul. Edited by David Ryazanov. London: Martin Lawrence.

"The Popular Magazines." 1910. *Saturday Evening Post*, December 10: 18.

Poulton, Edward Bagnall. 1909. "Darwin and His Modern Critics." *Quarterly Review* 211, no. 420: 1–38.

Preston, C. E. 1931. "The Science Column." Review of Wheat and Fitzpatrick, *Advanced Biology*. *High School Journal* 14, no. 4: 227.

Provine, William B. 1971. *The Origins of Theoretical Population Genetics*. Chicago History of Science and Medicine. Chicago: University of Chicago Press.

———. 1978. "The Role of Mathematical Population Geneticists in the Evolutionary Synthesis of the 1930s and 1940s." *Studies in the History of Biology* 2: 167–92.

———. 1980a (1998). "England." In *The Evolutionary Synthesis: Perspectives on the Unification of Biology*, edited by Ernst Mayr and William B. Provine, 329–34. Cambridge, MA: Harvard University Press.

———. 1980b (1998). "Genetics." In *The Evolutionary Synthesis: Perspectives on the Unification of Biology*, edited by Ernst Mayr and William B. Provine, 51–58. Cambridge, MA: Harvard University Press.

———. 1992. "Progress in Evolution and Meaning in Life." In *Julian Huxley: Biologist and Statesman of Science*, edited by C. Kenneth Waters and Albert Van Helden, 165–80. Houston, TX: Rice University Press.

Punnett, Reginald Crundall. 1905. *Mendelism*. London: Macmillan.

———. 1907. *Mendelism*. 2nd ed. Cambridge: Bowes and Parker.

———. 1909. *Mendelism (American Edition)*. 2nd ed. New York: Wilshire Book Company.

———. 1911. *Mendelism: Third Edition, Entirely Rewritten and Much Enlarged*. 3rd ed. New York: Macmillan.

Quelch, Tom. 1909. Review of Punnett, *Mendelism*. *Wilshire's Monthly* 13, no. 8: 11.

Quint, Howard H., and Gaylord Wilshire. 1969. *Wilshire's Monthly*. Radical Periodicals in the United States. Westport, CT: Greenwood Reprint Corp.

Qureshi, Sadiah. 2017. "Tipu's Tiger and Images of India, 1799–2009." In *Curating Empire: Museums and the British Imperial Experience*, edited by Sarah Longair and John McAleer, 207–24. Manchester: Manchester University Press.

Radick, Gregory. 2003. "Is the Theory of Natural Selection Independent of Its History?" In *The Cambridge Companion to Darwin*, edited by M. J. S. Hodge and G. Radick, 143–67. Cambridge: Cambridge University Press.

———. 2023. *Disputed Inheritance: The Battle over Mendel and the Future of Biology*. Chicago: University of Chicago Press.

Radin, Joanna. 2017. *Life on Ice: A History of New Uses for Cold Blood*. Chicago: University of Chicago Press.

Radin, Joanna, and Emma Kowal, eds. 2017. *Cryopolitics: Frozen Life in a Melting World*. Cambridge, MA: MIT Press.

Ramsden, George, and Hermione Lee, eds. 1999. *Edith Wharton's Library: A Catalogue*. Settrington, UK: Stone Trough Books.

Redvaldsen, David. 2017. "Eugenics, Socialists and the Labour Movement in Britain, 1865–1940." *Historical Research* 90, no. 250: 764–87.

Reed, David. 1997. *The Popular Magazine in Britain and the United States of America, 1880–1960*. London: British Library.

Rees, Amanda, and Iwan Rhys Morus. 2019. "Presenting Futures Past: Science Fiction and the History of Science." *Osiris* 34, no. 1: 1–15.

Reid, Constance. 2001. "The Alternative Life of E. T. Bell." *American Mathematical Monthly* 108, no. 5: 393–402.

Reingold, Nathan. 1962. "Jacques Loeb, the Scientist: His Papers and His Era." *Quarterly Journal of Current Acquisitions* 19, no. 3: 119–30.

———. 1979. "National Science Policy in a Private Foundation: The Carnegie Institution of Washington." In *The Organisation of Knowledge in America 1860–1920*, edited by Alexandra Oleson and John Voss, 313–41. Baltimore, MD: Johns Hopkins University Press.

Rembis, Michael A. 2006. "'Explaining Sexual Life to Your Daughter': Gender and Eugenic Education in the United States during the 1930s." In *Popular Eugenics: National Efficiency and American Mass Culture in the 1930s*, edited by Susan Currell and Christina Cogdell, 91–119. Athens: Ohio University Press.

Rémy, Catherine. 2014. "'Men Seeking Monkey-Glands': The Controversial Xenotransplantations of Doctor Voronoff, 1910–30." *French History* 28, no. 2: 226–40.

Review of *Biology in Education: A Handbook Based on the Proceedings of the National Conference on the Place of Biology in Education, Organised by the British Social Hygiene Council*. 1933. *Nature* 132, no. 3341: 729–30. https://doi.org/10 .1038/132729a0.

Review of de Vries, *Plant-Breeding*. 1907a. *Athenaeum*, August 31: 242–43.

———. 1907b. *Nation*, September 12: 238.

Review of Harwood, *New Creations in Plant Life*. 1906. *Athenaeum*, March 31: 395.

Review of Jordan and Kellogg, *Evolution and Animal Life*. 1907. *Independent* 68, no. 3070 (October 3): 818–19.

Review of Kellogg, *Darwinism To-Day*. 1907. *Nation*, November 21: 475–77.

———. 1908. *Athenaeum*, March 28, 388.

Review of Morgan, *Evolution and Adaptation*. 1904. *Manchester Guardian*, April 5: 7.

Review of Morgan, *Experimental Zoology*. 1907. *Scientific American* 96, no. 20: 418.

Review of Morgan, *The Physical Basis of Heredity*. 1920. *Nation*, June 19: 828–29.

Review of Osterhout, *Experiments with Plants*. 1905. *Athenaeum*, July 15: 84.

Review of Punnett, *Mendelism*. 1909. *Living Age*, August 7: 383–84.

Review of Wells, *Men Like Gods*. 1923. *Advocate of Peace through Justice* 85, no. 7: 279.

Review of Wells, *The Science of Life*. 1931. *Nature* 127, no. 3204: 477–79. https://doi .org/10.1038/127477a0.

"Reviews of New Books." 1903. *Washington Post*, November 21: 13.

Reynolds, Robert Dwight, Jr. 1974. "The Millionaire Socialists: J. G. Phelps Stokes and His Circle of Friends." PhD diss., University of South Carolina. https:// www.proquest.com/pqdtglobal1/dissertations-theses/millionaire-socialists-j-g -phelps-stokes-his/docview/302779669/sem-2.

Richards, Ellen Henrietta Swallow. 1905. *The Cost of Shelter*. New York: J. Wiley and Sons.

———. 1910. *Euthenics, the Science of Controllable Environment: A Plea for Better Living Conditions as a First Step toward Higher Human Efficiency*. Boston, MA: Whitcomb and Barrows.

Richards, Evelleen. 1983. "Darwin and the Descent of Women." In *The Wider Domain of Evolutionary Thought*, edited by David Oldroyd and Ian Langham, 57–111. Dordrecht: D. Reidel.

———. 1997. "Redrawing the Boundaries: Darwinian Science and Victorian Women Intellectuals." In *Victorian Science in Context*, edited by Bernard Lightman, 119–42. Chicago: University of Chicago Press.

———. 2017. *Darwin and the Making of Sexual Selection*. Chicago: University of Chicago Press.

Richards, Martin. 2004. "Perfecting People: Selective Breeding at the Oneida Community (1869–1879) and the Eugenics Movement." *New Genetics and Society* 23, no. 1: 47–71.

Richards, Robert J. 1987. *Darwin and the Emergence of Evolutionary Theories of Mind and Behavior*. Chicago: University of Chicago Press.

Richardson, Barbara. 2002. "Ellen Swallow Richards: 'Humanistic Oekologist,' 'Applied Sociologist,' and the Founding of Sociology." *American Sociologist* 33, no. 3: 21–57.

Ringel, Paul B. 2015. *Commercializing Childhood: Children's Magazines, Urban Gentility, and the Ideal of the American Child, 1823–1918*. Amherst: University of Massachusetts Press.

Rinn, Daniel. 2018. "Liberty Hyde Bailey: Pragmatic Naturalism in the Garden." *Environment and History* 24, no. 1: 121–38.

Roberts, Robin. 1993. *A New Species: Gender and Science in Science Fiction*. Urbana: University of Illinois Press.

Roemer, Kenneth M. 1976. *The Obsolete Necessity: America in Utopian Writings, 1888–1900*. Kent, OH: Kent State University Press.

———, ed. 1981. *America as Utopia*. New York: B. Franklin.

Roll-Hansen, Nils. 1978. "Drosophila Genetics: A Reductionist Research Program." *Journal of the History of Biology* 11, no. 1: 159–210.

Roosth, Sophia. 2017. *Synthetic: How Life Got Made*. Chicago: University of Chicago Press.

Ross, Andrew. 1991. *Strange Weather: Culture, Science and Technology in the Age of Limits*. London: Verso.

Rouyan, Anahita. 2015. "Radical Acts of Cultivation: Ecological Utopianism and Genetically Modified Organisms in Ruth Ozeki's *All Over Creation*." *Utopian Studies* 26, no. 1: 143–59.

———. 2017. "Shaping Public Discourses of Nature: Biological Mutation in the American Press, 1820–1945." PhD diss., History and Philosophy of Science, Università di Bologna. https://www.semanticscholar.org/paper/Shaping-Public-Discourses-of-Nature%3A-Biological-in-Rouyan/49f4cc917a82b2fbfb00cecf693749c01404376d?utm_source=direct_link.

Rudolph, John L. 2002. *Scientists in the Classroom: The Cold War Reconstruction of American Science Education*. New York: Palgrave Macmillan.

Runkle, Gerald. 1961. "Marxism and Charles Darwin." *Journal of Politics* 23, no. 1: 108–26.

Rupert, M. F. 1930. "Via the Hewitt Ray." *Science Wonder Quarterly* 1, no. 3: 370–83, 420.

Ruppert, Peter. 1986. *Reader in a Strange Land: The Activity of Reading Literary Utopias*. Athens: University of Georgia Press.

"The Sage of Santa Rosa." 1905. *Sunset* 14, no. 5: 541–42.

Saleeby, Caleb Williams. 1906. *Evolution, the Master-Key: A Discussion of the Principle of Evolution as Illustrated in Atoms, Stars, Organic Species, Mind, Society and Morals*. London: Harper and Brothers.

———. 1907. "[Comments on] Reid, G. Archdall, *The Biological Foundations of Sociology*." *Sociological Papers* 3: 3–37.

———. 1909. *Parenthood and Race Culture: An Outline of Eugenics*. New York: Moffat, Yard.

———. 1910. "Mendelism and Womanhood." *Forum* 44: 60–66.

———. 1914. "The Progress of Eugenics." *Forum* 51 (April): 542.

———. 1919. "The Creation of New Food Plants: A Study in the Eugenics of Wheat." *World's Work* 34, no. 202: 324–27.

Salvatore, Nick. 2007. *Eugene V. Debs: Citizen and Socialist*. 2nd ed. Urbana: University of Illinois Press.

Sapp, Jan. 2003. *Genesis: The Evolution of Biology*. Oxford: Oxford University Press.

Saunders, Judith P. 2009. *Reading Edith Wharton through a Darwinian Lens: Evolutionary Biological Issues in Her Fiction*. Jefferson, NC: McFarland.

Saunders, Max. 2019. *Imagined Futures: Writing, Science, and Modernity in the To-Day and To-Morrow Book Series, 1923–31*. Oxford: Oxford University Press.

Scharnhorst, Gary. 1985. *Charlotte Perkins Gilman*. Boston, MA: Twayne.

Schiebinger, Londa L. 1989. *The Mind Has No Sex? Women in the Origins of Modern Science*. Cambridge, MA: Harvard University Press.

———. 1993. *Nature's Body: Gender in the Making of Modern Science*. Boston, MA: Beacon Press.

Schivelbusch, Wolfgang. 2014. *The Railway Journey: The Industrialization of Time and Space in the Nineteenth Century*. Edited by Alan Trachtenberg. Berkeley: University of California Press.

"School Science." 1905. Review of Osterhout and others. *Speaker: The Liberal Review* 15, no. 386: 600–601.

Schurman, Jacob Gould. 1887. *The Ethical Import of Darwinism*. New York: Charles Scribner's Sons.

Schwartz, James. 2008. *In Pursuit of the Gene: From Darwin to DNA*. Cambridge, MA: Harvard University Press.

"Science [book reviews]." 1907. *Nation*, November 7: 426.

"Science News." 1929. *Science* 70, no. 1805: x–xiv.

Searle, Geoffrey R. 1976. *Eugenics and Politics in Britain, 1900–1914*. Leyden: Noordhoff International.

———. 1998. "Eugenics: The Early Years." In *Essays in the History of Eugenics*, edited by Robert A. Peel, 20–35. London: Galton Institute.

———. 2019. "Saleeby, Caleb Williams Elijah (1878–1940), Writer and Eugenist." In *Dictionary of National Biography*. Oxford: Oxford University Press.

Searles, A. Langley. 1947. Introduction to Garrett P. Serviss, *Edison's Conquest of Mars*, edited by A. Langley Searles. Los Angeles, CA: Carcosa House.

Secord, Anne. 1994. "Corresponding Interests—Artisans and Gentlemen in 19th-Century Natural History." *British Journal for the History of Science* 27, no. 95 (part 4): 383–408.

Secord, James A. 1985. "Darwin and the Breeders: A Social History." In *The Darwinian Heritage*, edited by David Kohn, 519–42. Princeton, NJ: Princeton University Press.

———. 2000. *Victorian Sensation: The Extraordinary Publication, Reception, and Secret Authorship of Vestiges of the Natural History of Creation.* Chicago: University of Chicago Press.

Segal, Howard P. 1994. *Future Imperfect: The Mixed Blessings of Technology in America.* Amherst: University of Massachusetts Press.

———. 2005. *Technological Utopianism in American Culture.* 2nd ed. Syracuse, NY: Syracuse University Press.

Serviss, Garrett P. 1898 (1947). *Edison's Conquest of Mars.* Edited by A. Langley Searles. Los Angeles, CA: Carcosa House.

———. 1905a. "Artificial Creation of Life." *Cosmopolitan: A Monthly Illustrated Magazine* 39, no. 5: 459–68.

———. 1905b. "How Burbank Produces New Flowers and Fruit: The Illimitable Field of Plant-Production Opened by Crossing and Selection." *Cosmopolitan: A Monthly Illustrated Magazine* 40, no. 2: 163–70.

———. 1905c. "Transforming the World of Plants: The Wonder-Work of Luther Burbank, Which Shows How Man Can Govern Evolution." *Cosmopolitan: A Monthly Illustrated Magazine* 40, no. 1: 63–70.

———. 1908. "Some Great American Scientists: IX. Luther Burbank." *Chautauquan: A Weekly Newsmagazine* 50, no. 3: 406–17.

"Set of Books on Burbank Work: Complete Scientific History of Eminent Scientist's Work in Realm of Nature." 1907. *Santa Rosa Press Democrat,* January 10: 5.

Setz, C. 2018. "'Apes Above Themselves': Evaluating Science in *The Dial* (1920–1929)." *Review of English Studies* 69, no. 291: 725–46.

Shapiro, Adam R. 2008. "Civic Biology and the Origin of the School Antievolution Movement." *Journal of the History of Biology* 41, no. 3: 409–33.

———. 2012. "Between Training and Popularization: Regulating Science Textbooks in Secondary Education." *Isis* 103, no. 1: 99–110.

———. 2013. *Trying Biology: The Scopes Trial, Textbooks, and the Antievolution Movement in American Schools.* Chicago: University of Chicago Press.

Shapiro, Michael. 2019. "The Legend of the Almost Lost." *Stanford* (July). https://stanfordmag.org/contents/the-legend-of-the-almost-lost.

Sharp, Patrick B. 2018. *Darwinian Feminism and Early Science Fiction: Angels, Amazons, and Women.* Cardiff, UK: University of Wales Press.

Sheehan, Helena. 1985 (1993). *Marxism and the Philosophy of Science: A Critical History.* 2nd ed. Atlantic Highlands, NJ: Humanities Press.

Sherborne, Michael. 2013. *H. G. Wells: Another Kind of Life.* London: Peter Owen.

Shinn, Charles Howard. 1901. "A Wizard of the Garden [reprinted as *Intensive Horticulture in California*]." *Land of Sunshine: The Magazine of California and the West* 14, no. 2: 3–39.

———. 1923. "Men Like Gods: Well's Utopian Novel of Earthlings Who Find a New World." *Overland Monthly and Out West* 81, no. 5: 44–45.

Shishin, Alex. 1998. "Gender and Industry in *Herland*: Trees as a Means of Production and Metaphor." In *A Very Different Story: Studies on the Fiction of Charlotte Perkins Gilman,* edited by Val Gough and Jill Rudd, 100–114. Liverpool: Liverpool University Press.

Shklar, Judith. 1965. "The Political Theory of Utopia: From Melancholy to Nostalgia." *Daedalus* 94, no. 2: 367–81.

Shpayer-Makov, Haia. 1987. "The Reception of Peter Kropotkin in Britain, 1886–1917." *Albion: A Quarterly Journal Concerned with British Studies* 19, no. 3: 373–90.

Shuttleworth, Sally. 2010. *The Mind of the Child: Child Development in Literature, Science, and Medicine, 1840–1900.* Oxford: Oxford University Press.

Silver, Queen. 1923 (2000). "Evolution from Monkey to Bryan." In *Queen Silver: The Godless Girl*, edited by Wendy McElroy, 175–91. Amherst, NY: Prometheus Books.

Silverberg, Robert, ed. 1974. *Mutants!* Ipswich, UK: Wildside Press.

Simons, Algie M. 1905. "Evolution by Mutation." *International Socialist Review* 6: 172–75.

[———]. 1906. Review of Harwood, *New Creations in Plant Life. International Socialist Review* 6, no. 8: 507–8.

Simpson, George Gaylord. 1944. *Tempo and Mode in Evolution.* New York: Columbia University Press.

Skinner, Quentin. 1979. "The Idea of a Cultural Lexicon." *Essays in Criticism* 29, no. 3: 205–24.

Sleigh, Charlotte. 2007. *Six Legs Better: A Cultural History of Myrmecology.* Baltimore, MD: Johns Hopkins University Press.

———. 2018. "'Come On You Demented Modernists, Let's Hear from You': Science Fans as Literary Critics in the 1930s." In *Being Modern: The Cultural Impact of Science in the Early Twentieth Century*, edited by Robert Bud, Paul Greenhalgh, Frank James, and Morag Shiach, 147–66. The Cultural Impact of Science in the Early Twentieth Century. London: UCL Press.

Slichter, Charles S. 1912. "Industrialism." *Popular Science Monthly* 81, no. 22: 355–63.

Smith, David C. 1986. *H. G. Wells: Desperately Mortal.* New Haven, CT: Yale University Press.

Smith, Harold Sherburn. 1957. "William James Ghent: Reformer and Historian." PhD diss., University of Wisconsin–Madison. https://www.proquest.com/pqdtglobal1/dissertations-theses/william-james-ghent-reformer-historian/docview/301916017/sem-2.

Smith, Jane S. 2009. *The Garden of Invention: Luther Burbank and the Business of Breeding Plants.* New York: Penguin.

———. 2010. "Luther Burbank's Spineless Cactus." *California History* 87, no. 4: 26–47, 66–68.

Smith, Pamela H. 2004. *The Body of the Artisan: Art and Experience in the Scientific Revolution.* Chicago: University of Chicago Press.

———. 2009. "Science on the Move: Recent Trends in the History of Early Modern Science." *Renaissance Quarterly* 62, no. 2: 345–75.

Smith, Scott Charles. 2012. "Edward J. Wickson's Quiet Voice for Change: The Origins of California's Secondary Agricultural Education Curriculum in the Early Twentieth Century." PhD diss., Graduate School of Education, University of California, Riverside. https://www.proquest.com/pqdtglobal1/dissertations-theses/edward-j-wicksons-quiet-voice-change-origins/docview/1033774121/sem-2.

Smocovitis, Vassiliki Betty. 1996. *Unifying Biology: The Evolutionary Synthesis and Evolutionary Biology.* Princeton, NJ: Princeton University Press.

———. 1999. "The 1959 Darwin Centennial Celebration in America." *Osiris* 14: 274–323.

———. 2009. "The 'Plant Drosophila': E. B. Babcock, the Genus Crepis, and the Evolution of a Genetics Research Program at Berkeley, 1915–1947." *Historical Studies in the Natural Sciences* 39, no. 3: 300–355.

Soloway, Richard A. 1995. *Demography and Degeneration: Eugenics and the Declining Birth Rate in Twentieth-Century Britain.* 2nd ed. Chapel Hill: University of North Carolina Press.

"Some Magazine Mysteries." 1895. *Nation*, November 14: 342–43.

Sorber, Nathan M. 2018. *Land-Grant Colleges and Popular Revolt: The Origins of the Morrill Act and the Reform of Higher Education.* Ithaca, NY: Cornell University Press.

Spargo, John. 1908. "Art and Literature." Review of London, *The Iron Heel. International Socialist Review* 8, no. 10: 628–29.

———. n.d. [c. 1908]. *Where We Stand.* Chicago: Charles H. Kerr.

———. 1909a. "Art and Literature." Review of Punnett, *Mendelism. International Socialist Review* 10, no. 1: 80–83.

———. 1909b. "Literature/Art." Review of Lewis, *Vital Problems in Social Evolution. International Socialist Review* 9, no. 9: 911–14.

———. 1912a. *Applied Socialism: A Study of the Application of Socialistic Principles to the State.* New York: B. W. Huebsch.

———. 1912b. *Karl Marx: His Life and Work.* New York: B. W. Huebsch.

Spargo, John, and George Louis Arner. 1912. *Elements of Socialism: A Text-Book.* New York: Macmillan.

Spary, Emma C. 2000. *Utopia's Garden: French Natural History from Old Regime to Revolution.* Chicago: University of Chicago Press.

"Speeds Breeding Types: University of Texas Savant Discovers How X-Ray Will Foster Heredity Traits." 1927. *Los Angeles Times*, October 2: 6.

Speer, John Bristol. 1944. "Mutant." *Fancyclopedia I.* Los Angeles: Los Angeles Science Fiction Society. https://fanac.org/Fannish_Reference_Works/Fancyclopedia/Fancyclopedia_I/.

Spengler, Joseph J. 1971. "Malthus on Godwin's 'Of Population.'" *Demography* 8, no. 1: 1–12.

Squier, Susan Merrill. 1994. *Babies in Bottles: Twentieth-Century Visions of Reproductive Technology.* New Brunswick, NJ: Rutgers University Press.

Stack, David. 2003. *The First Darwinian Left: Socialism and Darwinism, 1859–1914.* Cheltenham, UK: New Clarion Press.

Stamhuis, Ida H. 2015. "Why the Rediscoverer Ended Up on the Sidelines: Hugo De Vries's Theory of Inheritance and the Mendelian Laws." *Science and Education* 24: 29–49.

Stamhuis, Ida H., Onno G. Meijer, and Erik J. A. Zevenhuizen. 1999. "Hugo de Vries on Heredity, 1889–1903." *Isis* 90, no. 2: 238–67.

Stansfield, William D. 2006. "Luther Burbank: Honorary Member of the American Breeders' Association." *Journal of Heredity* 97, no. 2: 95–99.

Stapledon, Olaf. 1935 (1972). *Odd John and Sirius.* New York: Dover.

Star, Susan Leigh, and James R. Griesemer. 1989. "Institutional Ecology, 'Translations' and Boundary Objects: Amateurs and Professionals in Berkeley's Museum of Vertebrate Zoology, 1907–39." *Social Studies of Science* 19: 387–420.

Start the Boy Right. 1914. Monographs on the Improvement of Plant Life 2. Santa Rosa, CA: Luther Burbank Society.

Steen, Bart van der. 2019. "'A New Scientific Conception of the Human World': Anton Pannekoek's Understanding of Scientific Socialism." In *Anton Pannekoek: Ways of Viewing Science and Society*, edited by Chaokang Tai, Bart van der Steen, and Jeroen van Dongen, 137–55. Amsterdam: Amsterdam University Press.

Stern, Alexandra Minna. 2016. *Eugenic Nation: Faults and Frontiers of Better Breeding in Modern America*. 2nd ed. Oakland: University of California Press.

Stewart, Larry. 1992. *The Rise of Public Science: Rhetoric, Technology and Natural Philosophy in Newtonian Britain, 1660–1750*. Cambridge: Cambridge University Press.

Stocking, George W. 1987. *Victorian Anthropology*. New York: Free Press.

Stocks, J. L. 1936. "Eugenics of the Future?" Review of Muller, *Out of the Night*. *Manchester Guardian*, June 9: 7.

Stoltzfus, Arlin, and Kele Cable. 2014. "Mendelian-Mutationism: The Forgotten Evolutionary Synthesis." *Journal of the History of Biology* 47, no. 4: 501–46.

Stover, Leon. 1990. "Applied Natural History: Wells vs. Huxley." In *H. G. Wells under Revision*, edited by Patrick Parrinder and Christopher Rolfe, 125–33. London: Associated University Presses.

Suvin, Darko. 1979 (2016). *Metamorphoses of Science Fiction: On the Poetics and History of a Literary Genre*. Edited by Gerry Canavan. New ed. New Haven, CT: Yale University Press.

Swibold, Dennis L. 2006. *Copper Chorus: Mining, Politics, and the Montana Press, 1889–1959*. Helena: Montana Historical Society Press.

Swift, Jonathan. 1726 (2002). *Gulliver's Travels*. Norton Critical Editions, edited by Albert J. Rivero. New York: W. W. Norton.

Taine, John. 1931. "Seeds of Life." *Amazing Stories Quarterly* 4, no. 4: 434–505, 520.

Tattersdill, Will. 2016. *Science, Fiction, and the Fin-de-Siècle Periodical Press*. Cambridge Studies in Nineteenth-Century Literature and Culture. Cambridge: Cambridge University Press.

Taylor, H. J. 1941. "To Plant the Prairies and the Plains: The Life and Work of Niels Ebbesen Hansen." *Bios* 12, no. 1: 2–72.

Taylor, Michael. 2007. *The Philosophy of Herbert Spencer*. Continuum Studies in British Philosophy. London: Continuum.

Tennyson, Alfred. 1850. *In Memoriam*. Cambridge Library Collection. Fiction and Poetry. Cambridge: Cambridge University Press.

Theunissen, Bert. 1994a. "Closing the Door on Hugo de Vries's Mendelism." *Annals of Science* 51: 225–48.

———. 1994b. "Knowledge Is Power: Hugo de Vries on Science, Heredity and Social Progress." *British Journal for the History of Science* 27, no. 3: 475–89.

———. 1998. "The Scientific and Social Context of Hugo De Vries' *Mutationstheorie*." *Acta botanica neerlandica* 47, no. 4: 475–89.

Thompson, E. P. 1993. "The Moral Economy of the English Crowd in the Eighteenth Century." In *Customs in Common*, edited by E. P. Thompson, 185–258. New York: New Press.

Thompson, Graham. 2019. "William Dean Howells's Periodical Time." *Arizona Quarterly: A Journal of American Literature, Culture, and Theory* 75, no. 4: 77–106.

Thompson, Henry Clayton. 1907. *New Reading of Evolution: A Study Plan*. Chicago: New Reading.

Thomson, J. Arthur. 1908. *Heredity*. New York; London: G. P. Putnam's Sons; John Murray.

———, ed. 1922. *The Outline of Science: A Plain Story Simply Told*. 4 vols. New York: G. P. Putnam's Sons.

Thomson, J. Arthur, and Patrick Geddes. 1931. *Life: Outlines of General Biology*. 2 vols. London: Williams and Norgate.

Thone, Frank. 1927. "X-Rays Speed Up Evolution over 1,000 per Cent." *Science News-Letter*, October 15: 243–46.

Thurtle, Phillip. 2007. *The Emergence of Genetic Rationality: Space, Time, and Information in American Biological Science, 1870–1920*. Seattle: University of Washington Press.

Tichenor, Daniel J. 2002. *Dividing Lines: The Politics of Immigration Control in America*. Princeton, NJ: Princeton University Press.

Tomlinson, Stephen. 2005. *Head Masters: Phrenology, Secular Education, and Nineteenth-Century Social Thought*. Tuscaloosa: University of Alabama Press.

Toye, Richard. 2008. "H. G. Wells and Winston Churchill: A Reassessment." In *H. G. Wells: Interdisciplinary Essays*, edited by Steven McLean, 147–61. Newcastle, UK: Cambridge Scholars.

Trafton, Gilbert H. 1923. *Biology of Home and Community*. New York: Macmillan.

Trotter, David. 1992. *The English Novel in History, 1895–1920*. London: Routledge.

Turner, Frank Miller. 1993. *Contesting Cultural Authority: Essays in Victorian Intellectual Life*. Cambridge: Cambridge University Press.

Turney, Jon. 1998. *Frankenstein's Footsteps: Science, Genetics and Popular Culture*. New Haven, CT: Yale University Press.

Uncertain Commons. 2013. *Speculate This!* Durham, NC: Duke University Press.

Van Dijck, José. 1998. *Imagenation: Popular Images of Genetics*. London: Macmillan.

Van Wienen, Mark W. 2003. "A Rose by Any Other Name: Charlotte Perkins Stetson (Gilman) and the Case for American Reform Socialism." *American Quarterly* 55, no. 4: 603–34.

"Variation in Species: Scientific Investigation Pursued in a Washington Back Yard." 1902b. *Washington Post*, January 19: 29.

Vavilov, Nikolay Ivanovich. 1992. *Origin and Geography of Cultivated Plants*. Cambridge: Cambridge University Press.

Villiers de L'Isle-Adam, Auguste. 1886 (2001). *Tomorrow's Eve / L'Eve Future*. Edited by Robert Martin Adams. 1st paperback ed. Urbana: University of Illinois Press.

Vint, Sherryl. 2021. *Biopolitical Futures in Twenty-First-Century Speculative Fiction*. Cambridge Studies in Twenty-First-Century Literature and Culture. Cambridge: Cambridge University Press.

Vint, Sherryl, and Mark Bould. 2009. "There Is No Such Thing as Science Fiction." In *Reading Science Fiction*, edited by James E. Gunn, Marleen S. Barr, and Matthew Candelaria, 43–51. Houndmills, UK: Palgrave Macmillan.

———. 2011. *The Routledge Concise History of Science Fiction*. London: Routledge.

"Violent Science in State Legislatures: The Reaction of the Expert against Eugenics." 1914. *Current Opinion* 56, no. 2: 121–22.

Wagar, W. Warren. 2004. *H. G. Wells Traversing Time*. Middletown, CT: Wesleyan University Press.

Walker, Jeff. 2014. "'The Long Road': John Burroughs and Charles Darwin, 1862–1921." In *America's Darwin: Darwinian Theory and U.S. Literary Culture*, edited by Tina Gianquitto and Lydia Fisher, 40–58. Athens: University of Georgia Press.

Wallace, Alfred Russel. 1890. "Human Selection." *Fortnightly Review* 48, no. 285: 325–37.

———. 1908. "The Present Position of Darwinism." *Contemporary Review* 94: 129–41.

Walsh, James J. 1905. "The Present Position of Darwinism." *Catholic World: A Monthly Magazine of General Literature and Science* 80, no. 478: 499–511.

Walter, Herbert Eugene. 1914. *Genetics: An Introduction to the Study of Heredity*. New York: Macmillan.

Ward, Lester Frank. 1888 (2013). "Our Better Halves." In *Herland and Related Writings*, edited by Beth Sutton-Ramspeck. Peterborough, ON: Broadview Press.

———. 1913. "Eugenics, Euthenics, and Eudemics." *American Journal of Sociology* 18, no. 6: 737–54.

Watkins, John Elfreth, Jr. 1900. "What May Happen in the Next Hundred Years." *Ladies' Home Journal* 18, no. 1: 8.

———. 1906. "New Things to Eat: Fruits and Vegetables Which Are Strange to Most of Us." *Ladies' Home Journal* 23, no. 8: 23.

———. 1907a. "Man's Pedigree and Future." *Salt Lake Tribune*, June 16: 8.

———. 1907b. "New Species to Order." *Evening Star* (Washington, DC), February 23: 3 (part 3).

W.E.C. 1904. Review of Morgan, *Evolution and Adaptation*. *American Naturalist* 38, no. 449: 398–99.

———. 1905. Review of Bailey, *Plant-Breeding*. *American Naturalist* 39, no. 459: 174.

Weeks, Jeffrey. 2017. *Sex, Politics and Society: The Regulation of Sexuality since 1800*. 4th ed. Abingdon, UK: Routledge.

Weikart, Richard. 1993. "The Origins of Social Darwinism in Germany, 1859–1895." *Journal of the History of Ideas* 54, no. 3: 469–88.

———. 1998. *Socialist Darwinism: Evolution in German Socialist Thought from Marx to Bernstein*. San Francisco, CA: International Scholars Publications.

Weindling, Paul. 2012. "Julian Huxley and the Continuity of Eugenics in Twentieth-Century Britain." *Journal of Modern European History* 10, no. 4: 480–99.

Weismann, August. 1904. *The Evolution Theory*. Translated by J. Arthur Thomson and Margaret R. Thomson. 2 vols. London: Edward Arnold.

Wells, H. G. 1891. "Zoological Retrogression." *Gentleman's* 271: 246–53.

———. 1893. "The Man of the Year Million: A Scientific Forecast." *Pall Mall Gazette*, November 6: 3.

———. 1895 (2001). *The Time Machine*. Edited by Nicholas Ruddick. Peterborough, ON: Broadview Texts.

———. 1896a. "Human Evolution: An Artificial Process." *Fortnightly Review* 60, no. 358: 590–95.

———. 1896b (2009). *The Island of Doctor Moreau*. Edited by Mason Harris. Peterborough, ON: Broadview Press.

———. 1901. *The First Men in the Moon*. London: George Newnes.

———. 1902a. *Anticipations: Of the Reaction of Mechanical and Scientific Progress upon Human Life and Thought*. New York: Harper and Brothers.

———. 1902b. "The Discovery of the Future." *Nature* 65, no. 1684: 326–31. https://doi.org/10.1038/065326a0.

———. 1905. *A Modern Utopia.* London: Chapman and Hall.

———. 1921. *The Outline of History: Being a Plain History of Life and Mankind.* 3rd ed. New York: Macmillan.

———. 1923 (2002). *Men Like Gods.* Thirsk, UK: House of Stratus.

———. 1926. *Mr. Belloc Objects to "The Outline of History."* London: Watts.

———. 1937 (2006). *Star Begotten: A Biological Fantasia.* Middletown, CT: Wesleyan University Press.

———. 1939 (1982). "Utopias." *Science Fiction Studies* 9, no. 2: 117–21.

Wells, H. G., David Y. Hughes, and Harry M. Geduld. 1898 (1993). *A Critical Edition of "The War of the Worlds": H. G. Wells's Scientific Romance.* Bloomington: Indiana University Press.

Wells, H. G., Julian Huxley, and G. P. Wells. 1929–1930. *The Science of Life: A Summary of Contemporary Knowledge about Life and Its Possibilities.* 3 vols. London: Amalgamated Press / Waverley Book Company.

Wertheim, Margaret. 1997. *Pythagoras' Trousers: God, Physics and the Gender Wars.* London: Fourth Estate.

West, Rebecca. 1932 (1975). Review of Huxley, *Brave New World, Daily Telegraph,* February 5, 1932. In *Aldous Huxley: The Critical Heritage,* edited by Donald Watt, 197–202. London: Routledge and Kegan Paul.

Wharton, Edith. 1904. *The Descent of Man and Other Stories.* New York: Charles Scribner and Sons.

———. 1909. "The Debt." *Scribner's* 46, no. 23: 165–72.

Wheat, Frank M., and Elizabeth T. Fitzpatrick. 1929. *Advanced Biology.* New York: American Book Company.

White, Charles Abiathar. 1902a. *The Mutation Theory of Professor de Vries.* Washington, DC: Smithsonian Institution Press.

———. 1902b. "My Tomato Experiments." *Independent* 54, no. 2811 (October 16): 2460–65.

———. 1905. "The Mutations of Lycopersicum." *Popular Science Monthly* 66, no. 7: 151–61.

White, Paul. 2003. *Thomas Huxley: Making the "Man of Science."* Cambridge: Cambridge University Press.

Wickson, Edward James. 1902. *Luther Burbank, Man, Methods and Achievements: An Appreciation.* San Francisco, CA: Southern Pacific.

———. 1905. "The Real Luther Burbank." *Sunset* 15, no. 1: 2–16.

———. 1908. "Luther Burbank and His New Environment." *Sunset* 21, no. 2: 151–62.

Williams, Henry Smith. 1915. *Luther Burbank: His Life and Work.* New York: Hearst's International Library.

Williams, Raymond. 1961. *The Long Revolution.* London: Chatto.

———. 1976 (1983). *Keywords: A Vocabulary of Culture and Society.* London: Fourth Estate.

———. 1978. "Utopia and Science Fiction." *Science Fiction Studies* 5, no. 3: 203–14.

———. 1980. "Ideas of Nature." In *Problems in Materialism and Culture: Selected Essays,* 67–85. London: Verso.

Williamson, Jack. 1928. "The Metal Man." *Amazing Stories* 3, no. 9: 792–97.

Wilshire, Gaylord. 1905a. "How Man Paints the Butterfly." *Wilshire's Monthly* 8, no. 5: 8.

———. 1905b (1907). "The Mutation Theory Applied to Society." In *Socialism Inevitable (Wilshire Editorials)*, edited by Gaylord Wilshire. New York: Wilshire Book Company.

———, ed. 1907. *Socialism Inevitable (Wilshire Editorials)*. New York: Wilshire Book Company.

Wilson, J. Theo. 1904. "Evolution's Worst Knock." *San Francisco Chronicle*, October 2: 3, 7.

Windle, Bertram Coghill Alan. 1907. "Mendel and His Theory of Heredity." *Dublin Review* 141, no. 282: 345–56.

"Wizard of Horticulture: Luther Burbank, His Work and His Home in California." 1911. *New York Sun*, September 17: 7.

"Wizard's Wisdom." 1907. *Los Angeles Times*, September 6: 11.

Wohlsen, Marcus. 2012. *Biopunk: Solving Biotech's Biggest Problems in Kitchens and Garages*. New York: Penguin.

Woiak, Joanne Dawn. 1998. "Drunkenness, Degeneration, and Eugenics in Britain, 1900–1914." PhD diss., Institute for the History and Philosophy of Science and Technology, University of Toronto. https://www.proquest.com/pqdtglobal1/dissertations-theses/drunkenness-degeneration-eugenics-britain-1900/docview/304478430/sem-2.

Wolfe, Audra J. 2012. "The Cold War Context of the Golden Jubilee; or, Why We Think of Mendel as the Father of Genetics." *Journal of the History of Biology* 45, no. 3: 389–414.

Wolfreys, Julian. 2001. *Introducing Literary Theories*. Edinburgh: Edinburgh University Press.

"The Woman Vote in Russia." 1906. *Montana News*, June 14: 3.

Wood, Alison. 2009. "Darwinism, Biology, and Mythology in the To-Day and To-Morrow Series, 1923–1929." *Interdisciplinary Science Reviews* 34, no. 1: 22–31.

Woodbury, Charles H. 1910. "The Work of Luther Burbank." *Open Court* 24, no. 648: 299–309.

Woodward, Robert S. 1908. Confidential: Report [to trustees] on the history and the present status of the horticultural work in connection with Luther Burbank. Administrative Files. Carnegie Institution of Washington. Folder 4/25, part 3 of 8.

Woodward, Robert S., George Harrison Shull, and Luther Burbank. 1908. Confidential: Report [draft] on the history and the present status of the horticultural work in connection with Luther Burbank. Administrative Files. Carnegie Institution of Washington. Folder 4/25, part 3 of 8.

Wright, Hamilton. 1905. "The Spineless Cactus: The Latest Plant Marvel Originated by Luther Burbank." *World To-Day: A Monthly Record of Human Progress* (April): 381–85.

Yaszek, Lisa, ed. 2018. *The Future Is Female! 25 Classic Science Fiction Stories by Women, from Pulp Pioneers to Ursula K. Le Guin*. New York: Library of America.

Yaszek, Lisa, and Patrick B. Sharp. 2016. "Introduction: New Work for New Women." In *Sisters of Tomorrow: The First Women of Science Fiction*, xvii–xxv. Middletown, CT: Wesleyan University Press.

Yeo, Richard. 1985 (2001). "An Idol of the Marketplace: Baconianism in Nineteenth Century Science." In *Science in the Public Sphere: Natural Knowledge in British Culture, 1800–1860*, edited by Richard Yeo, 251–98. Aldershot: Ashgate/Variorum.

Young, Robert M. 1985. "Darwinism *Is* Social." In *The Darwinian Heritage*, edited by David Kohn, 609–38. Princeton, NJ: Princeton University Press.

Zammito, John. 2004. "Koselleck's Philosophy of Historical Time(s) and the Practice of History." Review of Koselleck, *Zeitschichten: Studien zur Historik (Mit einem Beitrag von Hans-Georg Gadamer)*. *History and Theory* 43, no. 1: 124–35.

Zevenhuizen, Erik. 1998. "Hugo de Vries: Life and Work." *Acta botanica neerlandica* 47, no. 4: 409–17.

Ziegler, Mary. 2008. "Eugenic Feminism: Mental Hygiene, the Women's Movement, and the Campaign for Eugenic Legal Reform, 1900–1935." *Harvard Journal of Law and Gender* 31: 211–35.

Zirkle, Conway. 1968. "The Role of Liberty Hyde Bailey and Hugo de Vries in the Rediscovery of Mendelism." *Journal of the History of Biology* 1, no. 2: 205–18.

Illustration Credits

Index

Page numbers in italics refer to figures and tables.

Bernal, J. D. (John Desmond), 173, 179, 180, 312, 345
Bernstein, Eduard, 240–41
Beverley, Robert, 81
Biffen, Roland, 320
Binner, Oscar, 105
biocentrism, 80, 200, 346
biological inventions, 31, 172, 174, 223, 268, 343, 348
biology: audiences for, 10–11, 21; and biotopianism, 330; civic, 195–98; "Excelsior biology," 164, 219; experimental, 104; genres, 7–8; and human nature, 11, 21; language for, 6–7; optimistic, 21; popularity of, 324–26; public, 315–17, 337; public prominence of, 3–4; as science of life, 7–8; synthetic, 348; teaching of, 194–95. *See also* science; textbooks
Biology of Home and Community (Trafton), 199
Biology of Man (Linville), 216
bionomics, 19n52. *See also* experimental evolution
biotechnology, 313–14
biotopia, 21–29
biotopianism: and biology, 330; criticism of, 344; and ethics, 189–90; and eugenics, 328–29, 341; and heredity, 28–30, 311; and human nature, 345–46; and nature, 268; and optimism, 29, 312, 347; and perverted futures, 154; resistance to, 347–48; and science fiction, 25; and socialism, 244–52; in textbooks, 352; in United States, 312–14
Blackwell, Antoinette Brown, 263
Blavatsky, Helena Petrovna, 48
Bliss, Arthur J., 67
Brave New World (*BNW*) (Huxley), 3–4, 177, 305–8, 311, 312–13, 332–34, 347
bricolage, 9, 19, 43, 46, 54, 315
British Broadcasting Corporation (BBC), 222
British Empire, 163
Brittain, Vera, 300–302, 304
Bryan, William Jennings, 193, 251
Burbank, Luther: and Liberty Hyde Bailey, 88–89, 204; and biotopianism, 313–14, 346–47; and cactus, spineless,

99–102, 150; and Carnegie Institution of Washington (CIW), 124–28, *126*, 133–36, 149, 150; celebrity endorsements, 106; and census of 1910, 150–51; and children, 108–10, 116–17; in *Civic Biology* (Hunter), 197; companies, 101–2; contradictions, 93–98; crop improvement, 320–21; data, 127–28, 135–36; desert wastes, reclamation of, 299; and Hugo de Vries, 33, 57, 59, 120, 121, 124, 138–45, 147; and Thomas Alva Edison, *307*; and experimental evolution, 24–25, 150–51; Experimental Grounds, 322; and Henry Ford, *307*; and heredity, 116–17; interpretation, 144, 149–50; inventions, 98–102; and kinetic theory of evolution, 132–33; and mass media, 139–42; monstrosities, 114, 116; and moralizing language, 115; and mutation theory, 56, 141–42, 143, 145; parodies, 113; and participatory culture, 19, 104, 293–94, 296; "plant freaks," 112–13, 114; publishing, 105–8; religious criticism, 114–15; and science fiction, 271–73; in *The Science of Life* (Wells, Huxley, and Wells), 225; as scientific mystic, 119–24, 137, 144, 148–49; as scientist, 137, 141, 149, 216; and George Shull, 134, *135*; and socialism, 230, 250–52; and speculative futures, 98–103; at Stanford University, 123; *Sunset* magazine, 120–21; in textbooks, 197, 198–99, 215–16, *352*; unnaturalness of work, 114–16; as unnatural parent, *97*, 108–17
Burroughs, John, 337–38
Butte Daily Bulletin, 231–33
butterflies, 269

cactus, spineless, 99–102, 150
California, 119–20
Callinicus (Haldane), 182
Carnegie Institution of Washington (CIW), 40, 60, 124–28, *126*, 133–36, 149, 150. *See also* Station for Experimental Evolution
cats, 259
Century Magazine, 99, 114, 120n3, 126, 142
chemical warfare, 182

Edge, H. T., 47n38
Edison, Thomas Alva, 44, 307
Edson, Milan C., 86–88
education, scientific, 87, 297–98, 316
elitism, 346
empire-building, 157–58
environmentalism, 22, 166, 315
Essay on Population (Malthus), 164
ethical codes, 153, 156–57, 159–60
ethical gardens, 154–63
ethical guidance, 153, 223, 260
The Ethical Import of Darwinism (Schurman), 17
ethical process, 153, 155, 159
ethical progress, 160–61
ethics: artificial, 181–85; naturalized, 24, 27, 28, 81, 183, 231, 261, 306, 314, 338, 343
eugenics: and arrival of the fittest, 323; and biotopianism, 27–28, 328–29, 341; and Luther Burbank, 116; criticism of, 324–27; and environmental effects, 319–20; and experimental evolution, 28, 326, 327; and Francis Galton, 4–5, 180; and heredity, 326; and Thomas Henry Huxley, 156, 158; left-wing, 328; legislation, 324; meaning of, 180, 329; and Mendelism, 336; misrepresentation of, 324; and Herman Muller, 177–79; and mutation theory, 323, 324; and natural selection, 317–18, 327; negative, 180, 328; popularity of, 324–26, 328; positive, 178, 186, 319, 323, 327–28; reform, 318, 323, 328; and science fiction, 330–32, 331; utopian, 179–80, 181; and white superiority, 72–73; and women, 330
Eugenics and Sex Harmony (Rubin), 332
euthenics, 319, 327–28
evening primrose. See *Oenothera lamarckiana*
Everybody's Magazine, 44–46, 64
"Every Man His Own Burbank" (Harwood), 294
evolution: backyard, 103, 293; concept of, 12; criticism of, 17–18; cyclical, 288n52; and Darwinism, 237; of eyes, 14; Charlotte Perkins Gilman, 260; human, 61, 88; human control over, 61; kinetic theory of, 128–29, 130–34, 137–38,

151–52; and Marxism, 237; meaning of, 151; and mutations, 290; and optimism, 63–64; and progress, 321; religious critique of, 47–49; and revolution, 16; and socialism, 235–37, 241–42, 243–44, 269; spiritualized, 123, 148; stadial, 16, 31, 236, 260–61; as violent, 153. See also Darwinism; experimental evolution; natural selection
Evolution (Lewis), 233, 242, 249
Evolution and Adaptation (Morgan), 54–55, 208–9
Evolution and Animal Life (Jordan and Kellogg), 135–36
Evolution and Ethics (Huxley), 153, 154–55, 197, 267
evolutionary ethics, 81
"The Evolutionary Monstrosity" (Harris), 299–300, 301
The Evolution of Sex (Geddes and Thomson), 261–62, 266
"Excelsior biology," 164, 219
experimental evolution: accessibility of, 296–97; and arrival of the fittest, 321; and Liberty Hyde Bailey, 204–5; and biological innovations, 31; and Luther Burbank, 150–51; cultural influence, 342; definition of, 4, 10; and Hugo de Vries, 70, 296; and eugenics, 28, 326, 327; evolution of, 34; and heredity, 309; interpretation, 315–17; and interpretation, act of, 33; and mass media, 19, 71, 139–40; and mutation theory, 5–6; and promised futures, 21, 24; and science fiction, 10, 289; and socialism, 244–45, 250–51; and speculative heredity, 29; Station for Experimental Evolution, 40; and textbooks, 208; as undisciplined science, 68. See also mutation theory; Station for Experimental Evolution
Experimental Zoology (Morgan), 210–12
Experiments with Plants (Osterhout), 46, 64, 205–6, 208
"extirpation," 157, 158
eyes, evolution of, 14

mutations: artificially induced, 57–58, 70–71, 220–21, 333; and bacteria, 300; and Luther Burbank, 143–44; and cosmic rays, 283–84, 285–86; and evolution, 290; human, 280, 283; meaning of, 144, 210–13, 214–15, 220, 339–40; parthenogenetic, 254–59, 267; and radiation, 220–21, 280–81, 285; and X-rays, 282–83, 290

Die Mutationstheorie (de Vries), 10, 13, 17, 39, 40, 50. *See also* mutation theory

mutation theory: afterlife of, 205–15; applications of, 56–62; and arrival of the fittest, 17, 20–21, 39, 131, 337–38; Liberty Hyde Bailey, 202–4; and Luther Burbank, 141–42, 143, 145; as candidate science, 337–40; and crop improvements, 339; cultural influence, 3, 340; and Darwinism, 13, 16–17, 51–56, 52; erasure from history, 38–39, 336–37; and eugenics, 323, 324; and *Everybody's Magazine*, 44–46; and evolution, 22; and experimental evolution, 5–6, 18; future implications, 39–40; and humans, 280; and kinetic theory of evolution, 151–52; and Marxism, 248–49; and mass media, 35, 39, 42–44, 62–63, 64–65, 139–42; and Mendelism, 26, 324, 340; and Thomas Hunt Morgan, 209, 210–11, 212; and Winthrop John Van Leuven Osterhout, 208; and public science, 40; religious critique of, 47–49, 51–54; and Scopes "Monkey Trial," 191–92; and socialism, 32, 229–30, 233, 235, 241, 242–43, 247–49; teaching of, 192–93; in textbooks, 199, 352, 364; and tipping-point analogy, 248n72; as undisciplined science, 142; *Youth's Companion* reports on, 41–43. *See also* experimental evolution

The Mutation Theory (de Vries). See *Die Mutationstheorie* (de Vries)

naturalistic fallacy, 24, 88, 156, 239, 269

naturalized ethics, 24, 27, 28, 81, 183, 231, 261, 306, 314, 338, 343

natural selection: and arrival of the fittest, 11, 129, 132, 289; and artificial selection, 13–14; challenges to, 1; and

eugenics, 317–18, 327; hostility of, 153; and Thomas Malthus, 167; and Thomas Hunt Morgan, 208–9, 212; and mutations, 290; negative value of, 208–9; in textbooks, 352, 364. *See also* Darwinism; evolution

nature. *See* human nature; Mother Nature; nature and nurture; science: and nature

Nature (journal), 43, 169

nature and nurture, 29, 180, 318, 320, 332–33

nature-study movement, 75–76

Needham, Joseph, 305

negative eugenics, 180, 328

neo-Lamarckianism, 129–30, 321. *See also* Lamarckian inheritance

New Age (periodical), 235

New Atlantis (Bacon), 22, 24, 154

New Civic Biology (Hunter), 339

New Creations in Plant Life (Harwood), 100, 105, 115, 199, 250

The New Earth (Harwood), 25, 251

"New Light on the Origin of Species," 35, 42–44

New Reading of Evolution (Thompson), 49–50, 64

News from Nowhere (Morris), 167

newspapers, 35–36, 36, 51–53, 57–59, 65. *See also* mass media

New Thought, 48–50

Nilsson, Hjalmar, 145

Nineteenth Century and After (journal), 65

Nordau, Max, 164

Nutting, Charles, 338

Odd John and Sirius (Stapledon), 279–80

Oenothera lamarckiana, 17, 38n12, 39, 48, 140, 191, 193, 211, 213–14, 220, 254, 296, 340

O'Hara, Edwin Vincent, 54–55, 56

Oklahoma territory, 93

"On a Piece of Chalk" (Huxley), 154–55

On the Origin of Species by Means of Natural Selection (Darwin), 1, 16, 311–12

Open Court (magazine), 50–51

original sin, 24, 87, 156n8, 162–63, 165, 167, 189, 306–7, 308–11, 327

orthogenesis, 129–30